SYSTEMATIC MODELING AND ANALYSIS OF TELECOM FRONTENDS AND THEIR BUILDING BLOCKS

THE KLUWER INTERNATIONAL SERIES IN ENGINEERING AND COMPUTER SCIENCE

ANALOG CIRCUITS AND SIGNAL PROCESSING

Consulting Editor: **Mohammed Ismail.** ***Ohio State University***

Related Titles:

LOW-POWER DEEP SUB-MICRON CMOS LOGIC SUB-THRESHOLD CURRENT REDUCTION
van der Meer, van Staveren, van Roermund
ISBN: 1-4020-2848-2

OPERATIONAL AMPLIFIER SPEED AND ACCURACY IMPROVEMENT
Ivanov and Filanovsky
ISBN: 1-4020-7772-6

STATIC AND DYNAMIC PERFORMANCE LIMITATIONS FOR HIGH SPEED D/A CONVERTERS
van den Bosch, Steyaert and Sansen
ISBN: 1-4020-7761-0

DESIGN AND ANALYSIS OF HIGH EFFICIENCY LINE DRIVERS FOR Xdsl
Piessens and Steyaert
ISBN: 1-4020-7727-0

LOW POWER ANALOG CMOS FOR CARDIAC PACEMAKERS
Silveira and Flandre
ISBN: 1-4020-7719-X

MIXED-SIGNAL LAYOUT GENERATION CONCEPTS
Lin, van Roermund, Leenaerts
ISBN: 1-4020-7598-7

HIGH-FREQUENCY OSCILLATOR DESIGN FOR INTEGRATED TRANSCEIVERS
Van der Tang, Kasperkovitz and van Roermund
ISBN: 1-4020-7564-2

CMOS INTEGRATION OF ANALOG CIRCUITS FOR HIGH DATA RATE TRANSMITTERS
DeRanter and Steyaert
ISBN: 1-4020-7545-6

SYSTEMATIC DESIGN OF ANALOG IP BLOCKS
Vandenbussche and Gielen
ISBN: 1-4020-7471-9

SYSTEMATIC DESIGN OF ANALOG IP BLOCKS
Cheung & Luong
ISBN: 1-4020-7466-2

LOW-VOLTAGE CMOS LOG COMPANDING ANALOG DESIGN
Serra-Graells, Rueda & Huertas
ISBN: 1-4020-7445-X

CIRCUIT DESIGN FOR WIRELESS COMMUNICATIONS
Pun, Franca & Leme
ISBN: 1-4020-7415-8

DESIGN OF LOW-PHASE CMOS FRACTIONAL-N SYNTHESIZERS
DeMuer & Steyaert
ISBN: 1-4020-7387-9

MODULAR LOW-POWER, HIGH SPEED CMOS ANALOG-TO-DIGITAL CONVERTER FOR EMBEDDED SYSTEMS
Lin, Kemna & Hosticka
ISBN: 1-4020-7380-1

DESIGN CRITERIA FOR LOW DISTORTION IN FEEDBACK OPAMP CIRCUITE
Hernes & Saether
ISBN: 1-4020-7356-9

CIRCUIT TECHNIQUES FOR LOW-VOLTAGE AND HIGH-SPEED A/D CONVERTERS
Walteri
ISBN: 1-4020-7244-9

DESIGN OF HIGH-PERFORMANCE CMOS VOLTAGE CONTROLLED OSCILLATORS
Dai and Harjani
ISBN: 1-4020-7238-4

CMOS CIRCUIT DESIGN FOR RF SENSORS
Gudnason and Bruun
ISBN: 1-4020-7127-2

SYSTEMATIC MODELING AND ANALYSIS OF TELECOM FRONTENDS AND THEIR BUILDING BLOCKS

by

Piet Vanassche

Katholieke Universiteit Leuven,
Heverlee, Belgium

Georges Gielen

Katholieke Universiteit Leuven,
Heverlee, Belgium

and

Willy Sansen

Katholieke Universiteit Leuven,
Heverlee, Belgium

Springer

A C.I.P. Catalogue record for this book is available from the Library of Congress.

ISBN 1-4020-3173-4 (HB)
ISBN 1-4020-3174-2 (e-book)

Published by Springer,
P.O. Box 17, 3300 AA Dordrecht, The Netherlands.

Sold and distributed in North, Central and South America
by Springer,
101 Philip Drive, Norwell, MA 02061, U.S.A.

In all other countries, sold and distributed
by Springer,
P.O. Box 322, 3300 AH Dordrecht, The Netherlands.

Printed on acid-free paper

Printed in the Netherlands.

to my Parents

Foreword

Analog circuits are fascinating artifacts. They manipulate signals whose information content is rich compared to digital signals that carry minimal amount of information; they are delicate in that any perturbation due to parasitic elements, to delays, to interactions with other elements and with the environment may cause a significant loss of information. The difficulty in dealing with these artifacts is to protect them from all possible attacks, even minor ones, from the physical world. The irony is that they are often used to funnel information from and to the physical world to and from the abstraction of the digital world and for this function, they are irreplaceable. No wonder then that analog designers form a club of extraordinary gentlemen where art (or magic?) rather than science is the shared trade. They are difficult to train since experience and intuition are the traits that characterize them. And they have difficulties in explaining what is the process they use to reach satisfactory results. Tools used for design (simulation) are mainly replacing the test benches of an experimental lab. However, the growing complexity of the integrated systems being designed today together with the increasing fragility of analog components brought about by shrinking geometries and reduced power consumption is posing severe challenges to traditional analog designers to produce satisfactory results in a short time. At the same time, the need for experienced analog designers has increased constantly since almost all designs, because of integration, do contain analog components. This situation has created a strong interest in developing design methodologies and supporting tools that are based on rigorous, mathematically literate, approaches. Doing so will make it possible to leverage the expertise of seasoned analog designers and to train new generations faster and better.

In the past, several attempts have been made in academia and industry to create these methodologies and to extend the set of tools available. They have had questionable acceptance in the analog design community. However, recently, a flurry of start-ups and increased investment by EDA companies in novel tools signal a significant change in market attention to the analog domain. I personally believe that to substantially improve quality and design time, tools are simply insufficient. A design methodology based on a hierarchy of abstraction layers, successive refinement between two adjacent layers, and extensive verification at every layer is necessary. To do so, we need to build theories and models that have strong mathematical foundations. The analog design technology community is as strong as it has ever been. Mathematically astute researchers are advancing the state of the art of simulation technology at a pace as fast as I have ever seen in my entire career. Reduced order modeling, relaxation-based techniques, Krylov subspace accelerations have made it possible to verify relatively large circuits with a degree of accuracy that was unthinkable only a few years ago. At the same time, (semi-) automatic layout and parametric optimization for analog circuits have made it in the commercial world. Yet, much research still needs to be done to bring productivity at the appropriate level. As we move towards submicron designs, there is an increasing fear that second-order physical effects can possibly not be compensated for challenging analog components. Hence, the benefits of integration may not be enough to warrant the increased design problems unless significant advances are made to build analog circuits that are robust with respect to submicron undesirable effects.

RF circuits are even more difficult to deal with. Next generation wireless devices are likely to be multi-standard (they are often referred to as software-defined radios) and pose two fundamental challenges: the design of the multi-frequency RF front-end and the reconfigurability needed at base-band. At this time, there has not been a satisfactory design that could yield a successful industrial product. It is safe to say that a good part of the reasons why this is so resides in the lack of methodologies and tools.

This book is about modeling, analysis and verification of RF circuit behavior addressed in a rigorous way. It is directed to a category of engineers who are between design and EDA. It has a good tutorial dimension so that it can be used in a course. In particular, the simple description in plain words of the essence of the different methods, this before diving into the mathematical details, makes reading this book a pleasant experience without sacrificing precision and rigor. Its strength is in its informed review of linear periodically time varying system analysis and the analysis of oscillator dynamics, including phase noise in oscillators. Both topics are essential in the design of analog and RF circuits.

Alberto Sangiovanni-Vincentelli

Contributing Authors

Piet Vanassche received the MSc and PhD degrees in Electrical Engineering from the Katholieke Universiteit Leuven in 1997 and 2003 respectively. His Ph.D. research dealt with techniques for high-level exploration and trade-off analysis for mixed signal telecommunication systems. For this work he obtained a fellowship from the Institute for Science and Technology (IWT). His research interests are in simulation and synthesis of telecommunication systems, network analysis, control theory and engineering education. Currently, he is working for Siemens on advanced process control. He is a member of IEEE.

Georges Gielen received the MSc and PhD degrees in Electrical Engineering from the Katholieke Universiteit Leuven in 1986 and 1990 respectively. He is now a full–time professor at the Katholieke Universiteit Leuven. His research interests are in the design of analog and mixed-signal integrated circuits, and especially in analog and mixed-signal CAD tools and design automation (modeling, simulation and symbolic analysis, analog synthesis, analog layout generation, analog and mixed-signal testing). He has authored or coauthored two books and more than 160 papers in edited books, international journals and conference proceedings. He regularly is a member of the Program Committees of international conferences (ICCAD, DAC, DATE, ...), he is a Deputy Editor of the IEEE Transactions on Circuits and Systems, part II, and is a member of the Editorial Board of the Kluwer international journal on Analog Integrated Circuits and Signal Processing. He is a member of the Board of Governors of the IEEE Circuits and Systems (CAS) Society, and is the Chairman of the IEEE Benelux CAS Chapter. He was the 1997 Laureate of the Belgian Royal Academy of Sciences, Literature and Arts, in the category of engineering sciences. He also received the 1995 Best Paper award of the John Wiley international journal on Circuit Theory and Applications. He is a Fellow of IEEE.

Willy Sansen has received the MSc degree in Electrical Engineering from the Katholieke Universiteit Leuven in 1967 and the PhD degree in Electronics from the University of California, Berkeley in 1972. In 1969 he received a BAEF fellowship. In 1972 he was appointed by the National Fund of Scientific Research (Belgium) at the ESAT laboratory of the K.U.Leuven, where he has been a full professor since 1980. During the period 1984-1990 he was the head of the Electrical Engineering Department. Since 1984 he has headed the ESAT-MICAS laboratory on analog design, which counts about fifty members and which is for 75 % active in research projects with industry. He is a fellow of the IEEE and is a member of several boards of directors. In 1978 he was a visiting professor at Stanford University, in 1981 at the EPFL Lausanne, in 1985 at the University of Pennsylvania, Philadelphia and in 1994 at the T.H. Ulm. Prof.Sansen is a member of several editorial and program committees of journals and conferences. He is cofounder and organizer of the workshops on Advances in Analog Circuit Design in Europe. He is a member of the executive and program committees of the IEEE ISSCC conference. He is program chair of the ISSCC-2002 conference. He has been involved in design automation and in numerous analogue integrated circuit designs for telecom, consumer electronics, medical applications and sensors. He has been supervisor of over forty PhD theses in these fields. He has authored and coauthored eleven books and more than 550 papers in international journals and conference proceedings.

Contents

Symbols and Abbreviations

Conventions

We use the following notations for the scalars, vectors and matrices:

x, X	scalar, slanted lower- and upper-case letters
$\mathbf{x}$	vector, bold lower-case letters
$\mathbf{X}$	matrix, bold upper-case letters. Sometimes, this is also used to denote a vector in its interpretation of a matrix with a single column.
$X_{i,j}$	matrix element located at the i-th row and the j-th column
$\tilde{\mathbf{X}}$	harmonic transfer matrix, bold upper-case letters with a tilde on top
$\overline{x}$, $\overline{\mathbf{x}}$	averaged scalar- or vector-quantity corresponding to x, $\mathbf{x}$
$x^*, \mathbf{x}^*, \mathbf{X}^*$	complex conjugate of a scalar, vector or matrix

Operators

$\lvert\cdot\rvert$	Absolute value of a real number or modulus of a complex number
$\lVert\cdot\rVert_p$	p-norm of a vector or matrix
$\lVert\cdot\rVert$	Shorthand notation for the two-norm of a vector or matrix
$\mathcal{F}\{\cdot\}$	Fourier-transform operator
$\mathcal{L}\{\cdot\}$	Laplace-transform operator
$\mathrm{Im}\{\cdot\}$	Imaginary part of a complex number
$\mathrm{Re}\{\cdot\}$	Real part of a complex number

Symbols

$[a,b]$	interval of real numbers located between a and b.
A	oscillating signal's amplitude
C	symbol for a capacitance
f	frequency in [Hz]
G	symbol for a conductance
$\mathbf{I}$	unity matrix
$\mathbf{I}_N$	unity matrix belonging to $\mathbb{R}^{N\times N}$
j	complex number that equals $\sqrt{-1}$
L	symbol for an inductance
$O(\varepsilon^n)$	the Landau symbol also called big-O. A function $f(\varepsilon)$ is said to be $O(\varepsilon^n)$ if $\exists K > 0 : \lvert f(\varepsilon)\rvert < A\,\lvert\varepsilon^n\rvert$
$\mathbf{p}$	state vector of an oscillator's core system

R	symbol for a resistance
s	Laplace transform variable (complex frequency variable)
t	time
ε	perturbation variable
$\Phi(t,\tau)$	autocorrelation function of a non-stationary stochastic process $n(t) \in \mathbb{R} \to \mathbb{R}$. It is defined as $\Phi(\tau) = E\{n(t)n(t-\tau)\}$. If $n(t)$ is stationary, then $\Phi(t,\tau) = \Phi(\tau)$ does not depend on t.
θ	oscillating signal's phase
σ	Standard deviation of a stochastic variable
τ	normalized time
ω	(angular) frequency in [rad/sec]
$\mathbb{C}$	set of complex numbers
$\mathbb{R}$	set of real numbers
$\mathbb{Z}$	set of integer numbers

Abbreviations

CAD	Computer-Aided Design
CMOS	Complementary Metal Oxide Semiconductor
CT-LTI	Continuous-Time Linear Time-Invariant
DAE	Differential Algebraic equation
DCS	Digital Cellular System
DT-LTI	Discrete-Time Linear Time-Invariant
EDA	Electronic Design Automation
FPGA	Field Programmable Gate Array
HTF	Harmonic Transfer Function
HTM	Harmonic Transfer Matrix
HPSD	Harmonic Power Spectral Density
IC	Integrated Circuit
IF	Intermediate-Frequent
ISF	Impulse Sensitivity Function
LNA	Low-Noise Amplifier
LPTV	Linear Periodically Time-Varying
LTI	Linear Time-Invariant
LTV	Linear Time-Varying
LO	Local Oscillator
MIMO	Multi-Input Multi-Output
MOS	Metal Oxide Semiconductor
MOST	MOS Transistor
ODE	Ordinary differential equation
PAM	Pulse Amplitude Modulation
PCA	Principal Component Analysis

PDF	Probability Density Function
PFD	Phase-Frequency Detector
PLL	Phase-Locked Loop
PPV	Perturbation Projection Vector
PSK	Pulse Shift Keying
PSD	Power Spectral Density
QPSK	Quadrature Pulse Shift Keying
RF	Radio Frequent
RMS	Root Mean Square
SAW	Surface Acoustic Wave
SISO	Single-Input Single-Output
SPICE	Name of a circuit simulator originally developed at Berkeley
SPICE-like	In this text, the word "SPICE-like" is used to indicate simulation algorithms that either build on a Runge-Kutta method or on numerical differentiation formulas.
VCO	Voltage-Controlled Oscillator
Verilog	A language to describe the operation of digital electronic systems and circuits.
Verilog-AMS	A language to describe the operation of mixed-signal electronic systems and circuits
VHDL	Very high speed integrated circuit Hardware Description Language. A language to describe the operation of digital electronic systems and circuits.
VHDL-AMS	Very high speed integrated circuit Hardware Description Language with Analog and Mixed-Signal extensions. A language to describe the operation of mixed-signal electronic systems and circuits.
WLAN	Wireless Local Area Network

Chapter 1

Introduction

To those who do not know mathematics it is difficult to get across a real feeling as to the beauty, the deepest beauty, of nature ... If you want to learn about nature, to appreciate nature, it is necessary to understand the language that she speaks in. —*Richard Feynmann*

Mathematics is too important to be left to the mathematicians.
—*David Hestenes*

The ability to analyze system or circuit behavior is one of the key requirements for successful design. To put an idea to work, a designer needs both the knowledge and tools for analyzing the behavior of that new system architecture or that experimental circuit topology. Design decisions are grounded on the results obtained from analysis.

Producing dedicated methods for analyzing each particular problem at hand is of course inefficient. It's like reinventing the wheel time and again. Therefore, successful methods should be applicable to large classes of system and circuit behavior. This implies a classification that makes abstraction of the underlying implementation details. This process of abstraction can be considered as a formalization of design knowledge. This formalization is important for two reasons: in the short run, it speeds up the future design of similar systems by enabling us to reuse exsiting methods; in the long run, it eases a transfer of knowledge to generations to come.

This book reports on research in the field of methods for modeling and analysis of telecommunication frontends and their building blocks. In doing so, it deals with fundamental theory and algorithms for behavioral model extraction.

1.1 Structured analysis, a key to successful design

1.1.1 Electronics, a competitive market

Since the birth of the first transistor at Bell-Labs (1947) over the creation of the first integrated circuit (IC) at Texas Instruments (1958), electronics has experienced a tremendous growth, both technologically and economically. Fig. 1.1 shows both the evolution of the transistor dimensions and the transistor count of the Intel processors over the last couple of decades. This gives a clue as to the tremendous pace with which technology evolves. With the latest Intel Pentium 4 containing over 100 million transistors

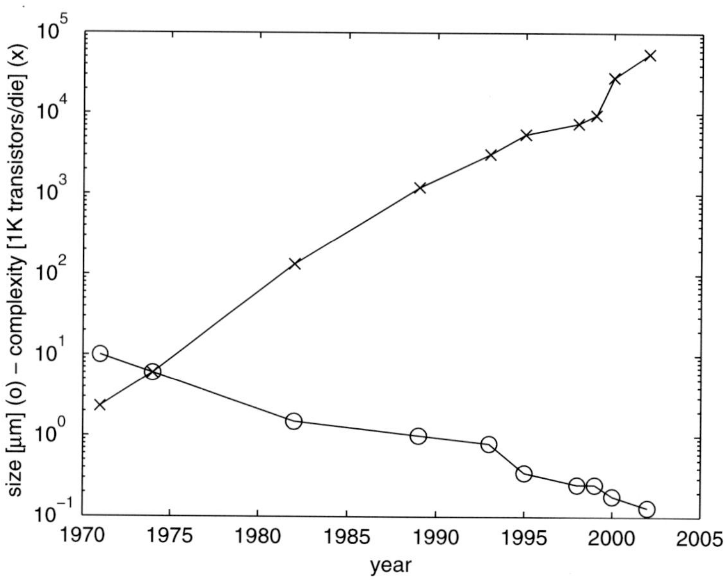

Figure 1.1: Evolution of transistor sizes and number of transistors per die for the Intel processor [source: Intel].

and with the first 1 billion transistor processor expected by 2007, integrating complex functionalities on a single die is becoming reality.

Packing more transistors onto a single chip has resulted into a dramatic cost reduction. Fig. 1.2 shows that the average transistor cost drops exponentially with time. This price evolution drives a huge number of commercial applications, making them affordable for mass markets. These applications have pervaded almost all aspects of our daily lives: computers are used to run complex administrations; electronic control systems are omnipresent, from chemical plants to automobiles; electronic signal processing has made global communication a reality.

As a consequence of these successes and of many years of high-paced growth, the electronics industry has become a highly competitive business. There are a lot of companies that want market share. Often, the first company to offer a particular product at a reasonable price acquires a substantial share in overall sales. A short time-to-market is therefore of vital importance in being successful. The electronics industry is a business where complex systems need to be developed in a minimum of time.

1.1.2 Analog design: A potential bottleneck

In order to cope with the requirements of a demanding market, systems are made highly programmable. This is accomplished using field programmable gate arrays (FPGAs), micro-processors and reconfigurable logic. Introducing software components on chip

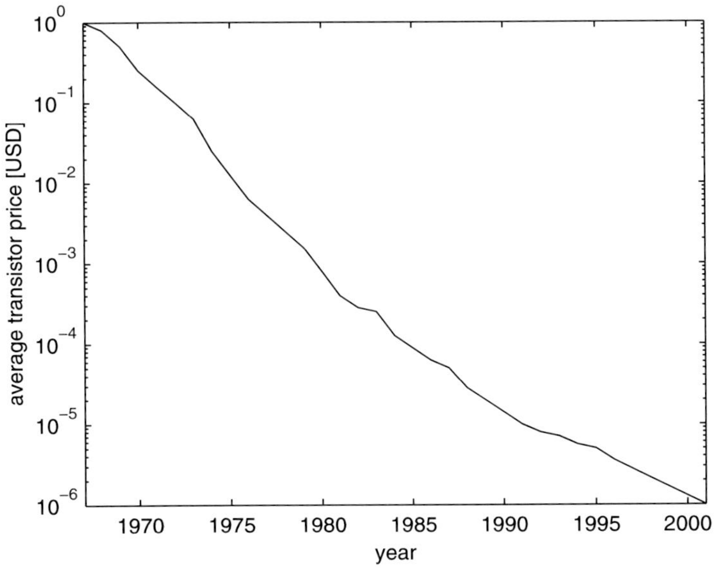

Figure 1.2: Evolution of the cost of a single transistor [source: Intel].

allows us to make last-minute changes, promotes reuse and provides a lot of flexibility in system debugging. This trend is supported by the growing speeds and decreasing prices of these programmable devices. However, in each system, there are building blocks that are hard to make highly programmable. Analog circuits are typical examples of such blocks. They are small but important parts that are present in almost any system.

Systems or subsystems are said to be analog if it is not possible to make abstraction of the continuous nature of the system's input, output and internal signals. Within the overall application, the analog part typically constitutes the interface with the physical world, e.g. the frontend of a telecommunication system. Due to their continuous nature, analog systems are highly complex in their design. As such, they require a disproportional fraction of the overall design time as compared to the complexity of the signal processing operations they implement. Moreover, they tend to be highly sensitive to all kinds of process parasitics, like substrate couplings and mismatch. Still, these analog subsystems must be made first time right. If not, they become bottlenecks in getting the overall system to market in time. Productivity and reliability demands like this press for structured analog design methods.

1.1.3 Structured analog design

As is the case for any design process, analog design will always require creative inputs, e.g. some ground-breaking idea for a new system or circuit topology. However, this does not imply that the design process can not be structured. A structured design flow

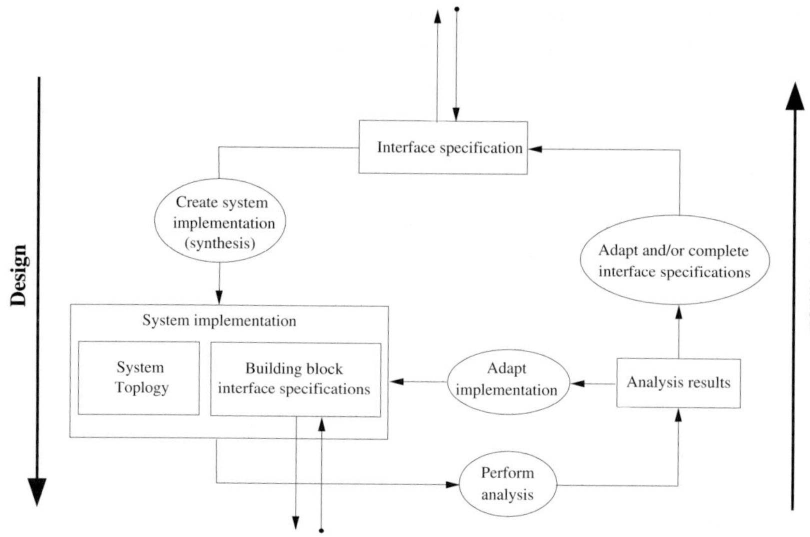

Figure 1.3: A structured design flow implements system realization in a hierarchical manner. At each stage, the initial specifications are converted to a candidate implementation. An implementation consists an interconnection of building blocks, each implementing part of the overall functionality. Next, the performance of the implementation is evaluated. The results may lead to adjusting the system implementation or even the initial set of system specifications.

implements system realization in a hierarchical manner. The problem statement is gradually refined by decomposing it into more manageable subproblems.

During each stage in the design process, this decomposition proceeds along the lines illustrated in Fig. 1.3. The starting point is a set of system specifications that represent the required functionality, performance and cost constraints. To meet these specifications, a designer suggests a number of candidate implementations[1]. An implementation involves a set of building block specifications —each intended to realize part of the overall functionality— and a system topology, i.e. the way in which the building blocks are interconnected. By means of techniques for modeling and analysis, the resulting system performance is evaluated. If specifications are met, one can proceed with the implementation of the building blocks. If not, the current system implementation needs to be adjusted. If it seems impossible to meet specifications, one can request to relax them. This, however, may impact implementations at previous stages in the design hierarchy.

[1] This —synthesis— step is the one that mainly requires a designer's creative input. Of course, automated synthesis is possible, but it will always be based on filling out the degrees of freedom of some template solution devised by a —human— designer. The art of automated synthesis is making the template as general as possible.

Structured analog design requires designers to have both the knowledge and tools to tackle the problems of analysis and synthesis that occur. For this, we need

- to formalize the knowledge available on analysis and synthesis of analog building blocks. This implies embedding that knowledge within a global theoretical framework.
- algorithms and tools to support the process of analysis and design. This part relates to the areas of computer-aided design (CAD) and electronic design automation (EDA).

Having theory and tools available improves understanding, speeds up future design of similar systems and eases transfer of knowledge and experience to trainee designers.

The history of the feedback amplifier offers us an excellent example as to the gains of a structured design process [Mind02]. First suggested by Harold Black, these amplifiers often showed undesired oscillatory behavior. Understanding and predicting this behavior has been the driver for Harry Nyquist and Hendrik Bode to develop the theory of feedback. This theory embeds feedback amplifiers within a global mathematical framework. It was a major step towards a better understanding and a more systematic design of feedback systems. Feedback theory has resulted in shorter design times by ensuring *a priori* the absence of oscillatory behavior. Nowadays, application of the theory is supported by numerous toolboxes.

1.1.4 Structured analysis

Embedding knowledge within a global theoretical framework requires a structured and hierarchical approach towards analysis. Rather than having to reinvent the wheel time and again, theory and methods for modeling and analysis should apply to all systems exhibiting similar behavior. This implies a classification of systems according to their behavioral properties. It induces a tree-like hierarchy whereby each class is partitioned into further subclasses. This is accomplished by refining the behavioral properties that define (sub)class members. The different levels of hierarchy correspond to different levels of abstraction that either ignore or account for specific details of the system's behavior and/or implementation. Based on the set of properties that define a particular class of systems, methods for analysis —and hence design— are developed that apply to all systems of a that class.

In Fig. 1.4 we consider the class of linear systems. Linear systems are defined by the requirement that the principle of superposition holds. This is a general and abstract requirement that is approximately satisfied by systems ranging from a single-stage amplifier to a complete receiver frontend. However, it can be exploited to construct some dedicated methods useful for the analysis linear system behavior. All these methods have in common that they only rely on superposition for their results to be valid. They can be further refined by taking even more system-specific information into account, e.g time-invariance, as in filters, or periodic time-variance, as in mixers. The extra information can be exploited to speed up analysis. Moreover, by exploiting system-specific information, results can often be presented in a way that is simpler to interpret.

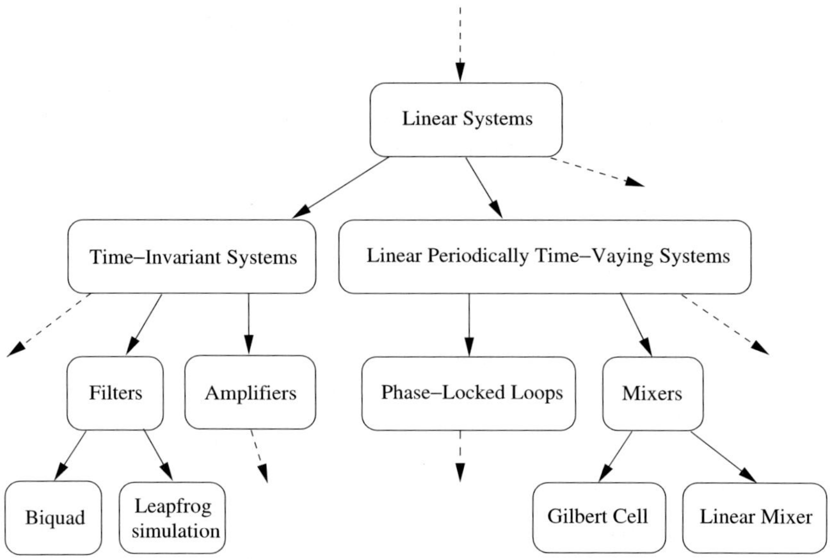

Figure 1.4: Structured analysis requires a hierarchical classification of systems according to their behavioral properties and/or implementation.

However, results obtained in this manner are only valid for a (more) limited class of systems.

The starting point for a structured approach towards analysis requires gathering all *a priori* information that one has on the application at hand[2]. This information is used to locate the application's position within our classification of systems and circuits, e.g. the tree in Fig. 1.4. This way, suitable theory and methods for analyzing the application are identified as they are tied to the class or classes of systems to which the application belongs. Often, methods can be further refined by taking more application-specific information into account. This approach promotes the analysis of a particular system by using techniques that, initially, were developed to deal with a completely different application. Furthermore, it supports a transfer of knowledge between different areas of science by making people recognize the properties that problems have in common. Last but not least, for generations to come, it provides a well-organized gateway to acquire and expand knowledge and experience.

1.2 This work

This work focuses on theory and methods for modeling and analysis of telecommunication frontends and their building blocks. The theory builds on a classification of building blocks according to their behavioral properties. These properties are exploited to

[2]There is no shame in using previous experience to make life a little easier. However, it is important to embed this experience within a structured framework.

construct methods that render analysis as efficient as possible. Both theory and methods apply to large classes of systems. Their use often extends beyond telecommunications. In summary, this work contributes to a structured approach towards telecommunication system analysis and provides a well-organized gateway to methods that may be of use in capturing system and/or building block behavior.

1.2.1 Main contributions

This book reports work on two particular subjects:

Linear periodically time-varying (LPTV) systems: LPTV behavior arises when system or circuit behavior is linearized in the neighborhood of a periodic (time-varying) operating point. This is comparable to traditional linear time-invariant (LTI) small-signal analysis where the operating point is constant. LPTV behavior is typically observed in systems driven by a large periodic signal, e.g. mixers and phase-locked loops (PLLs). It is characterized by the up- and downconversion of signal content.

A first part develops frequency-domain methods that extend traditional time-invariant techniques to cope with LPTV behavior. The starting point is the Harmonic Transfer Matrix (HTM) representation of an LPTV system, a concept borrowed from power electronics and microwave theory. HTMs allow us to handle LPTV systems in a manner that is similar to dealing with LTI systems using (Laplace- or frequency-domain) transfer functions. Unfortunately, the —in principle infinite-dimensional— size of the HTMs tends to make computations unwieldy, hampering their introduction into the circuit design community.

This work introduces powerful techniques that render HTM-based analysis useful for design practice. Analysis of up- and downconversion behavior, time-varying noise analysis and stability analysis are made practical. The tehniques allow us to compute both numerical results and symbolic expressions. By exploiting the properties of typical LPTV circuits, we obtain compact approximations that cope with the HTM inversions induced by feedback loops. Furthermore, it is shown that HTMs make up a very natural representation for frequency-domain modeling of LPTV behavior. Their elements have a well-defined physical meaning that is in close agreement with intuition on the matter. All of this establishes HTMs as a powerful and practical framework for capturing time-varying behavior.

Theory on HTMs is applied to several examples. Most notable is the HTM-based analysis of PLL behavior. PLLs allow an easy and exact description in terms of HTMs. For slow PLL feedback loops, this exact —time-varying— description reduces to the well-known time-invariant feedback model. For the first time, this time-invariant model is given solid mathematical underpinnings with its shortcomings clearly identified. Finally, it is also shown how a HTM-based analysis unifies the derivation of both continuous-time and discrete-time PLL models.

HTMs, together with the methods developed to perform computations with them, are well suited as a framework to teach structured analysis of LPTV system behavior. They extend intuitions on traditional LTI systems to the more general class of LPTV systems. The methods presented to manipulate them can equally be applied to applications as different as mixers and PLLs. They are suited to obtain both symbolic and numerical results.

Oscillator behavior: Oscillators are key building blocks in many telecommunication frontends. They are, amongst others, needed for channel selection and clock and data recovery, etc. Their behavior is often characterized by the presence of signal components that vary at widely different rates. This causes simulation times to soar when using traditional algorithms, e.g. SPICE. This poses a problem, especially when running repetitive or lengthy system-level simulations.

A second part of this work presents methods to extract compact models that capture an oscillator's dynamics. To do so, these methods exploit the widely spaced time constants that characterize the oscillator's behavior. While these time constants often poses a bottleneck to traditional simulation algorithms, they enable the methods here presented to explicitly separate the oscillator's slow- and fast-varying signal components. This results in models that are readily solved using multi-rate simulation techniques whichallows us to greatly boost simulation speed. The modeling strategy is solidly grounded on the theory of dynamical systems, perturbation analysis and averaging. It yields clear insights into the mechanisms that govern oscillator dynamic behavior.

Applications of the theory involve oscillator phase noise analysis and modeling the settling behavior of harmonic oscillators. As phase noise analysis is concerned, theory is greatly simplified as compared to the current approach based on stochastic differential equations. The harmonic oscillator models provide a means for efficient simulation.

The theoretical foundations presented in this work provide a sound basis for understanding oscillator dynamics. It can be used to teach analysis of oscillator stability, settling and noise behavior. The algorithms for harmonic oscillator behavioral modeling are readily implemented on top of existing harmonic balance or shooting algorithms as found in commercial simulators.

1.2.2 Math, it's a language

Formalizing knowledge requires a language that is accurate enough to describe that knowledge. We must be able to express conditions, perform analyses and write down algorithms in a manner that does not suffer from the ambiguities common to everyday languages like English or Dutch. On the other hand, there is always the need for intuitive understanding of descriptions written down in some abstract language. Whatever you do, you must always understand what you are doing.

Over centuries, mathematics has developed as the mainstream language of science. It is conceived to provide means for a consistent, accurate and quantitative description of our observations of the world around us. Since this text aims at consistency and preciseness in its description of the different topics being treated, it has a highly mathematical content. For example, the theory of LPTV systems is grounded on concepts of integral equations and functional analysis while oscillator modeling is build on the theory of stable manifolds and averaging. However, the reader should not be deterred by this. It has been tried to clarify all mathematics in this text by means of intuitive (geometrical) interpretations.

Especially among engineers, mathematics often has a reputation for getting too abstract. Concepts like manifolds and integral equations seem like nice toys for a mathematician, but too abstract and complicated to be useful in (circuit) engineering design practice[3]. This is reinforced by a mathematician's love for abstraction. They carefully avoid physical interpretation. The latter is considered as compromising the universal abstract character of mathematics. However, although abstraction can help linking topics as diverse as climate dynamics and oscillator phase noise, it goes past the essence of mathematics: a language for describing our "everyday" observations. Every mathematical expression should therefore have meaning to those who read it. It should invoke images and should help us in our quest for intuitive understanding.

Decoupling meaning from formalism, semantics from syntax, reduces mathematics to a set of rules for deducing new statements from old ones. One stops wondering whether those new statements make sense, what their connection to reality is. This results in a collection of statements, a story, that is no longer easily understood. So, people stop trying to understand and focus on results only. If results are all right, this validates the mathematical procedure. This instrumentalist point of view [Pop34] reduces mathematics to a toolbox. As a consequence, mathematics often is more likely to conceal than to clarify the nature of things. It is no longer telling us a story and hence looses its attractiveness towards many —young— people. People are fascinated by stories, and telling stories is what languages are all about[4].

In this text, we've tried to keep the story we tell in mathematics consistent with the one we tell in English. It is attempted to clarify mathematical derivations as much as possible through intuitive (geometrical) interpretations. These interpretations are summarized in a section called *"The story behind the math"* that comes at the beginning of most chapters. In this way, we aim for mathematical accuracy while avoiding that the basic message gets lost in a jungle of equations.

1.3 Outline of this book

This book is subdivided into six chapters. Each chapter starts with a brief summary followed by an introduction that reviews existing state of the art methods to solve similar problems. Chapters 3 to 5 contain the technical core. When relevant, it contains a section *"The story behind the math"* that attempts to provide a clear and intuitive

[3]Of course, this perception strongly depends upon the engineering discipline being considered. Mechanical engineering, for example, has a long-standing tradition which dates back to way before the advent of computers. Lack of computers forced people to develop, for example, sophisticated mathematical approximation strategies. On the other hand, development of micro-electronics went in parallel with that of computing devices. As a consequence, circuit design practice heavily relies on virtual prototyping using computer-aided design tools.

[4]To this account, it is recommended to read the work of David Hestenes [Hest87] and Edwin Jaynes [Jayn03]. They provide splendid examples on how mathematics can be used to tell fascinating stories about the world we live in. David Hestenes is concerned with a clear and consistent algebra for capturing nature's geometry. His *geometric algebra* corresponds more closely to our intuitive notions on the matter than do the often artificial constructions of traditional vector and matrix algebra. Edwin Jaynes' account on probability shows it to be the only consistent framework for logic inference. His narrative approach is in great contrast to, for example, the abstract and axiomatic one of Kolmogorov [Kolm92].

perspective on their contents. All mathematics is introduced when needed. The outline of the subsequent chapters goes as follows:

- **Chapter 2** (***Basic concepts***) outlines basic concepts on telecommunication systems and methods for modeling and analyzing them. This chapter provides a conceptual framework that puts the methods developed in further chapters into the perspective of the art on modeling and analysis of telecommunication systems.

- **Chapter 3** (***Linear periodically time-varying system analysis***) elaborates methods to deal with LPTV systems. It provides a brief overview of existing methods and presents a coherent framework of frequency-domain techniques based on the HTM formalism.

- **Chapter 4** (***Applied LPTV system analysis***) demonstrates the power and elegance of the framework presented in chapter 3 through applications like PLLs and mixers. Furthermore, we develop an algorithm for automated symbolic analysis of LPTV systems.

- **Chapter 5** (***Modeling oscillator dynamic behavior***) treats the behavior of perturbed oscillators and its applications to circuit analysis. It is demonstrated how small disturbances cause slow-varying processes to occur on top of the rapid oscillations. Examples of such slow-varying processes are an oscillator's phase noise behavior or the settling behavior of high-Q harmonic oscillators. A general framework is presented to deal with this kind of behavior together with several examples.

- **Chapter 6** (***Conclusions***), draws conclusions and provides directions for researchers who would like to further explore the tracks outlined in this text.

Chapter 2

Modeling and analysis of telecom frontends: basic concepts

You insist that there is something a machine cannot do. If you tell me precisely what it is that a machine cannot do, then I can always make a machine which will do just that.
—John von Neumann

Good models and efficient methods for constructing and evaluating them are of utmost importance in making true top-down design of telecom frontends a reality. This book intends to contribute to the work in this area. Firstly, however, we must clearly define "models", "good models", "modeling" and "analysis". You cannot realize something, e.g. a good model, if you cannot tell what it is you want to realize. Therefore, section 2.1 spends some time to elaborate these concepts.

Incorporating all relevant prior knowledge and experience is one of the main properties that characterize good (telecom building block) models. It makes models reliable in a sense that their behavior can be trusted to correspond with that of physical implementations. This is of great importance in avoiding redesigns. Furthermore, exploiting prior knowledge helps to improve, for example, simulation efficiency. It allows us to use "shortcuts" in capturing a system's behavior. As this book mainly deals with telecom frontends and their building blocks, section 2.2 reviews some typical telecom frontend architectures and summarizes our prior knowledge of their behavior. Further chapters will often exploit this knowledge to simplify analysis.

In communications, the relevance of the different aspects of a building block's behavior —and therefore the need to incorporate them into (good) models— is measured by their impact on the system's overall performance in transmitting information. Physically, information is stored by modulating the properties of the waveforms that are transmitted, e.g. their amplitude and phase. Information is lost due to distortion of a waveform's shape as it travels from sender to receiver. These shape distortions are, for example, due to linear, nonlinear and stochastic building block behavior. Predicting them is of great importance in estimating the probability of transmission errors to occur (bit error rate). Hence our need for accurate building block models.

2.1 Models, modeling and analysis

A major part of this text is about models, modeling and analysis. Before focusing on particular applications, it is useful to spend some time to clarify these concepts.

In what follows, "models", "good models", "modeling" and "analysis" are given a precise meaning. This provides the context for the methods presented in this text. We also stress the importance of having good models available to make top-down design truly possible.

2.1.1 Models: what you want or what you have

Modeling and analysis is all about manipulating models. So, the first topic to be addressed sounds: what is a model? Almost any introductory textbook on physics provides a definition. A fairly thorough and consistent treatment on the matter can be found in [Hest87]. There, the author defines:

Definition (Model–Physics): *A model is a conceptual representation of a real object.*

The means used for representing an object can be quite general: it may involve a verbal description, a set of mathematical equations, a collection of measurement data or combinations of the aforementioned. Note that for engineering purposes, the word "object" is often too general and abstract. It is often better to use "system" or "circuit" instead.

From an engineering point of view, the definition above does not satisfy. It highlights the fact that physics is mainly concerned with description and analysis of systems (objects) that exist. Engineers, on the other hand, are also concerned with the creation of new systems. Most of the time, they describe systems that they would like to have but that are not there yet. In [Hest87], such descriptions would be called *fictitious models*. Since this viewpoint puts unequal emphasis on objects that already exist, this text prefers a different, more balanced definition:

Definition (Model–Engineering): *A model is a conceptual representation of a system that you want to realize (specification model) or that you have realized (implementation model).*

This definition puts equal weight on both reasons for using models in engineering practice: the specification of a system's behavior involves constructing a model for the system as we would like to have it; the verification of a particular implementation involves the construction of a model for a physically realized system as it is given to us. The first kind of models occur as we go down in the design hierarchy while the second kind occurs when going up.

Example (Low-noise amplifiers models): Consider the example in Fig. 2.1 where two people construct models for a low-noise amplifier (LNA). The system designer is responsible for the realization of a complete frontend architecture. He or she will construct a specification model that describes the LNA behavior as (s)he wants it. The circuit designer, on the other hand, realizes an LNA in terms of transistors, coils, capacitors and resistors. The resulting netlist represents an implementation model that describes the LNA as it is or will be physically available. Of course, the goal is to realize an LNA that meets the specifications. Phrased in the terminology introduced above: it must be possible to map the implementation model realized by the circuit designer on the specification model provided by the system designer. ▲

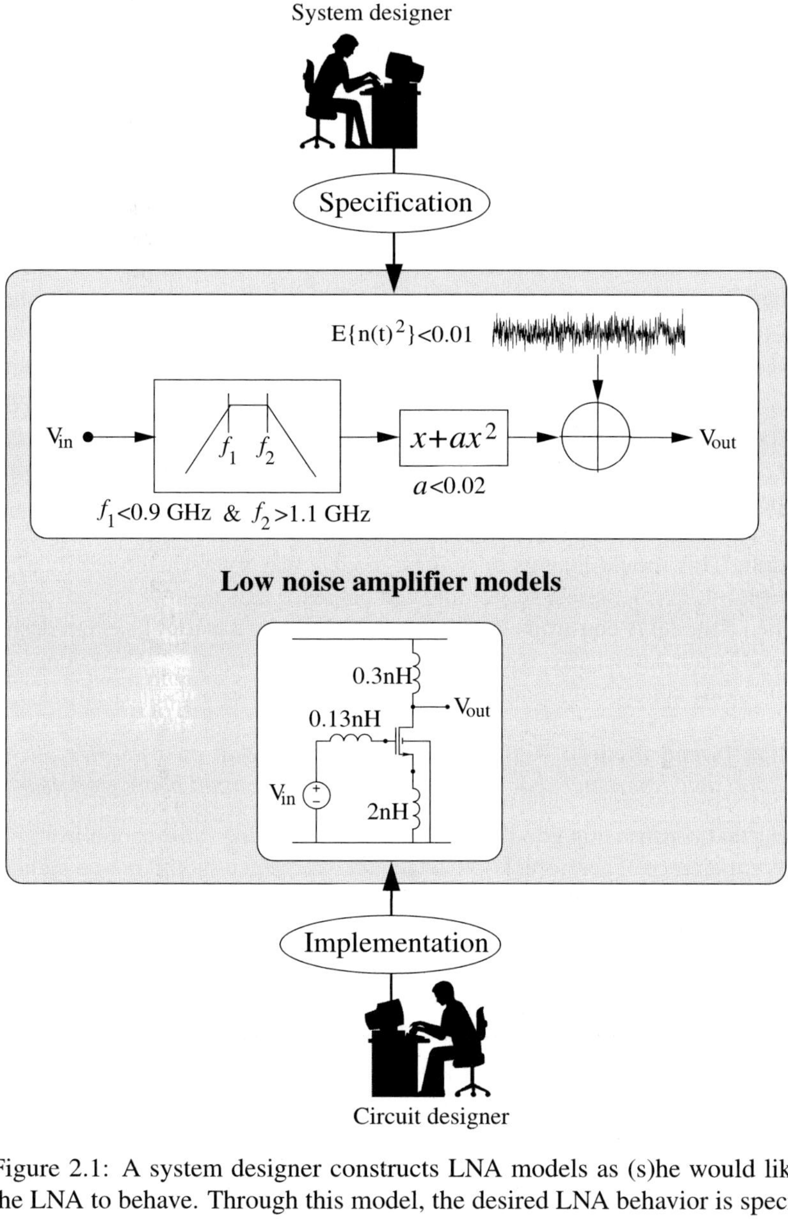

Figure 2.1: A system designer constructs LNA models as (s)he would like the LNA to behave. Through this model, the desired LNA behavior is specified, i.e. it represents the LNA as we would want it to be. A circuit designer realizes an LNA in terms of transistors, coils, etc. The resulting netlist represents a model for the LNA that is physically available, i.e. it represents the LNA as we have it.

2.1.2 Good models

In the example in Fig. 2.1, one may wonder why the system designer makes the LNA model so complicated. Basically, what (s)he really wants is represented by the first block: a simple bandpass filtering operation with some gain. No system designer wants nonlinear distortion or noise. So, why mention it in the specification model? The reason for this is that there is no use in living in utopia. If a system designer makes decisions based on building block models that do not correspond to reality, these decisions often result in implementations that fail to meet the overall performance requirements. A model is said to correspond to reality if it captures all (relevant) behavior exhibited by an actual implementation. Stated differently, it is possible to map the entire behavior of a physical implementation to the (specification) model. If there is no such correspondence, system performance might get ruined by (unwanted) behavior that was not accounted for. This results in costly redesigns.

Clearly, there is a catch in the previous discussion. How is it possible to construct (specification) models that account for the entire behavior of implementations *that may not yet exist*? Often, we don't even have a clue on how the implementation will look like. The answer, of course, is that capturing the *entire* behavior is impossible. The only thing we can do is bundling all our knowledge, gained from similar design experiences in the past. This should help us to suggest models that approximate reality as closely as possible. It is impossible to account for behavior that nobody is expecting at the time when a model is constructed. However, it would be a waste if system design fails because all prior knowledge was not exploited in constructing proper models. This brings us to the concept of a *good model*.

Definition (Good model): *A good model is a model that incorporates all relevant information and experience that we have on the system/circuit being modeled.*

It is clear that constructing good models requires retrieving information from previous (design) experience. Therefore, knowledge and experience should not be gathered and stored in an ad hoc manner. In order to avoid looking for a needle in a haystack of experience, knowledge should be structured and compacted. As outlined in chapter 1, this promotes a hierarchical classification of systems according to their characteristics and properties.

One way to ensure the "goodness" of —especially system-level— models is to show that, at least in theory, they can be derived from existing circuit-level implementations by means of a number of approximations. This links the model to reality while reducing model complexity. Moreover, it is well controlled which behavior is taken into account and which behavior is neglected. Transfer functions, for example, derive their status as a powerful concept to model and specify circuit behavior from the fact that they adequately describe the behavior of a circuit's small-signal approximation. This small-signal approximation linearizes circuit behavior in the neighborhood of a constant (DC) operating point. However, it neglects all nonlinearities. Hence, as illustrated in Fig. 2.2, transfer-function-based (specification) models link to (possible) circuit implementations through a number of well-controlled approximations. A likewise approach towards constructing good models is pursued in chapters 3 and 4.

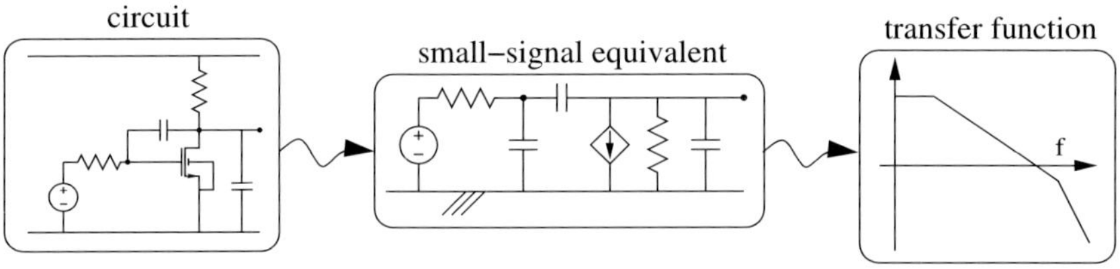

Figure 2.2: Transfer functions derive their status as a powerful concept to model and specify circuit behavior from the fact that they adequately describe the behavior of a circuit's small-signal approximation.

As a final note, it should be stressed that all good models should be as compact as possible. Overly detailed descriptions bear tedious computations and lengthy simulations with them. They do not contribute much to design efficiency. Irrelevant model behavior, for instance, involves non-dominant nonidealities[1] or high-frequency transients. Removing irrelevant details from a model is very important, e.g. for efficient optimization-based design.

2.1.3 The importance of good models in top-down design

As discussed in chapter 1, minimizing the time necessary to design analog frontends requires the introduction of hierarchy in the design process[2]. As illustrated in Fig. 2.3, a hierarchical, top-down method initially tackles frontend design in terms of low-noise amplifiers, filters, mixers, oscillators, A/D converters, etc. In turn, these blocks are implemented using integrators, operational amplifiers, etc. In a next step, circuit- and transistor-level details are filled out. Beyond the circuit level, there is layout and manufacturing.

For CAD support of such a hierarchical flow, good models are of utmost importance. They are among the cornerstones in realizing each of the steps in the design flow hierarchy. The models are used as interfaces to communicate design decisions between the different levels in the design tree. Many actions in a top-down design flow result in specification and/or implementation models that are passed down/up the design hierarchy. As illustrated in Fig. 2.1, system engineers, responsible for the overall frontend design, communicate their desires by means of specification models for the behavior of the frontend building blocks. Creating these models involves synthesis, specification translation and system exploration. Circuit engineers create a transistor netlist. This netlist acts both as a specification model for layout and as an implementation model for building block verification at the architectural level. Here, in a verification step, one checks whether the implementation model can be mapped on the specification model that was initially passed down.

[1] A nonideality is said to be non-dominant if there is no frequency band of interest in which the nonideality contributes the major part of the unwanted signal energy.

[2] For a more complete introduction to hierarchical design methods, see [Donn98].

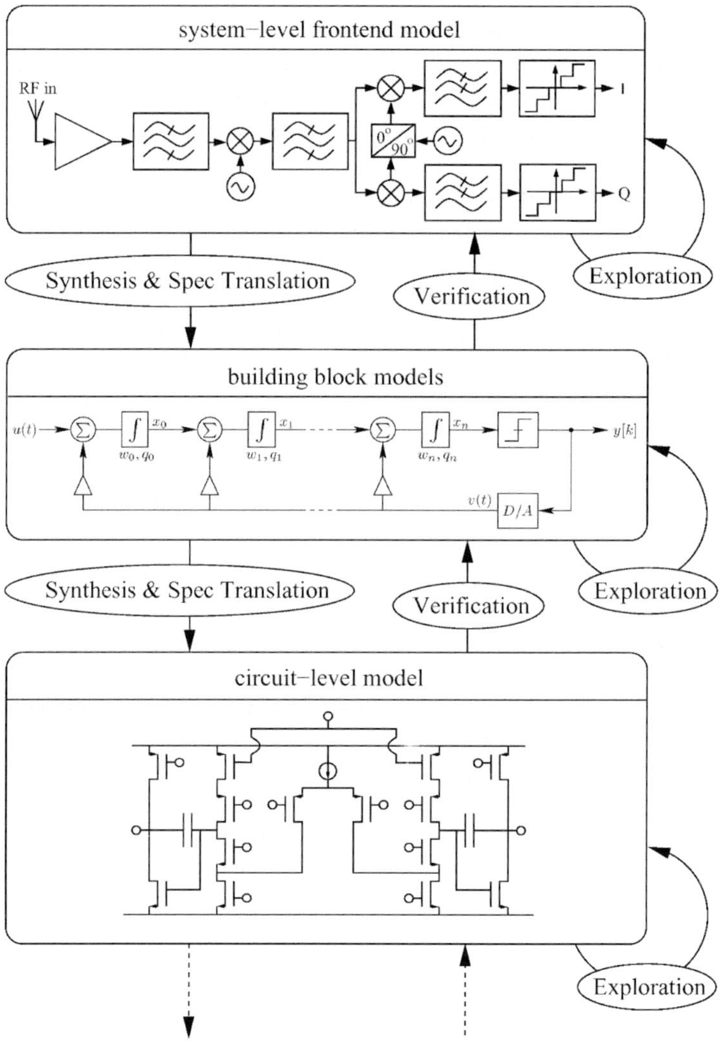

Figure 2.3: Minimizing design time requires the introduction of hierarchy in the design process. Frontend design is done in terms of LNAs, filters, mixers, A/D converters, etc. These building blocks (e.g. a $\Delta\Sigma$ A/D converter) are in turn implemented using integrators, operational amplifiers, etc. In a next step, transistor-level details are filled out. To support this hierarchical design process, good models are of utmost importance. They serve as interfaces used to communicate design decisions between the different levels in the design tree.

All models used in a top-down design flow need to be "good models": they must accurately represent real-life system behavior. For example, a system-level mixer model should capture all of the relevant behavior that characterizes a mixer's transistor-level implementation. Otherwise, wrong design decisions will be taken at the architectural level, them being based on analysis using a defective or incomplete mixer model. This often results in costly redesigns. As mentioned before, since we do not know all implementation details in advance, we must rely on prior knowledge and experience to construct realistic model templates. This especially holds for models constructed during the early stages of a design.

Finally, in realizing efficient (partially automated) hierarchical design methods, it is very important that models should be kept as simple as possible. Complex and overly detailed models compromise efficiency of analysis as they result in tedious computations and lengthy simulations. This in turn hampers architecture exploration or optimization-based synthesis and/or specification translation. Since these steps often require numerous evaluations of the building block models, complex models render it slow and sometimes even infeasible. This especially holds at the architectural level where a great number of building block models need to be evaluated simultaneously. Low-complexity models are therefore among the cornerstones of (semi-)automated design. In summary, capturing complex building block behavior in a manner that is as simple as possible is one of the great challenges in CAD.

2.1.4 Modeling languages

Describing system behavior requires a proper formalism, i.e. a language, to do so. Choice of the language is driven by the nature of the system we want to describe. Human behavior, for one, is best described in a spoken language like English. On the other hand, the quantitative nature of engineering problems renders mathematics a natural choice. Often, new language constructs (syntaxes) need to be developed to capture newly encountered objects and their behavior. For example, electrical netlists have driven the creation of the SPICE input syntax [Vlad94]. More recently, describing mixed-signal systems has brought about languages like VHDL-AMS [VHDL] and VERILOG-AMS [VERI]. It is, however, not the intention of this text to give a complete overview of such languages nor to discuss their strengths and shortcomings. In what follows, mathematics will be the main language of choice.

2.1.5 Modeling and analysis: model creation, transformation and interpretation

Models are almost never given to us in a manner that directly suits our needs. In the very beginning, there might even be no model available at all. There are only our observations or a vague idea that we have in mind. Therefore, we need techniques to create and manipulate models. This is where modeling and analysis come into play.

Definition (Modeling and Analysis): *Modeling and analysis concern the acts of creating, manipulating and interpreting models. Hereby, the aim of modeling is to create a model, i.e. models are considered a result of the operation. In carrying out analysis,* *models are just a means to gain the information necessary to make design decisions. Analysis involves interpreting models.*

The main difference between both concepts comes down to whether models are considered a result of the act or a means to obtain results. Looked upon in this manner, analysis always involves modeling steps followed by the interpretation of the resulting models and the consequences they imply towards design.

One of the hardest parts of the modeling process is the creation of a starting point. We need to construct an initial specification or implementation model. This requires a formalized description of ideas or observations. There is often nothing to guide us but our past experience and our capability to detect similarities between different systems and problems. Here, again, a hierarchical classification of systems and their behavior can be of great help (see chapter 1).

Once an initial model is created, we can proceed by transforming one model into another. A circuit netlist, for example, can be transformed into a small-signal model. Symbolic modeling techniques [Fern98, Gie91, Gie94, Lin91, Wamb98b] in turn transform this small-signal model into a set of transfer functions. The oscillator modeling methods presented in chapter 5 proceed by gradually transforming a set of circuit equations into a more compact description. Even techniques for numerical simulation can be considered as model transformations. Here, the key observation is to recognize that simulation methods produce models that capture input-output behavior by means of (*input signal,output signal*) tuples. Hence, the models that result from numerical simulation can be described as

$$M_{sim} = \{(\mathbf{x}_1(t), \mathbf{y}_1(t)); (\mathbf{x}_2(t), \mathbf{y}_2(t)); \ldots\} \tag{2.1}$$

were the $\mathbf{x}_k(t)$ represent vectors of input signals and the $\mathbf{y}_k(t)$ vectors of corresponding output signals. The example at the end of this section illustrates this model transformation process for a single-stage amplifier.

At each stage in the process of modeling and analysis, we often have numerous possible transformations to proceed with. Which one to choose? Choice of a proper model transformation should be driven by the target we have in mind. Which part of the system's behavior is of greatest interest to us? Which kind of input signals are we dealing with? For example, if we want the resulting model to be parameterized, e.g. to be of use for trade-off analysis, the transformations should —at least partially— be symbolic in nature. If we are interested in the response to a limited set of input signals, numerical simulation might be the way to go. Note that in almost all cases, we want the resulting model to be as simple as possible.

Example (Analysis of a single-stage amplifier): Fig. 2.4 illustrates modeling and analysis for a single-stage amplifier. The starting point is a parameterized netlist. New models are derived by transforming this netlist model. A typical transformation, for instance, involves dumping the modified nodal equations. These equations in turn serve as a starting point for further transformations.

The transformations that we select should be driven by the interest in mind. This interest corresponds to what we want to know about the amplifier and its behavior. When

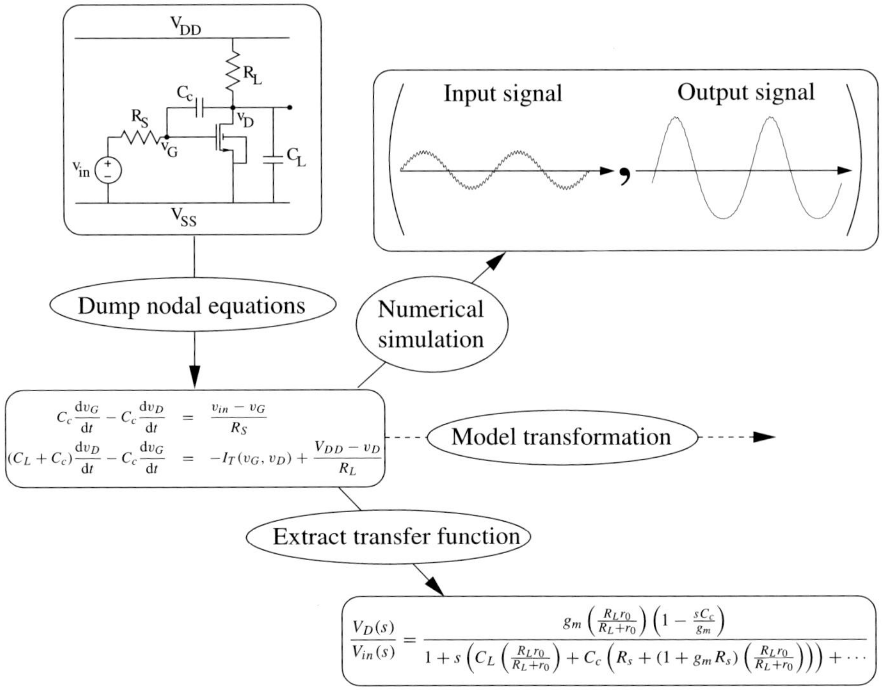

Figure 2.4: Analyzing the behavior of a single-stage amplifier involves transforming one model for the amplifier into another one. The idea is to select those transformations that yield the most simple results and/or that emphasize that part of the amplifier's behavior in which we are interested.

interested in the response to a particular input signal, a transformation involving numerical simulation is the way to go. When interested in the impact of the load capacitance C_L on the gain-bandwidth product, symbolic transformations present themselves as suitable candidates. Of course, when the symbolic expressions become too complicated, results are useless and we need to try a different transformation. In short, the art of system and circuit analysis comes down to selecting the proper sequence of model transformations. ▲

As a final note, we address automated modeling and analysis. Creating the initial model is often hard to automate. Coming up with a suitable circuit topology or an adequate elementary building block (transistor) model will always require some ingenuity. Transforming the models, however, can be automated if the transformations are formalized to the point that they can be written down as computer algorithms. This requires clear specification of the structure of the models that serve as an input. Furthermore, we must also state the conditions for the transformation to yield reliable results. Algorithms for symbolic analysis [Fern98], model-order reduction [Odab97] and numerical simulation [Kund97, Nag75] represent some well-known examples of automated model transformations.

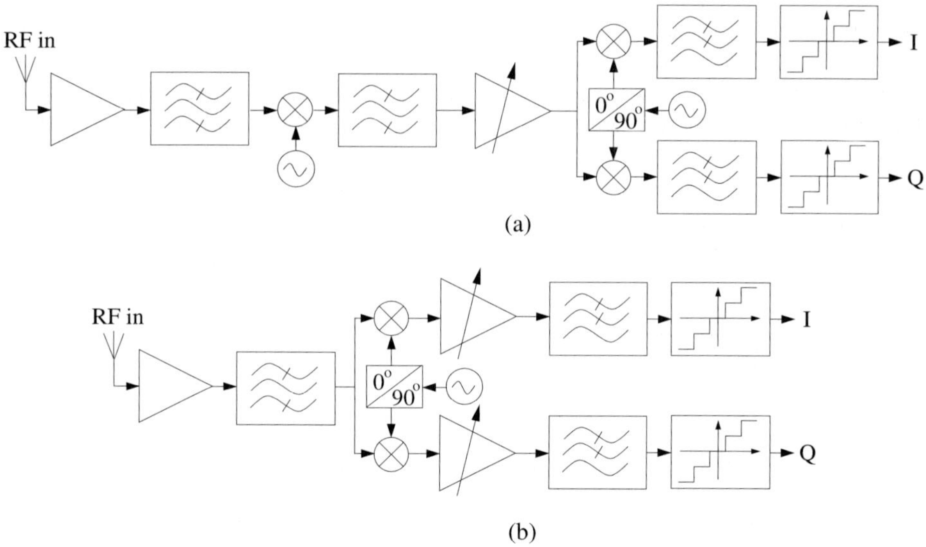

Figure 2.5: Frontend architectures of (a) a heterodyne receiver and (b) a low-IF receiver.

2.2 Good models for telecommunication frontends: Architectures and their behavioral properties

Having considered models, modeling and analysis from a global perspective, the main topic of this book can be phrased as: *the construction of good models for telecommunication systems and their building blocks.* As was outlined in section 2.1.2, the construction of good models requires us to incorporate all relevant prior knowledge and experience. This section summarizes this knowledge. It presents a brief review of common frontend architectures, their building blocks and their behavioral properties. It attempts to answer the question: what has experience taught us on the behavior of telecommunication frontends and their building blocks?

2.2.1 Frontend architectures and their building blocks

Fig. 2.5 depicts two commonly used receiver frontend architectures. The heterodyne receiver on top provides good performance in terms of channel selectivity and sensitivity, but typically requires surface-acoustic wave (SAW) bandpass filters in combination with additional IF circuitry. The low-IF architecture on the bottom requires less parts, but exhibits inherent problems such as self-mixing, $1/f$ noise and sensitivity. Self-mixing comes from the local oscillator (LO) signal making its way to the input of the mixer. This generates a DC component at the mixer output, possibly saturating the filters and gain amplifiers that follow. In [Crol97] a more rigorous overview of receiver (and transmitter) architectures is presented.

Observing both architectures, it is seen that their functioning relies on similar building blocks: low-noise amplifiers (LNAs), automatic gain control (AGC), filters, mixers, analog-to-digital converters (ADCs) and local oscillator (LO) signals. The latter are typically derived from a reference signal using a voltage-controlled oscillator (VCO) embedded in a phase-locked loop (PLL). Modeling an entire receiver frontend therefore requires us to have models for each of these building blocks. This book focuses on techniques that can be used to model the behavior of mixers, oscillators and PLLs.

Note that Fig. 2.5 only considers receiver architectures. Transmitters, however, have similar structures. In this case, the I- and Q-channels serve as inputs that are fed to digital-to-analog converters (DACs). The signals are then combined, upconverted to RF and transmitted through the antenna which is driven by a power amplifier (PA). Modeling DACs and PAs is, however, not a topic in this book[3].

2.2.2 Properties of frontend building block behavior

The construction of good frontend building block models requires us to incorporate all relevant prior knowledge. As will be illustrated in subsequent chapters, exploiting this knowledge helps us to improve model quality. This, for instance, makes simulations run faster. In what follows, we give a brief overview of some properties in common to many building blocks that occur in telecommunication frontend architectures: their almost linear nature, the presence of widely spaced time constants and the presence of stochastic (noisy) components in their behavior. Note that subsequent chapters also exploit other properties. However, these properties are often tied to a single building block. It is therefore not relevant to discuss them here.

2.2.2.1 Almost linear

Most building blocks in telecommunication frontends are, by design, intended to behave linearly. Here, *linear* should be interpreted in its most general, time-varying setting. As will be discussed in depth in chapter 3, up- and downconversions, induced by a multiplication $y(t) = u(t) \times \cos(\omega_0 t)$ of an input signal $u(t)$ with some periodic carrier signal $\cos(\omega_0 t)$, are linear operations: the principle of superposition still holds. This stands contrary to popular belief that considers multiplying two non-constant signals as a nonlinear operation. The key issue to observe is that one of the signals in the product —the carrier signal— is *known at the time when the mixer is designed*: It does not depend on the the information (the data bits) that is transmitted or received. In chapter 3 and 4, it will be shown how linearity can be exploited to significantly facilitate analysis of up- and downconversion behavior.

However, in almost any real system, it also occurs that two information-carrying (data) signals $u_1(t)$ and $u_2(t)$ intermodulate. This happens when the system's output depends, for instance, on the product $u_1(t) \times u_2(t)$ of the two data signals. The intermodulation of two data signals is truly nonlinear: the principle of superposition no longer holds. Fig. 2.6 illustrates the difference between the linear intermodulation of a data signal

[3] With regard to PAs, it is possible to handle them using the HTM-formalism in chapter 3 as well as the separation of time constants methods in chapter 5.

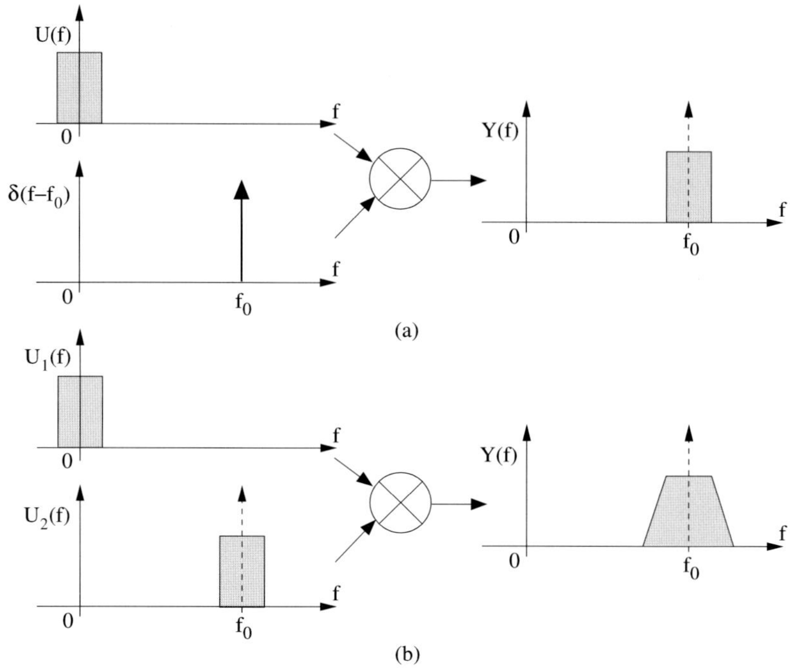

Figure 2.6: Difference between linear and nonlinear intermodulation. (a) Linear intermodulation involves the product of a (a priori unknown) data signal $u(t)$ and a (known) carrier. As a result, the information contained in $u(t)$ is shifted along the frequency axis. (b) Nonlinear intermodulation involves the product of two data signals $u_1(t)$ and $u_2(t)$. As a result, $u_1(t)$ is upconverted and, on top of that, its content is spread out over the frequency axis. Typically, this spectral spreading is not desired.

and a known carrier and the nonlinear intermodulation of two data signals.

Nonlinear intermodulation is in most cases undesired. Too large a contribution to the overall signal content may very well ruin the system's performance. In order to prevent this from happening, a good design should keep the signal components due to nonlinear intermodulation as small as possible. This means that their magnitude is well below that of the signal components generated by linear system behavior. Good design therefore requires most telecom building blocks to behave *almost linearly*. In traditional literature [Wamb98a], it is more common to say that the system behaves in a *weakly nonlinear* manner.

2.2.2.2 Widely spaced time constants

Many wireless applications transmit their information by upconverting it to the GHz frequency range or beyond. For example, GSM operates in the 900 MHz frequency

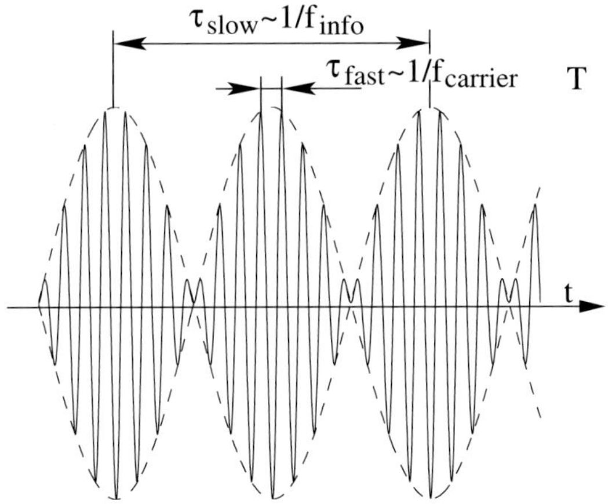

Figure 2.7: Many wireless applications transmit information by upconverting it to the GHz range. This results in signals containing widely spaced time constants. The carrier introduce a fast time constant τ_{fast} while the upconverted information induces a slow time constant τ_{slow}.

band, DCS1800 in the 1.8 GHz band while wireless LAN (WLAN) applications operate near 2.4 GHz (ISM band) or 5 GHz (military applications). Signal bandwidths, however, are often orders of magnitude smaller. GSM bands are, for instance, 200 kHz wide while WLAN bands occupy a few MHz. Hence, the rate at which information is transmitted is well below the carrier frequency, or

$$f_{info} \ll f_{carrier} \quad . \tag{2.2}$$

The resulting signals, illustrated in fig. 2.7, can be described as rapid oscillations with a slowly modulated amplitude and phase. This kind of behavior can also be generated by a building block's internal transient dynamics. For example, both high-Q harmonic oscillators and PLLs settle at a rate that is much slower than that of their steady-state output oscillation.

Traditional SPICE-like simulators [Nag75] experience a great deal of trouble in evaluating this kind of behavior. As illustrated in Fig. 2.8, the time step τ_{sim} with which simulations proceed is inversely proportional to the highest signal frequency, or

$$\tau_{sim} \sim \frac{1}{f_{carrier}} \quad . \tag{2.3}$$

The total time interval T_{sim} over which simulations are performed is typically a multiple of the length of a single information symbol, or

$$T_{sim} \sim \frac{N}{f_{info}} \quad . \tag{2.4}$$

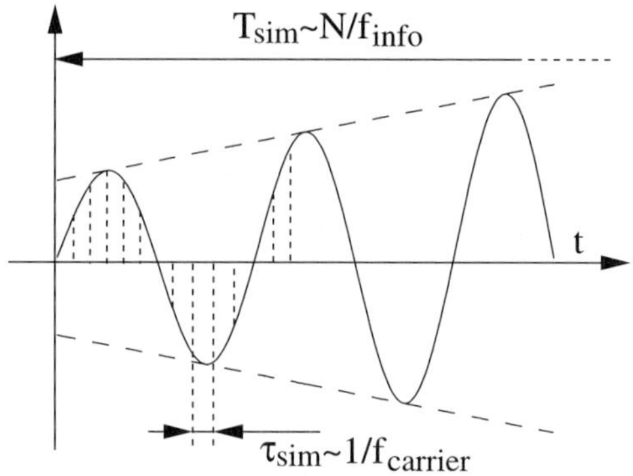

Figure 2.8: The time step τ_{sim} with which Spice-like simulations proceed is inversely proportional to the highest signal frequency. This time step is much smaller than the length T_{sim} of the entire simulation interval. As a consequence, the presence of widely spaced time constants forces SPICE-like simulation algorithms to take an enormous number of simulation steps.

Hence, to cover the entire simulation interval, a SPICE-like simulator needs to take

$$S = \frac{T_{sim}}{\tau_{sim}} \sim N \frac{f_{carrier}}{f_{info}} \tag{2.5}$$

steps. With $f_{carrier}/f_{info}$ often being over 1000 and with, for bit error rate simulations, N being a million in order of magnitude, SPICE-like simulators are forced to take over a billion steps to complete a single simulation run. As a result, the simulation of a complete system architectures may take days or even weeks to complete.

Both chapters 3 and 5 present methods that cope with this problem by means of models that keep slow- and fast-varying behavior explicitly separated. In this way, the situation in Fig. 2.8 is avoided. Wide separation of carrier and settling time constants is even a necessary condition for the oscillator modeling methods in chapter 5 to work properly. It is an excellent example on how building block characteristics that at first seem to hamper efficient analysis, can be exploited to speed up simulation.

2.2.2.3 Stochastic behavior

All physical systems, and, therefore, all telecom systems and their building blocks, have in common that they produce noise. For example, both thermal and $1/f$ noise occur in a wide variety of applications. Noisy signals, also called *stochastic processes*, are disturbances about which little information is available. Hence, they cannot be compensated for and, as a consequence, they cause data transmission errors. As it is the case for signal components due to nonlinear intermodulation, it is important to keep their magnitude small.

Contrary to deterministic signals, the shape of a stochastic process $n(t)$ is never known exactly. There is only a certain probability that a particular shape will occur. Stochastic processes are therefore characterized by probability density functions that quantify the likelihood that the process assumes a particular shape. However, it is often more practical to characterize signal stochastics by means of its moments or, equivalently, its cumulants [Mid87]. If we are mainly interested in a stochastic signal's energy content, then it is sufficient to know all moments up to second order, i.e. the signal's expected value $\mu(t) = E\{n(t)\}$ and autocorrelation $\Phi(t, \tau) = E\{n(t+\tau/2)n(t-\tau/2)\}$.

In order to cope with noise, any modeling framework for a particular class of building blocks must at least be able to characterize stochastic input-output behavior by means of the moments up to second order. Chapters 3, 4 and 5 outline how to perform noise computations for, respectively, LPTV systems and oscillators.

2.3 Conclusions

Models are descriptions of systems that we want to realize or that we have realized. In a top-down design flow they are used to communicate design decisions between the different levels of hierarchy. For example, a frontend system engineer will pass an LNA specification model to a circuit engineer. It describes the LNA as the system engineer would like it to behave. For the circuit engineer, it makes up the starting point for circuit-level synthesis, resulting in an implementation model, e.g. a Spice netlist. This model describes the LNA as it is realized. It is used by the system engineer for frontend-level verification.

In order for a top-down design flow to be successful, both specification and implementation models must be as accurate as possible in capturing the behavior of actual on-chip implementations. Defective and incomplete models are one of the main reasons for a redesign to be required. Of course, at any given stage of the design process, one only disposes of the knowledge gathered up to that point in time. This includes experience obtained from similar designs in the past. Models including all relevant prior knowledge and experience are called good models. The target of this work can therefore be phrased as "constructing good models for telecom frontends and their building blocks".

Past designs of telecommunication frontends show that their behavior and that of their building blocks all have some important properties in common:

- Their behavior is almost linear (weakly nonlinear). Here, linearity must be interpreted in its most general, time-varying setting.
- They produce signals containing widely spaced time constants. This especially holds for the RF sections of, for example, wireless applications.
- They all produce noise, i.e. they produce signals containing random elements.

Subsequent chapters in this book exploit these and other properties to construct compact models that allow accurate and efficient evaluation of telecom frontends and their building blocks. This assists us in their design and in the identification of performance bottlenecks.

Finally, models should only capture relevant behavior. Irrelevant details only burden analysis and simulation without deepening a designer's understanding of the system's behavior. In communications the relevance of building-block behavioral characteristics —and therefore the need to incorporate them into (good) models— is measured by their impact on the system's ability to transmit information without great loss. Unfortunately, it turns out that few building block behavior can be neglected. This especially holds when pushing the limits of performance. Hence, one of the great challenges for CAD is to capture complex building block behavior in a manner that is as compact as possible.

Chapter 3

A framework for frequency-domain analysis of linear periodically time-varying systems

You may always depend on it that algebra, which cannot be translated into good English and sound common sense, is bad algebra. —*William Clifford*

Telecom frontend building blocks like mixers and PLLs are known to involve up- and downconversion of signal content. This means that signal content is transferred from one periodic carrier wave (frequency band) to another. Often, these transfers are linear in nature: the principle of superposition still holds. Systems that exhibit this kind of behavior are termed *linear periodically time-varying* (LPTV). Handling LPTV system behavior requires us to extend traditional techniques used for linear time-invariant (LTI) system analysis. A single scalar transfer function is unable to capture signal transfers from one frequency band to another. This chapter elaborates a framework for frequency-domain LPTV system analysis based on harmonic transfer matrices. Applications of the framework are presented in the next chapter.

A HTM-based representation considers the signal content in each frequency band as separate and independent inputs. In this way, a single-input single-output (SISO) LPTV system can be modeled as a multi-input multi-output (MIMO) LTI system. Hereby, the contents of the different frequency bands are treated as different inputs. Hence, LPTV behavior can be captured using a matrix of transfer functions, called a *harmonic transfer matrix* (HTM). Each HTM element models the transfer of signal content from a particular input frequency band to a particular output frequency band. The basic concepts underlying the HTM framework are outlined in section 3.1. The mathematical details are discussed in section 3.3.

HTMs allow us to adapt many techniques for frequency-domain analysis of LTI systems to handle LPTV systems as well. With regard to this matter, section 3.4 discusses the construction of HTMs for elementary LPTV systems. It also outlines how to compute HTMs of composed systems. Finally, section 3.5 treats stability and noise analysis of LPTV systems.

A major problem in performing HTM-based computations lies in a HTM's infinite dimensional matrix structure. Prior art (see section 3.2) has dealt with this problem by truncating the HTM to some large but finite size. However, when this size grows large, matrix computations become very tedious. This especially holds for HTM inversions —needed for feedback computations— and eigenvalue computations —needed for stability analysis. Symbolic computations become intractable. Sections 3.4 and 3.5

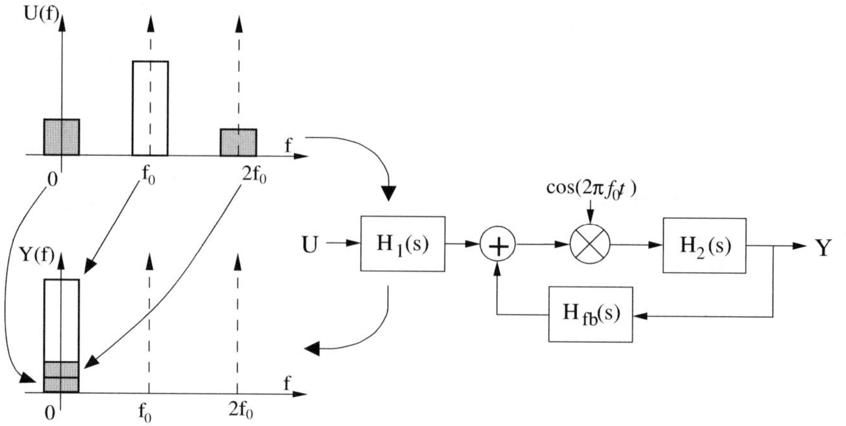

Figure 3.1: The characterization the behavior of a downconversion stage with (parasitic) feedback requires understanding and quantifying all signal transfers that occur, both wanted and unwanted. The wanted signal components are indicated in white, the unwanted ones are in gray. Only the signal transfers to the output frequency band of interest (at DC) are drawn.

therefore elaborate approximation strategies that render HTM-based computations efficient. As will be illustrated in this and the next chapter, these strategies allow us to obtain both numerical results and symbolic expressions.

3.1 The story behind the math

Telecommunication frontend building blocks like mixers and PLLs involve up- and downconversion of signal content. A GSM transmitter frontend, for example, is required to shift the input signal content, located at baseband, to the 900 MHz band. This is accomplished using one or more mixing stages. Such a mixing stage is a typical example of an LPTV system. In what follows, we present the basic ideas behind the HTM framework that is introduced as a means to model and analyze all aspects of LPTV system behavior. Firstly, however, we consider a simple example that helps us to identify the kind of information a designer needs when dealing with this type of system behavior.

3.1.1 What's of interest: A designer's point of view

Let us consider the (LPTV) downconversion stage illustrated on the right-hand side of Fig. 3.1. It consists of a bandpass pre-filtering stage, a mixer responsible for the downconversion and a lowpass filter that suppresses high-frequent residues. Furthermore, we've added a (parasitic) feedback from the output to the mixer's input. Such parasitic couplings are, for instance, due to transistor gate-source or gate-drain capacitances and are typically highpass in nature. This downconversion stage is fed with an incoming signal $U(f)$ that contains a wanted component near the carrier frequency and unwanted components near DC and twice the carrier frequency. These unwanted signal components are, for example, caused by nonlinearities in previous stages.

As depicted in the left-hand side of Fig. 3.1, the downconversion stage is designed with the intent to transfer the input information content around the carrier frequency to baseband. However, due to the parasitic feedback, unwanted input signal content at DC and twice the carrier frequency also manages to make its way to the output frequency band at DC. The latter transfers are highly unwanted and must be suppressed beneath a certain level. So, in judging performance, a designer is interested to know both whether the downconversion stage properly passes the wanted signal component at the carrier frequency to baseband and whether it sufficiently suppresses unwanted signal transfers.

Any framework that aims to assist designers in their task to develop the type of systems discussed above, should therefore be able to quantify all signal transfers that occur. Furthermore, all methods preferably operate in the frequency domain: this most closely corresponds to the way designers reason on mixers and PLLs.

3.1.2 Using harmonic transfer matrices to characterize LPTV behavior

Among all representations of LPTV behavior, harmonic transfer matrices [Maas88, Moll00, Vana02c, Were91a] are the only ones that directly meet the requirements stated above. They operate in the frequency domain and directly describe the signal transfers that occur between the different frequency bands.

HTMs build on the observation that, as signal content can get transfered from one frequency band to another, a single scalar spectrum $U(s)$ is no longer efficient to describe the content of a signal $u(t)$. As illustrated in Fig. 3.2, it is natural to subdivide a signal's overall spectrum into components contained within evenly spaced frequency bands centered around multiples of f_0. Here, f_0 represents the rate of the oscillator signal that drives the LPTV system. This suggests to represent a signal $u(t)$ by means of a vector of spectra

$$\mathbf{U}(s) = \begin{bmatrix} \vdots \\ U_0(s) \\ U_1(s) \\ U_2(s) \\ \vdots \end{bmatrix} \tag{3.1}$$

instead of a single scalar spectrum. In (3.1), $U_m(s)$ models the signal content of $u(t)$ in the frequency band centered around mf_0. Although not strictly necessary, the spectra $U_m(s)$ are typically chosen as bandlimited to the frequency band $[-f_0/2, f_0/2]$.

A HTM-based description of LPTV behavior relies on the property that, given the vectorized signal model (3.1), input-output behavior can be captured by means of the

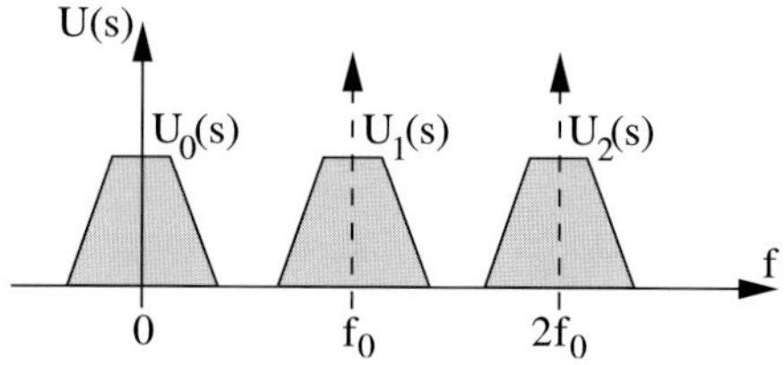

Figure 3.2: As information can get transfered from one frequency band to another, it is natural to subdivide a signal's overall spectrum into components contained within evenly spaced frequency bands. These frequency bands are centered around multiples of f_0.

simple matrix-vector multiplication

$$\begin{bmatrix} \vdots \\ Y_0(s) \\ Y_1(s) \\ Y_2(s) \\ \vdots \end{bmatrix} = \begin{bmatrix} & \vdots & \vdots & \vdots & \\ \cdots & \tilde{H}_{0,0}(s) & \tilde{H}_{0,1}(s) & \tilde{H}_{0,2}(s) & \cdots \\ \cdots & \tilde{H}_{1,0}(s) & \tilde{H}_{1,1}(s) & \tilde{H}_{1,2}(s) & \cdots \\ \cdots & \tilde{H}_{2,0}(s) & \tilde{H}_{2,1}(s) & \tilde{H}_{2,2}(s) & \cdots \\ & \vdots & \vdots & \vdots & \end{bmatrix} \begin{bmatrix} \vdots \\ U_0(s) \\ U_1(s) \\ U_2(s) \\ \vdots \end{bmatrix} . \tag{3.2}$$

The matrix is called a *harmonic transfer matrix* (HTM). The elements of a HTM can be any scalar transfer function. If we consider a single row of (3.2), for example

$$Y_1(s) = \sum_{m=-\infty}^{+\infty} \tilde{H}_{1,m}(s) U_m(s) \ , \tag{3.3}$$

then it is noted that the spectrum $Y_1(s)$ is a linear combination of all input frequency band spectra $U_m(s)$. The matrix elements $\tilde{H}_{1,m}(s)$ represent the corresponding input-output weight factors. Hence, $\tilde{H}_{1,m}(s)$ characterizes the transfer of the signal content from the input signal frequency band around mf_0 to the output signal band at carrier frequency. This process is illustrated in Fig. 3.3. A HTM-based description of LPTV behavior therefore directly captures the signal transfers that take place between the different frequency bands. It is therefore very useful to help us understand and quantify the behavior of LPTV systems like mixers and PLLs.

A different point of view on (3.2) considers each $U_m(s)$ as a separate input variable to the system and each $Y_n(s)$ as a separate output variable. The relation (3.2) can then be interpreted as the description of a multi-input multi-output LTI system. Hence, use of HTMs allows us to consider a single-input single-output LPTV system as a multivariable LTI system. Hereby, the signal content in each frequency band is to be treated as a separate variable.

3.1.3 LPTV behavior and circuit small-signal analysis

Up to now, it was taken for granted that linear periodically time-varying systems, and the HTMs associated with them, are relevant to circuit design practice. But, do there

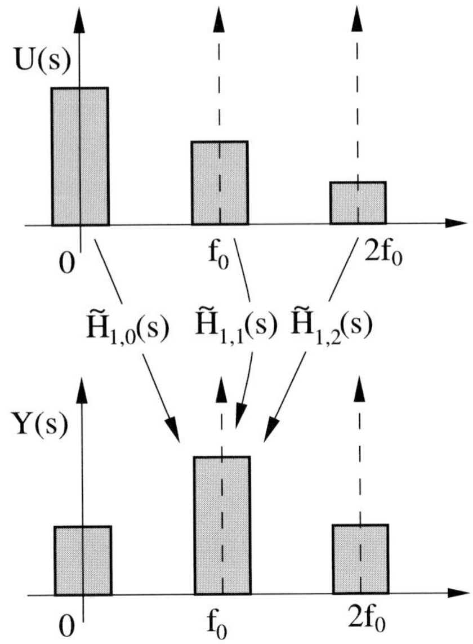

Figure 3.3: The elements of the harmonic transfer matrix directly characterize the signal transfers from the different input frequency bands to the different output frequency bands.

exist circuits exhibiting this type of behavior? If so, how are their HTMs related to the circuit equations?

It turns out that HTMs are related to mixer and PLL circuits in much the same way as traditional (scalar) transfer functions are related to opamps and filters. More precisely, they arise by linearizing a set of circuit equations near a periodically time-varying operating point.

Example (Linearized mixer behavior): Fig. 3.4 shows an elementary mixing stage. As inputs, we apply

$$V_{osc}(t) = \cos(\omega_0 t) \tag{3.4}$$

$$u(t) = u_0 + \Delta u(t) \ , \tag{3.5}$$

respectively a (large) oscillator signal and an input signal $u(t)$ consisting of a DC bias u_0 and a small-signal variation $\Delta u(t)$.

If we use a first-order transistor model and if we assume the transistor to operate in the linear region, then we find for the current $I(t)$ flowing through the transistor

$$I(t) = \beta(u(t) - V_{th})V_{osc}(t) - \frac{\beta}{2}V_{osc}(t)^2 \tag{3.6}$$

$$= \beta(u_0 + \Delta u(t) - V_{th})\cos(\omega_0 t) - \frac{\beta}{2}\cos(\omega_0 t)^2 \ . \tag{3.7}$$

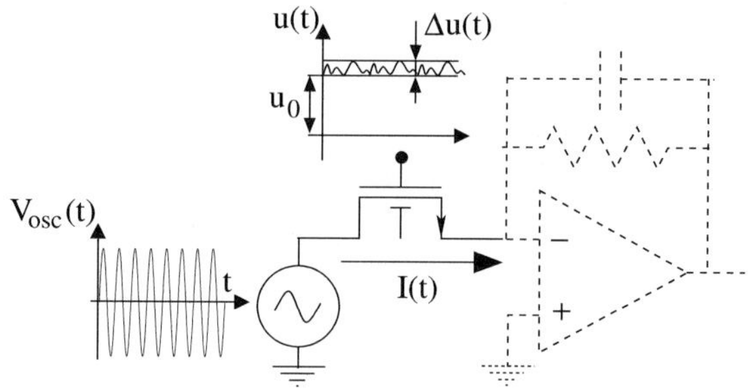

Figure 3.4: Elementary mixing stage to which we apply a large oscillator signal $V_{osc}(t)$ and an input signal $u(t)$. The latter consists of a DC bias u_0 and a small-signal variation $\Delta u(t)$.

Since $\Delta u(t)$ is small, we can linearize (3.7) around $\Delta u(t) = 0$. In this way, we obtain

$$I(t) = I_{op}(t) + \Delta I(t) \tag{3.8}$$

with

$$I_{op}(t) = \beta(u_0 - V_{th})\cos(\omega_0 t) - \frac{\beta}{2}\cos(\omega_0 t)^2 \tag{3.9}$$
$$\Delta I(t) = (\beta\cos(\omega_0 t))\,\Delta u(t) \; . \tag{3.10}$$

Here, $I_{op}(t)$ is the operating point current. This is the current that flows through the transistor when all small-signal input components are set to 0. Note that it varies periodically with time. $\Delta I(t)$ is the small-signal current that comes on top of $I_{op}(t)$. It is due to a non-zero small-signal input $\Delta u(t)$. This small-signal current is proportional to $\Delta u(t)$ with a proportionality constant that varies periodically with time. For this reason, (3.10) is said to model the LPTV system that describes the small-signal current flowing through the mixing transistor.

As will be proven in section 3.4.1, the HTM that corresponds to the multiplication in (3.10) equals

$$\tilde{\mathbf{H}}_{mixer}(s) = \begin{bmatrix} \ddots & \ddots & & & 0 \\ \ddots & 0 & \beta/2 & & \\ & \beta/2 & 0 & \beta/2 & \\ & & \beta/2 & 0 & \ddots \\ 0 & & & \ddots & \ddots \end{bmatrix} \; . \tag{3.11}$$

This (frequency-independent) HTM relates the small-signal transistor current variation ΔI to the small-signal gate voltage variation Δu. ▲

The example above illustrates the procedure used to extract a HTM from a set of circuit equations. This procedure, also depicted in Fig. 3.5, consists of the following steps:

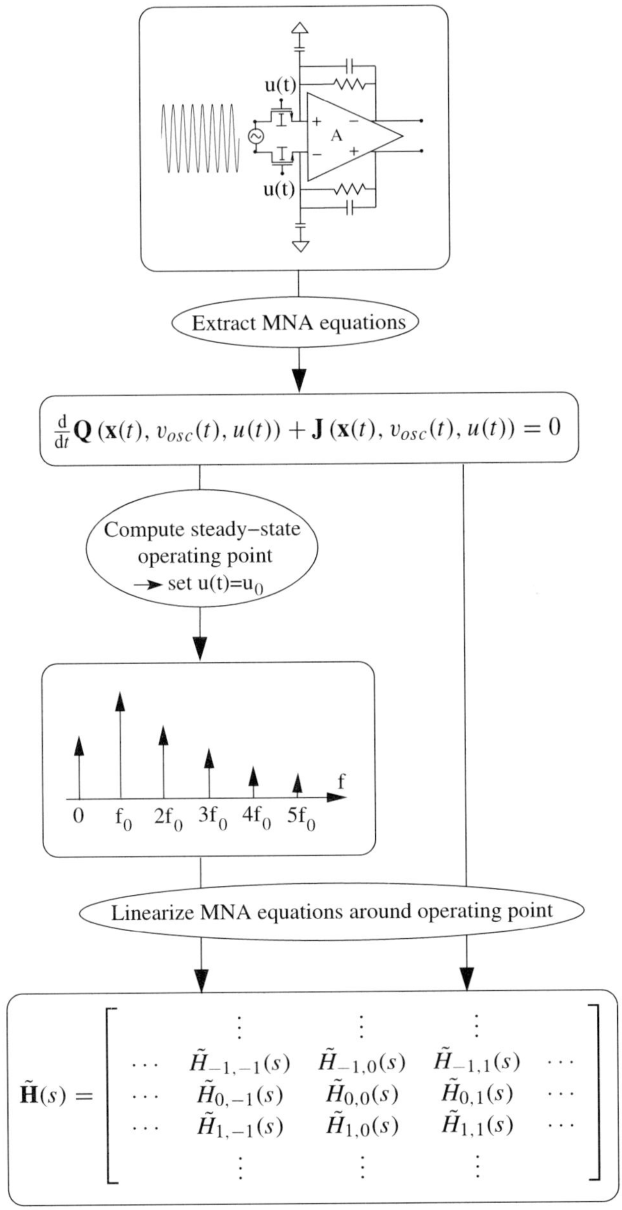

Figure 3.5: HTMs are derived from circuit behavior by linearizing the circuit equations in the neighborhood of a periodically time-varying operating point. The operating point is obtained by neglecting the small-signal component of the input signal $u(t)$. However, in computing the operating point, the large periodic oscillator signal that drives the mixer is not neglected. This results in periodic operating point waveforms with spectra that contain discrete tones at multiples of the oscillation frequency.

1. Construct the circuit's MNA equations. This can, for example, be accomplished using the algorithm described in [Freu95].

2. Compute the operating point voltages and currents. In doing so, input small-signal variations are to be neglected, i.e. they are set to 0. However, the periodic large-signal oscillations that drive the circuit are *not* neglected during operating point computations. Hence, operating point voltages and currents may vary periodically with time.

3. Linearize the circuit equations in the neighborhood of the periodic operating point. For stable systems, input small-signal variations cause the circuit's voltages and currents to deviate slightly from their operating point values. The smallness of these deviations justifies linearization. This yields a system of linear equations that relates the small-signal variations of the voltages and currents to the input small-signal variations. This system of equations describes an LPTV system.

4. Using the results described in section 3.4.4, the linearized circuit equations can be converted into HTMs.

The procedure above is very similar to the traditional small-signal analysis of an opamp. There, one computes a DC operating point by discarding all small input signals. Linearizing the opamp's behavior in the neighborhood of this operating point then yields a small-signal model. The behavior of this small-signal model can be captured by scalar transfer functions. As far as LPTV systems are concerned, e.g. mixers and PLLs, the main difference with opamp small-signal modeling lies in the fact that the operating point is periodically time-varying rather than constant. Hence, we need a HTM, i.e. a matrix of transfer functions, to describe their small-signal behavior rather than a scalar transfer function. In this sense, HTMs generalize LTI transfer functions to handle LPTV systems as well.

3.2 Prior art

LPTV system analysis dates back as far as the 19th century. In 1883, G. Floquet [Floq83] published his study on linear differential equations with periodically time-varying coefficients. Since then, his approach was further elaborated while alternative techniques have emerged. This section briefly introduces prior art on LPTV system analysis.

3.2.1 Floquet theory

Originally, Floquet developed his theory to study the solutions an n-th order single-input single-output differential equation

$$\frac{\mathrm{d}^n}{\mathrm{d}t^n}y(t) + b_{n-1}(t)\frac{\mathrm{d}^{n-1}}{\mathrm{d}t^{n-1}}y(t) + \ldots + b_0(t)y(t) = u(t) \ . \tag{3.12}$$

Here, all coefficients $b_k(t) \in \mathbb{R} \to \mathbb{R}$ are T-periodic in t. In a modern setting [Lam97], this is rephrased as studying the solutions of a system of n first-order differential equations

$$\frac{\mathrm{d}}{\mathrm{d}t}\mathbf{x}(t) = \mathbf{A}(t)\mathbf{x}(t) + \mathbf{b}(t)u(t) \tag{3.13}$$

$$y(t) = \mathbf{c}(t)^T\mathbf{x}(t) + d(t)u(t) \tag{3.14}$$

where $\mathbf{A}(t) \in \mathbb{R} \to \mathbb{R}^{n\times n}$, $\mathbf{b}(t), \mathbf{c}(t) \in \mathbb{R} \to \mathbb{R}^n$, and $d(t) \in \mathbb{R} \to \mathbb{R}$ are T-periodic matrix-valued functions. The formulation (3.13)-(3.14) is called a state-space representation of (3.12).

Floquet theory proves that it is possible to determine a T-periodic transformation of variables

$$\mathbf{z}(t) = \mathbf{P}(t)\mathbf{x}(t) \ , \tag{3.15}$$

with $\mathbf{P}(t) \in \mathbb{R} \to \mathbb{R}^{n\times n}$, that transforms (3.13)-(3.14) into a state-space representation with *constant* matrix $\mathbf{A}'$, or

$$\frac{\mathrm{d}}{\mathrm{d}t}\mathbf{z}(t) = \mathbf{A}'\mathbf{z}(t) + \mathbf{b}'(t)u(t) \tag{3.16}$$

$$y(t) = \mathbf{c}'(t)^T\mathbf{z}(t) + d(t)u(t) \ . \tag{3.17}$$

The dynamic behavior, e.g. settling times and stability, of the LPTV system is therefore completely determined by the eigenvalues of $\mathbf{A}'$. These eigenvalues are called the *Floquet exponents* of the LPTV system.

The Floquet exponents and the transformation matrix $\mathbf{P}(t)$ can be obtained from the fundamental matrix $\Phi(t,\tau) \in \mathbb{R} \times \mathbb{R} \to \mathbb{R}^{n\times n}$ satisfying

$$\frac{\mathrm{d}}{\mathrm{d}t}\Phi(t,\tau) = \mathbf{A}(t)\Phi(t,\tau), \ \ \Phi(t,t) = \mathbf{I} \ . \tag{3.18}$$

Here, $\mathbf{I} \in \mathbb{R}^{n\times n}$ is the unity matrix. Given $\Phi(t,\tau)$, we find

$$\mathbf{A}' = \ln(\Phi(T,0)) \tag{3.19}$$

$$\mathbf{P}(t)^{-1} = \Phi(t,0)\exp(-\mathbf{A}'t) \ . \tag{3.20}$$

The matrix $\Phi(T,0)$ is called the *monodromy matrix*. Note that $\Phi(t,0)$ can be found by integrating (3.18) over the interval $[0,T]$. $\mathbf{A}'$ and $\mathbf{P}(t)$ can also be computed using frequency-domain techniques [Dem03]. These techniques solve the time-varying eigenvalue problem

$$\left(\frac{\mathrm{d}}{\mathrm{d}t} + \mathbf{A}(t)\right)\mathbf{u}(t) = \lambda\mathbf{u}(t), \mathbf{u}(T) = \mathbf{u}(0) \ . \tag{3.21}$$

Numerically, the latter method is better conditioned than the direct computation of $\Phi(t,0)$.

3.2.2 Lifting

A second way to deal with LPTV behavior involves converting (3.13)-(3.14) to a time-invariant discrete-time system [Bam92]. This is called *lifting*. If we introduce the processes $u_k(t) = u(t+kT)$ for $t \in [0,T]$, then we find

$$\mathbf{x}_{k+1} = \hat{\mathbf{A}}\mathbf{x}_k + \int_0^T \Phi(t,\tau)\mathbf{b}(\tau)u_k(\tau)d\tau \; . \tag{3.22}$$

Here, $x_k = x(kT)$ and $\hat{\mathbf{A}} = \Phi(T,0)$ with $\Phi(t,\tau)$ satisfying (3.18). The properties of continuous-time LPTV systems can now be analyzed using time-invariant discrete-time techniques. For example, stability of the LPTV system (3.13)-(3.14) is equivalent to the stability of (3.22). This requires the magnitude of all eigenvalues of $\hat{\mathbf{A}}$ to be smaller than one. This in turn is equivalent with the Floquet exponents —the eigenvalues of $\ln(\Phi(T,0)) = \ln(\hat{\mathbf{A}})$— having negative real parts.

3.2.3 Frequency-domain approaches

Both methods discussed above are time-domain methods. They are well suited for constructing compact models that can be used for simulation (verification). However, they provide little direct insight into the signal transfers induced by an LPTV system. As was outlined in section 3.1.1, these signal transfers are most naturally described in the frequency domain as this closely corresponds with design practice.

Different authors in different fields of engineering have made the same observation. This all lead them to introduce a similar matrix-based frequency-domain description for LPTV systems. In [Maas88], analysis of nonlinear microwave mixers drove the idea the *harmonic conversion matrix* as a means to capture a mixer's small-signal behavior. This harmonic conversion matrix amounts to the harmonic transfer matrix evaluated at the imaginary axis. In order to deal with the ∞-dimensional nature of the matrix, the author proposed to truncate it to finite dimensions. However, the treatment in [Maas88] is somewhat incomplete. The structural properties characterizing the conversion matrix are not noticed. Furthermore, topics like noise and stability analysis are not addressed.

Almost at the same time and independently, a similar representation was developed in the field of control theory [Were91a, Were91b]. Here, developments were driven by LPTV stability analysis. The representation is called the *harmonic transfer function* or also the *harmonic transfer matrix* characterizing an LPTV system. It is derived starting from an integral representation for LPTV operators. Theory is developed in the Laplace domain with frequency-domain analysis being a limiting case as the Laplace variable approaches the imaginary axis. In this way, the Nyquist stability criterion for linear time-invariant systems can be extended to handle periodically time-varying ones as well. In [Moll00] this representation was adopted to analyze the stability of power supply networks. However, although the treatments in [Were91a, Moll00] overcome many of the shortcomings in [Maas88], dealing with the ∞-dimensional nature of the matrix is still inefficient. Also, little attention is given to the structural properties that characterize the matrix representation.

Finally, although not always mentioned explicitly, this matrix formulation also underlies many other approaches for frequency-domain analysis of LPTV system behavior [Phil98, Royc98, Royc99, Teli96]. As most of these methods target numerical analysis, they deal with the matrix inversions through truncation and Krylov subspace techniques.

3.2.4 Contributions of this work

This chapter adopts the HTM framework for Laplace-domain (frequency-domain) analysis of LPTV system behavior. It contributes to the theory on HTMs in the following ways:

- As far as theory is concerned, results previously scattered over different publications and application areas [Maas88, Moll00, Royc98, Were91a] are collected in a coherent framework. To the author's knowledge, this treatment is the first of its kind in elaborating the HTM framework in its full strength. Existing results are made consistent with each other. Additionally, the way in which results are derived and synthesized is original. The derivations emphasize clear-cut physical interpretations while avoiding too high a degree of mathematical abstraction. Moreover, they lend themselves to extend the HTM concept towards arbitrary classes of (non-periodic) LTV systems.
- Even more than prior treatments, emphasis is placed on making it practical to apply the HTM framework. We develop methods for approximate computations that render HTM-based analysis of many practical systems and circuits better feasible. It is shown how complex operations, like the inversion of ∞-dimensional HTMs needed to solve feedback loops, can be tackled efficiently by exploiting the properties of LPTV systems as they occur in design practice. These methods bridge the gap between HTMs as an abstract theoretical framework and HTMs as a powerful tool for both numerical and symbolic computations.

In the next chapter, the theory will be illustrated for different telecommunication-related examples like mixers and PLLs. The intent is to demonstrate the practical use of HTMs for analyzing telecommunication systems and their building blocks. Firstly, however, we discuss the theoretical foundations.

3.3 Laplace-domain modeling of LPTV systems using Harmonic Transfer Matrices

This chapter builds on HTMs to model and analyze different aspects of LPTV system behavior. Hence, in a first step, we spend some time to elaborate this concept in depth. The approach here presented in obtaining the HTM representation is to some extent similar to the ones in [Moll00, Were91a]. However, although starting from the same integral representation, it stays closer to the idea of an LPTV system as a system that transfers signal content from one frequency band to another.

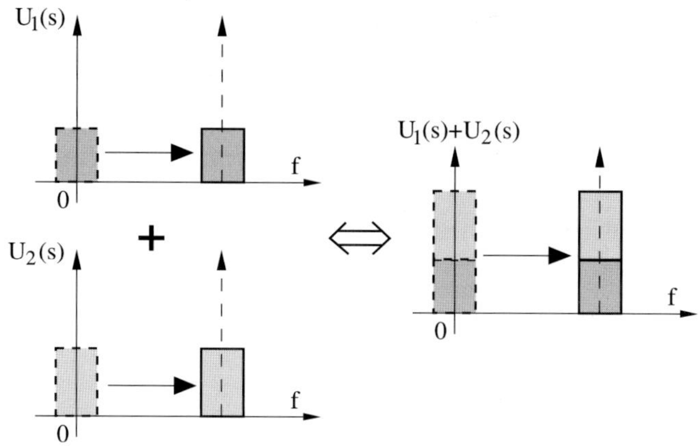

Figure 3.6: Frequency translations are linear operations as the principle of superposition holds. Taking the sum of two frequency-shifted signals is equivalent with frequency shifting the sum of the original signals.

Firstly, section 3.3.1 stresses the linear nature of the frequency translations induced by LPTV systems. This allows us to describe LPTV behavior using an integral representation. Next, section 3.3.2 formalizes the idea of "the signal content contained in a particular frequency band". In section 3.3.3 both results are combined, yielding a matrix-based representation for capturing LPTV system behavior. Section 3.3.4 discusses some of the structural properties that characterize this matrix representation. Finally, section 3.3.5 comments on the ∞-dimensional nature of the matrix while section 3.3.6 indicates how to extend the idea of matrix-based representations to arbitrary (non-periodic) LTV systems.

3.3.1 LPTV systems: implications of linearity and periodicity

LPTV systems are characterized by signal content that is translated along the frequency axis. As illustrated in Fig. 3.6, frequency translations are *linear* operations[1]. They satisfy the principle of superposition, the basic requirement for linearity. Taking the sum of two frequency-shifted signals is equivalent with frequency shifting the sum of

[1]This linearity is seldom stressed. Since frequency-translating circuits, like mixers, implement frequency translations by means of a multiplication of an input signal with a carrier signal, e.g. a cosine, they are often thought of as nonlinear. This is a misconception that complicates the way we reason on these type of circuits. The key issue to observe is that one of the signals involved in the multiplication, the carrier, is known *at the time when the circuit is implemented.* It can therefore be considered as part of the system and not as an external input. Hence, with regard to the other inputs, the multiplications that give rise to upconversions are linear time-varying operations and not nonlinear ones.

the original signals. Stated in mathematical terms, we find

$$\begin{aligned} L[\alpha u_1(t)+\beta u_2(t)] &= e^{j2\pi k f_0 t}(\alpha u_1(t)+\beta u_2(t)) \\ &= \alpha e^{j2\pi k f_0 t}u_1(t)+\beta e^{j2\pi k f_0 t}u_2(t) \\ &= \alpha L[u_1(t)]+\beta L[u_2(t)] \ . \end{aligned} \tag{3.23}$$

Furthermore, in LPTV systems, frequency translations are not arbitrary but occur over distances that are a multiple of a fundamental frequency f_0. The corresponding carrier waves $e^{j2\pi k f_0 t}$ in (3.23) are therefore all T-periodic with $T = 1/f_0$. This results in a type of behavior that is termed *periodically time-varying*. In this section, we discuss the implications of both linearity and periodic time variance on the mathematical representation of LPTV systems.

3.3.1.1 Linearity

With LPTV systems being linear, we can use linear time-varying (LTV) system theory [Zad61] to obtain a template that characterizes LPTV input-output behavior. Furthermore, we are only interested in *causal* systems, i.e. systems that have no knowledge about the future. The behavior of a causal linear time-varying system is captured by the following definition:

Definition (causal linear time-varying system): *A causal linear time-varying system $H[u(t)]$ computes the output at a time t as a linear combination of the input samples $u(t)$ that it has received until that time. The coefficients used for computing this linear combination may depend upon the time instance t at which the output is computed.*

The most natural way to translate this definition into mathematics goes as

$$y(t) = H[u(t)] = \int_0^\infty h(t,\tau)u(t-\tau)d\tau \ . \tag{3.24}$$

Here, $u(t)$ represents the input signal(s) and $y(t)$ the output signal(s)[2]. In (3.24) the integration indicates a summation over a continuous range of (past) input samples. The quantity $h(t,\tau)$ is called the *kernel* characterizing the linear system. This kernel can be interpreted as a (possibly time-varying) impulse response. Since causal LPTV systems are a subclass of causal LTV systems, linearity has taught us that their behavior can be captured using (3.24).

3.3.1.2 Periodic time variance

Not all kernels $h(t,\tau)$ represent periodically time-varying systems. To do so, the input-output behavior described by (3.24) must satisfy:

Definition (Periodically time-varying): *A T-periodic time-varying system is a system whose response depends in a periodic manner on time. More precisely, equal input signals applied at instances of time that are a period T apart result in equal output signals.*

---

[2]For notational convenience, we denote $u(t)$ and $y(t)$ as scalars; they can however also be considered as vectors in which case $h(t,\tau)$ is a matrix.

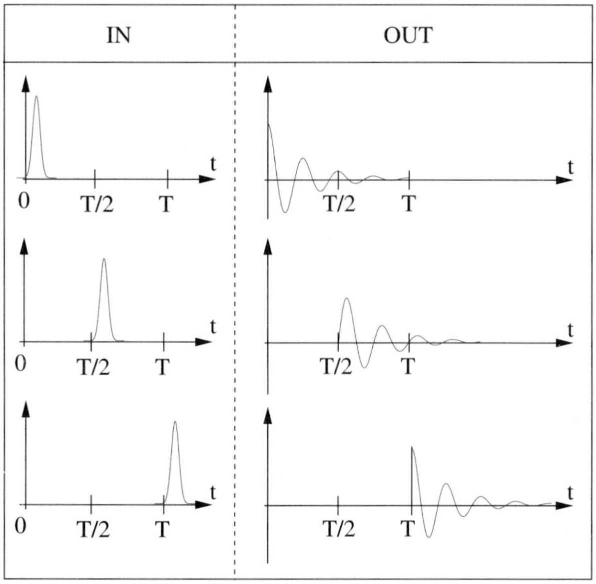

Figure 3.7: A T-periodic time-varying system produces similarly shaped output signals when a Dirac impulse is applied at time instances that are a period T apart (compare first and last input-output combination). However, when the Dirac impulses are applied at time instances that are not a period T apart, e.g. when they are half a period apart, output signals may have different shapes (compare first and second input-output combination).

Fig 3.7 illustrates this definition for Dirac impulses applied at several instances in time. As is observed, if an input signal $u(t)$ is applied at time T, the system produces the same output $y(t)$ as if $u(t)$ were applied at time 0. The only difference of course is that the output is delayed in time by the period T. Note that this invariance does not necessarily hold when the same $u(t)$ is applied at time instances that are not an integer multiple of T apart.

To understand the implications of the definition above on the properties of a periodically time-varying linear system with kernel $h(t,\tau)$, we consider a simple Dirac pulse being applied at a time instance t_0, i.e.

$$u(t) = \delta(t - t_0) \quad . \tag{3.25}$$

If the system is linear, it immediately follows from (3.24) that

$$y(t) = H[\delta(t - t_0)] = h(t, t - t_0) \quad . \tag{3.26}$$

The definition of periodic time variance in turn implies that

$$\begin{aligned} h(t-T,t-t_0-T) &= D_T\left[H[\delta(t-t_0)]\right] \\ &= H\left[D_T[\delta(t-t_0)]\right] \\ &= H[\delta(t-t_0-T)] \\ &= h(t,t-t_0-T) \ . \end{aligned} \quad (3.27)$$

Hereby, $D_T[\cdot]$ denotes the delay operator $D_T[u(t)] = u(t-T)$. Since t_0 is arbitrary, (3.27) implies

$$h(t-T,\tau) = h(t,\tau) \ , \quad (3.28)$$

i.e. the kernel $h(t,\tau)$ is T-periodic in t. Hence, by pursuing both the implications of the system's linearity and periodicity, we have obtained a first, integral, representation characterizing LPTV system behavior.

3.3.2 Linear periodically modulated signal models

The LPTV system representation derived above does not tell us much about their signal processing characteristics. More specifically, it does not clearly reveal how LPTV systems transfer signal content from one frequency band to another. However, in order to describe this kind of LPTV system behavior, we must first formalize the notion of "the signal content that is stored in a particular frequency band".

3.3.2.1 Storing signal content on carriers

Storing signal content in a particular frequency band relates to the more general question on how information is stored on waveforms. As outlined in chapter 2, the information contained in a set of signals $\{u_m(t) \in \mathbb{R} \to \mathbb{C}\}$ is stored by using them to modulate the properties of a set of carrier waveforms. The most straightforward way to do so is by means of a linear modulation scheme [Proa01]. Here, the overall signal $u(t)$ is composed as the sum of a set of weighted carriers $\psi_m(t) \in \mathbb{R} \to \mathbb{C}$. The information signals $u_m(t)$ are used as the weights, or

$$u(t) = \sum_m u_m(t)\psi_m(t) \ . \quad (3.29)$$

The principles of this linear modulation scheme are illustrated in Fig. 3.8. It can be used to transmit several information signals $u_m(t)$ simultaneously. Hereby, each carrier wave $\psi_m(t)$ can be considered as a separate transmission channel[3]. The signals $u_m(t)$ are called the *envelopes* or also the *equivalent baseband components* associated with the carriers $\psi_m(t)$.

The harmonic functions

$$\psi_m(t) = e^{jm\omega_0 t} \ , \quad (3.30)$$

with $\omega_0 = 2\pi f_0$, constitute a set of carrier waves that frequently occur in practice. For (3.30), the transmission channels correspond with evenly spaced frequency bands

[3] Of course, in order to avoid excessive interchannel interference, the $u_m(t)$ must be restricted to vary on a time scale that is much larger than the time scale over which the different carriers $\psi_m(t)$ are correlated.

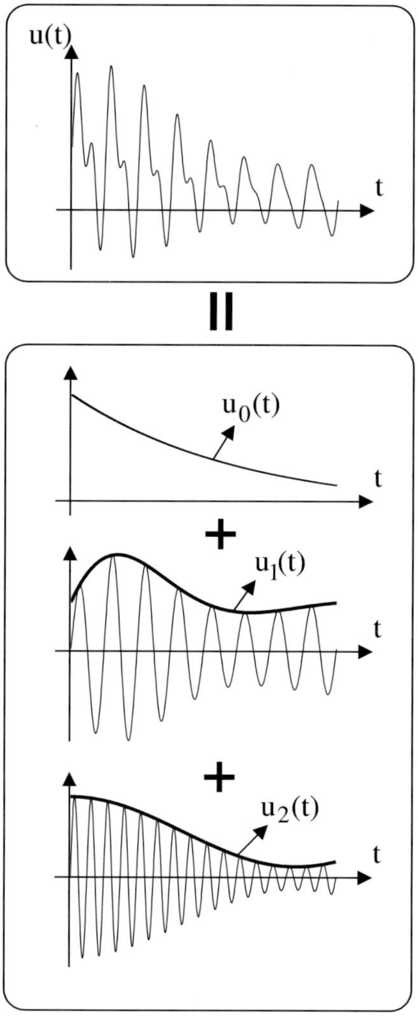

Figure 3.8: the information contained in a set of signals $\{u_m(t)\}$ (fat solid lines) is stored by using them to modulate the properties of a set of carrier waveforms $\{\psi_m(t)\}$ (fast-oscillating lines).

centered around multiples of f_0. Hence, given a set of envelopes $u_m(t)$, the overall waveform $u(t)$ becomes

$$u(t) = \sum_{m=-\infty}^{+\infty} u_m(t) e^{jm\omega_0 t} \ . \tag{3.31}$$

With $U_m(s) = \mathcal{L}\{u_m(t)\}$ the Laplace transform of the envelope $u_m(t)$, the Laplace transform of the overall signal then equals

$$U(s) = \mathcal{L}\{u(t)\} = \sum_{m=-\infty}^{+\infty} U_m(s - j\omega_0) \tag{3.32}$$

As was illustrated in Fig. 3.2, the component $u_m(t)$ is upconverted to the frequency band centered around mf_0. The notion of the signal content contained in a particular frequency band is therefore formalized as the envelope (equivalent baseband component) $u_m(t)$ used to modulate the carrier wave $\psi_m(t) = e^{jm\omega_0 t}$. Due to the periodic nature of the carrier waves, waveforms obtained using (3.31) are called *linear periodically modulated waveforms*. Also important is the fact that, for any signal $u(t)$, it is always possible to construct a set of envelopes such that (3.31) is satisfied. Such a set constitutes a *linear periodically modulated signal model.*

Example (Modeling a sinewave): Consider the signal

$$u(t) = \sin(\omega_0(t)) \ . \tag{3.33}$$

For this simple case, the most obvious linear periodically modulated signal model is specified as

$$\left\{u_{-1}(t) = -\frac{1}{2j},\, u_0(t) = 0,\, u_1(t) = +\frac{1}{2j}\right\} \ , \tag{3.34}$$

i.e. a set of constant envelopes. This model is illustrated on the left-hand side in Fig. 3.9.

A second linear periodically modulated signal model for $u(t)$ is depicted on the right-hand side in Fig. 3.9. Here, we select

$$\{u'_{-1}(t) = 0,\, u'_0(t) = \sin(t),\, u'_1(t) = 0\} \ . \tag{3.35}$$

Admittedly, this second solution is counterintuitive. Nevertheless, it is mathematically correct since (3.31) is satisfied. ▲

3.3.2.2 On the non-uniqueness of linear periodically modulated signal models

The example above illustrates that, for a given signal $u(t)$, the set of envelopes $\{u_m(t)\}$ for which (3.31) holds *is not unique*. There are always an infinite number of possibilities for which (3.31) is satisfied. However, if we look at Fig. 3.9, we see only one set of envelopes that intuitively seems "right". This solution is the one on the left with all envelopes slow-varying. When talking about modulated sinewaves, it seems natural

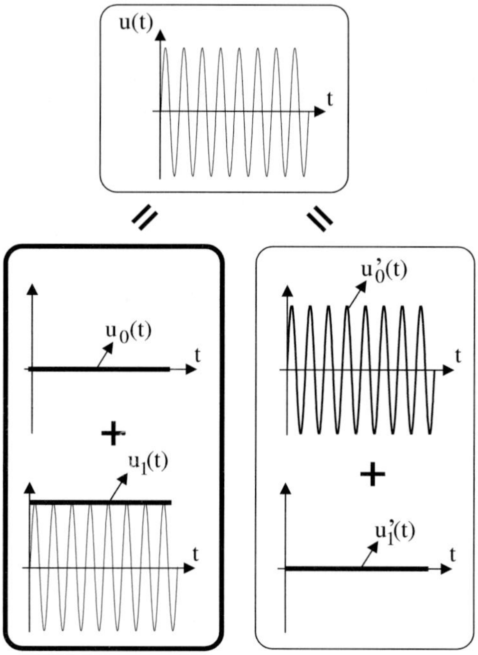

Figure 3.9: A sinewave $u(t) = sin(\omega_0 t)$ can be represented using several different linear periodically modulated signal models. The most obvious model is the one on the left encircled in bold. Here, all envelopes are slow-varying. Technically, however, the model on the right is also correct although it is somewhat counterintuitive.

to demand that an envelope $u_m(t)$ varies at a rate that is slower than that of the corresponding (sinusoidal) carrier. Modulating a 1 MHz carrier with an envelope that is oscillating at 1 GHz does not seem a sensible thing to do.

In the frequency-domain, slow-varying $u_m(t)$ correspond to bandlimited spectra $U_m(s)$. More specifically, one could require that

$$U_m(j\omega) = 0, \text{ for } |\omega| > \frac{\omega_0}{2} \ . \tag{3.36}$$

In words, the equivalent baseband spectra $U_m(j\omega)$ are strictly bandlimited to the frequency band $[-\omega_0/2, \omega_0/2]$. The advantages of requiring (3.36) to hold are twofold:

1. The condition (3.36) fixes all degrees of freedom in determining a linear periodically modulated signal model for a given signal $u(t)$. There is only one solution for the envelopes $\{u_m(t)\}$ that satisfies both (3.31) and (3.36). This is of great help in automating computations.
2. With (3.36) being satisfied, the spectra $U_m(j\omega)$ have a clear cut physical interpretation: they represent that part of the signal content of $u(t)$ located in the frequency band centered around $m\omega_0$. They do not contribute to the signal content in other frequency bands. It is this interpretation that has inspired the alternate name "equivalent baseband components".

It should, however, be stressed that (3.36) is a constraint that *can* be imposed. From a mathematical point of view, one is not required to do so. As presented, the HTM framework also allows us to deal with envelopes $u_m(t)$ that contain more rapid variations, i.e. the spectra $U_m(j\omega)$ may have content outside $[-\omega_0/2, \omega_0/2]$. But what would be the advantages in selecting these "weird" kind of envelopes? Again, there are two reasons:

1. For a single wideband input signal, it is often artificial to break it up into different components. For example, for the signal in Fig. 3.10, a decomposition of the overall spectrum $U(s)$ into the components $U_m(s)$ depicted on the left-hand side is counterintuitive. A single logical entity is split up in an artificial manner. Selecting the envelope spectra as on the right-hand side of Fig. 3.10, i.e. $U_0(s) = U(s)$ with the other $U_m(s)$ set to 0, seems more natural. Note that, since most of the signal energy is contained in $[-\omega_0/2, \omega_0/2]$, one can still stick to the interpretation of $U_0(s)$ as being (approximately) the signal content at baseband.
2. Sometimes, it helps to simplify computations if we choose $U_m(s)$ that do not strictly satisfy (3.36). This is especially true for symbolic computations. In such a case, the need to simplify the problem setup is more important than strict adherence to (3.36).

For these reasons, in what follows, we develop methods that can deal with arbitrary envelope spectra $U_m(s)$, i.e. *the envelope spectra can have non-zero signal content over the entire frequency axis*. However, in gaining understanding and giving interpretation to the theory that is developed, it is often helpful to think of the $U_m(s)$ as being bandlimited.

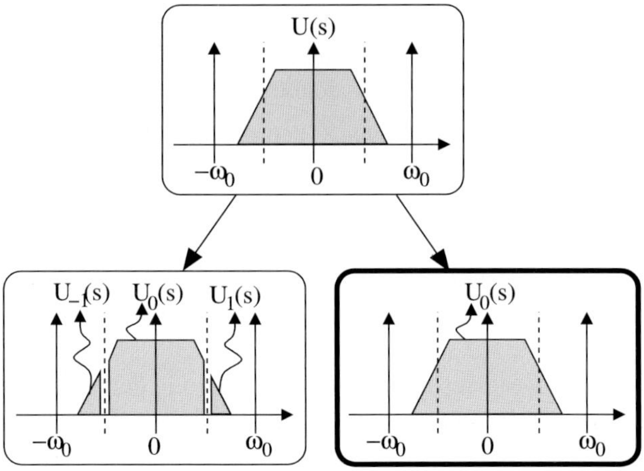

Figure 3.10: For a single wideband input signal, it is often artificial to break it up into different components. In this case, it is more convenient to choose $U_0(s)$ as on the right, i.e. in a way that does not strictly adhere to (3.36).

3.3.3 Harmonic transfer matrices: capturing transfer of signal content between carrier waves

Linear periodically modulated waveforms capture the idea of different signal contents that are stored in different frequency bands. LPTV systems are claimed to transfer signal content from one frequency band to another. Hence, t is natural to ask how LPTV systems act upon linear periodically modulated waveforms: what kind of output signals result and how can we compute them?

To answer this question, we substitute (3.31) into the input-output relation (3.24). This yields

$$\begin{aligned} y(t) &= \int_0^\infty h(t,\tau)u(t-\tau)d\tau \\ &= \sum_{m=-\infty}^{+\infty} \int_0^\infty h(t,\tau)e^{jm\omega_0(t-\tau)}u_m(t-\tau)d\tau \ . \end{aligned} \tag{3.37}$$

Since $\forall\tau$, $h(t,\tau)$ is T-periodic in t, with $T = 2\pi/\omega_0$, we can expand the kernel using a Fourier series, or

$$h(t,\tau) = \sum_{k=-\infty}^{+\infty} h_k(\tau)e^{jk\omega_0 t} \ . \tag{3.38}$$

Substituting (3.38) into (3.37), we obtain

$$\begin{aligned} y(t) &= \sum_{m=-\infty}^{+\infty} \sum_{k=-\infty}^{+\infty} e^{j(m+k)\omega_0 t} \int_0^\infty h_k(\tau) e^{-jm\omega_0\tau} u_m(t-\tau)d\tau \\ &= \sum_{n=-\infty}^{+\infty} e^{jn\omega_0 t} \left(\sum_{m=-\infty}^{+\infty} \int_0^\infty h_{n-m}(\tau) e^{-jm\omega_0\tau} u_m(t-\tau)d\tau \right) \\ &= \sum_{n=-\infty}^{+\infty} y_n(t) e^{jn\omega_0 t} \quad . \end{aligned} \tag{3.39}$$

In obtaining the second equality, we changed summation indices using $n = m+k$. The result shows that an LPTV system acting on a linear periodically modulated signal produces another linear periodically modulated signal with equivalent baseband components that equal

$$y_n(t) = \sum_{m=-\infty}^{+\infty} \int_0^\infty h_{n-m}(\tau) e^{-jm\omega_0\tau} u_m(t-\tau)d\tau \quad . \tag{3.40}$$

Each term in the right-hand side of (3.40) represents a linear time-invariant filtering operation. Hence, (3.40) is readily rewritten in the Laplace domain as

$$Y_n(s) = \sum_{m=-\infty}^{+\infty} H_{n-m}(s + jm\omega_0) U_m(s), \forall n \in \mathbb{Z} \tag{3.41}$$

where $U_m(s) = \mathcal{L}\{u_m(t)\}$, $Y_n(s) = \mathcal{L}\{y_n(t)\}$ and $H_k(s) = \mathcal{L}\{h_k(t)\}$.
By organizing the $U_m(s)$ in an ∞-dimensional column vector

$$\mathbf{U}(s) = \begin{bmatrix} \cdots & U_{-2}(s) & U_{-1}(s) & U_0(s) & U_1(s) & U_2(s) & \cdots \end{bmatrix}^T , \tag{3.42}$$

and using a similar definition for $\mathbf{Y}(s)$, (3.41) can be rewritten as a matrix-vector multiplication

$$\mathbf{Y}(s) = \tilde{\mathbf{H}}(s)\mathbf{U}(s) \quad . \tag{3.43}$$

Here, $\tilde{\mathbf{H}}(s)$ is a doubly ∞-dimensional matrix

$$\tilde{\mathbf{H}}(s) = \begin{bmatrix} & \vdots & \vdots & \vdots & \\ \cdots & \tilde{H}_{-1,-1}(s) & \tilde{H}_{-1,0}(s) & \tilde{H}_{-1,1}(s) & \cdots \\ \cdots & \tilde{H}_{0,-1}(s) & \tilde{H}_{0,0}(s) & \tilde{H}_{0,1}(s) & \cdots \\ \cdots & \tilde{H}_{1,-1}(s) & \tilde{H}_{1,0}(s) & \tilde{H}_{1,1}(s) & \cdots \\ & \vdots & \vdots & \vdots & \end{bmatrix} \tag{3.44}$$

with its elements equal to

$$\tilde{H}_{n,m}(s) = H_{n-m}(s + jm\omega_0) \quad . \tag{3.45}$$

The $H_k(s)$ represent the Laplace transforms of the signals $h_k(\tau)$ obtained from the Fourier series expansion (3.38) of the kernel $h(t,\tau)$.

At this point, we introduce some names and notational conventions that will be used throughout this and the next chapter. The matrix $\tilde{\mathbf{H}}(s)$ is called a *harmonic transfer matrix* (HTM) [Moll00, Were91a, Vana02c]. Its elements $\tilde{H}_{n,m}(s)$ are termed *harmonic transfer functions* (HTFs). The latter choice differs, for example, from [Royc98] where the name harmonic transfer function is reserved for the $H_k(s)$. However, the quantities $\tilde{H}_{n,m}(s)$ are the ones that directly model signal transfers from the m-th input to the n-th output frequency band. Hence, they are the ones having physical meaning and therefore deserve the name harmonic transfer functions. Furthermore, a tilde on top of a quantity denotes that that quantity is a structured matrix or an element of a structured matrix. In a structured matrix, the elements cannot be chosen independently of each other. As follows from (3.45), HTMs are structured with their elements related as $\tilde{H}_{n+k,m+k}(s) = \tilde{H}_{n,m}(s+jk\omega_0)$. We refer to the next section for a more detailed discussion on this structure. The elements of matrices that have no tilde on top of them can freely be chosen.

The physical meaning of HTMs and their elements follows directly from (3.41). As was outlined in section 3.1.2 and as illustrated in Fig. 3.3, the harmonic transfer function $\tilde{H}_{n,m}(s)$ directly captures the transfer of signal content from the input frequency band centered around mf_0 to the output frequency band centered around nf_0. Stated in a more abstract manner: $\tilde{H}_{n,m}(s)$ models the transfer of signal content from the m-th carrier wave $\psi_m(t) = e^{jm\omega_0 t}$ to the n-th carrier wave $\psi_n(t) = e^{jn\omega_0 t}$. In a lot of applications, this is the knowledge that is of greatest interest.

3.3.4 Structural properties of HTMs

As is observed from (3.45), the elements of a HTM $\tilde{\mathbf{H}}(s)$ can not be chosen independently. These structural properties of HTMs arise due to non-uniqueness of the signal model (3.31). This non-uniqueness manifests itself as

$$\begin{aligned} u(t) &= \sum_{m=-\infty}^{+\infty} u_m(t)e^{jm\omega_0 t} \\ &= \sum_{m=-\infty}^{+\infty} \left(u_{m-1}(t)e^{-j\omega_0 t}\right) e^{jm\omega_0 t} \\ &= \sum_{m=-\infty}^{+\infty} u'_m(t)e^{jm\omega_0(t)} \quad . \end{aligned} \tag{3.46}$$

As illustrated in Fig. 3.11, a tone located at a frequency $f_0+\Delta f$ in the spectrum of $u_{m-1}(t)$ can equally well be captured as a tone at Δf in the spectrum of $u'_m(t)$. This implies that the vectors

$$\mathbf{U}(s) = \begin{bmatrix} \vdots \\ U_0(s) \\ U_1(s) \\ U_2(s) \\ U_3(s) \\ \vdots \end{bmatrix} \Leftrightarrow \begin{bmatrix} \vdots \\ U_{-1}(s+j\omega_0) \\ U_0(s+j\omega_0) \\ U_1(s+j\omega_0) \\ U_2(s+j\omega_0) \\ \vdots \end{bmatrix} = \mathbf{P}\mathbf{U}(s+j\omega_0) = \mathbf{U}'(s) \tag{3.47}$$

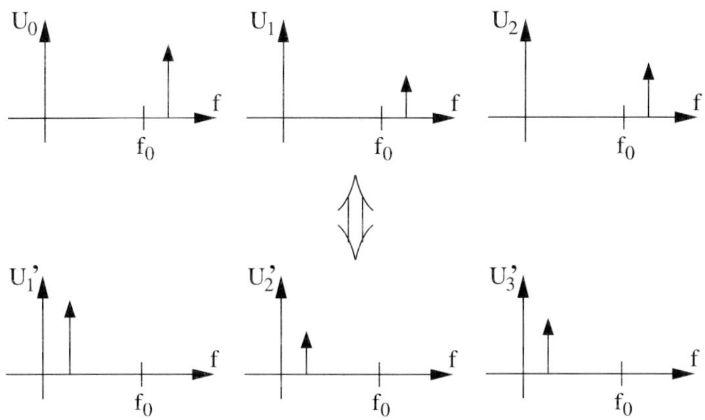

Figure 3.11: A linear periodically modulated signal model is non-unique. A tone located at a frequency $f_0 + \Delta f$ in the spectrum of $u_{m-1}(t)$ can equally well be captured as a tone at Δf in the spectrum of $u'_m(t)$.

represent the same signal $u(t)$. Here, $\mathbf{P}$ denotes the permutation operator shifting the elements of a matrix downwards, i.e. $x'_m = x_{m-1}$. It can be represented as a matrix with elements

$$P_{n,m} = \delta(m - n + 1) \tag{3.48}$$

where $\delta(k) \in \mathbb{Z} \to \{0, 1\}$ is the Kronecker delta function. It equals 1 for $k = 0$ and 0 otherwise.

If both vectors in (3.47) represent the same (input) signal, the same must hold for the (output) signal models obtained by multiplying $\mathbf{U}(s)$ and $\mathbf{U}'(s)$ with $\tilde{\mathbf{H}}(s)$, the HTM corresponding to some arbitrary LPTV operator. If one applies the same input signal to a given system, the same output signal will result, no matter the representation that is used for either signal. As a consequence, we obtain

$$\begin{aligned} \mathbf{P}\tilde{\mathbf{H}}(s + j\omega_0)\mathbf{U}(s + j\omega_0) &= \mathbf{P}\mathbf{Y}(s + j\omega_0) && (3.49) \\ &= \mathbf{Y}'(s) && (3.50) \\ &= \tilde{\mathbf{H}}(s)\mathbf{U}'(s) && (3.51) \\ &= \tilde{\mathbf{H}}(s)\mathbf{P}\mathbf{U}(s + j\omega_0) \ . && (3.52) \end{aligned}$$

Since $\mathbf{U}(s)$ is arbitrary, this implies that

$$\tilde{\mathbf{H}}(s + j\omega_0) = \mathbf{P}^{-1}\tilde{\mathbf{H}}(s)\mathbf{P} \ . \tag{3.53}$$

In terms of the HTM elements $\tilde{H}_{n,m}(s)$, this yields

$$\tilde{H}_{n,m}(s + j\omega_0) = \tilde{H}_{n+1,m+1}(s) \ . \tag{3.54}$$

Hence, all elements on a same HTM diagonal are frequency-shifted copies of each other. Although (3.54) can also be obtained from (3.45), the derivation above is a more fundamental one: it relates the structure of HTMs to the waveform's non-unique linear periodically modulated signal model. This point of view is much more fruitful in extending LPTV theory to other classes of LTV systems.

As a consequence of (3.53), it is sufficient to know $\tilde{\mathbf{H}}(s)$ for all $s = \sigma + j\omega$ with $\sigma \in \mathbb{R}$ and $\omega \in [-\omega_0/2, \omega_0/2]$. Furthermore, if $\tilde{\mathbf{H}}(s)$ is analytic in the right-hand complex plane, i.e. the LPTV system is stable and contains no right-hand plane poles, it is even sufficient to know $\tilde{\mathbf{H}}(j\omega)$ for $\omega \in [-\omega_0/2, \omega_0/2]$. Using Cauchy's integral theorem and analytic continuation [Arf85], it is possible to find $\tilde{\mathbf{H}}(s), \forall s \in \mathbb{C}$.

Another important consequence of the structural relation (3.53) is the repetitive structure of the poles and eigenvalues of $\tilde{\mathbf{H}}(s)$. As will be seen in section 3.5.2, this is important for LPTV stability analysis. If $p \in \mathbb{C}$ is a pole of $\tilde{\mathbf{H}}(s)$, i.e.

$$\lim_{s \to p} \left\| \tilde{\mathbf{H}}(s) \right\| = \infty \tag{3.55}$$

with $\left\| \tilde{\mathbf{H}}(s) \right\|$ the HTM's two-norm evaluated at $s \in \mathbb{C}$, then (3.53) immediately implies that $p + j\omega_0$ is also a pole of $\tilde{\mathbf{H}}(s)$. Furthermore, given a complex number λ and a vector $\mathbf{U}$ such that for a given s

$$\tilde{\mathbf{H}}(s)\mathbf{U} = \lambda \mathbf{U} \ , \tag{3.56}$$

we find, again using (3.53), that

$$\begin{aligned} \tilde{\mathbf{H}}(s + j\omega_0)\left(\mathbf{P}^{-1}\mathbf{U}\right) &= \mathbf{P}^{-1}\tilde{\mathbf{H}}(s)\mathbf{U} \\ &= \lambda\left(\mathbf{P}^{-1}\mathbf{U}\right) \ . \end{aligned} \tag{3.57}$$

This means that if λ is an eigenvalue of $\tilde{\mathbf{H}}(s)$, it is also an eigenvalue of $\tilde{\mathbf{H}}(s + j\omega_0)$. Hence, poles repeat themselves in the direction of the imaginary axis while, in that same direction, $\tilde{\mathbf{H}}(s)$ is characterized by the periodicity of its eigenvalues.

3.3.5 On the ∞-dimensional nature of HTMs

One of the main problems in dealing with HTMs is their ∞-dimensional matrix structure. As always, care must be taken in dealing with ∞ in a sense that a proper limiting procedure is required: how is infinity being approached? As different paths to ∞ may lead to different results, many paradoxes in mathematics can be traced down to careless specification of a proper limiting procedure. In this text, any ∞-dimensional HTM $\tilde{\mathbf{H}}(s)$ should be considered as the limit of a sequence of square K-dimensional submatrices $\tilde{\mathbf{H}}_K(s)$. Written formally:

$$\tilde{\mathbf{H}}(s) = \lim_{K \to \infty} \tilde{\mathbf{H}}_K(s) \ . \tag{3.58}$$

Moreover, $\tilde{\mathbf{H}}_K(s)$ is a submatrix of $\tilde{\mathbf{H}}_{K+1}(s)$. With mild conditions on the behavior of the LPTV operator captured by $\tilde{\mathbf{H}}(s)$, use of any such sequence $\left\{\tilde{\mathbf{H}}_K(s)\right\}$, in the limit for $K \to \infty$, yields the same results. The basic reason for this is that, mostly, the signal energy located beyond a particular finite frequency is of negligible importance. In what follows, we deal with HTMs as if they were finite-dimensional matrices. Unless when relevant, we do not explicitly refer to their ∞-dimensional structure as this would only complicate expositions.

3.3.6 Matrix-based descriptions for arbitrary LTV behavior

We conclude this section with some notes on the generalization of the HTM concept to handle other, more general classes of LTV systems. Just as a matrix representation is developed for capturing signal transfers between the harmonic carriers $e^{jm\omega_0 t}$, it is possible to do so for systems that transfer signal content between carrier waves $\psi_m(t)$ other than the harmonic ones. As such, we create a class of LTV systems that can be associated with the carriers $\psi_m(t)$, just as T-periodic LPTV systems are associated with the harmonic carriers $e^{jm\omega_0 t}$. Developing the matrix representation for such systems and unraveling its structural properties proceeds along the same lines as outlined above for LPTV systems. Moreover, although further discussion in this chapter is restricted to linear T-periodic systems, it is worth indicating that most of the analysis that follows is straightforward to generalize.

3.4 LPTV system manipulation using HTMs

Having discussed HTMs and their basic properties in depth, we now turn to their use for the manipulation and analysis of LPTV system behavior. Use of HTMs allows us to extend techniques for manipulating LTI systems to handle LPTV ones as well. Section 3.4.1 introduces HTM representations for two types of elementary LPTV systems. The basic apparatus for computing the HTMs of composed systems is outlined in sections 3.4.2 and 3.4.3. Together, these sections provide an elegant framework to manipulate LPTV systems and to compute the overall input/output behavior, given the system's building block characteristics. As an application of the material, section 3.4.4 shows how the HTM representation of an LPTV system is related to its state-space representation.

3.4.1 HTMs of elementary systems

Two fundamental subclasses of LPTV systems are LTI systems and multiplications with a periodic signal $p(t)$. Any other LPTV system can always be modeled as an interconnection of blocks belonging to these basic subclasses. The procedure to determine the HTM representations for these special types of LPTV systems is straightforward. In a first step, we find the kernels $h(t,\tau)$ associated with either an LTI system or a multiplication. The HTM representation then follows from (3.38) and (3.45).

3.4.1.1 LTI systems

For LTI systems, it holds that $h(t,\tau) = h(\tau)$. This $h(\tau)$ is the LTI system's impulse response. Note that it does not depend on the variable t (hence the label *time-invariant*). As a consequence, in the series expansion (3.38), all $h_k(\tau) = 0$ except for $k = 0$. Using (3.45), we then obtain that the HTM corresponding to an LTI operator with transfer characteristic $H(s) = \mathcal{L}\{h(\tau)\}$ is determined by

$$\begin{cases} \tilde{H}_{n,m}(s) = H(s + jm\omega_0) & m = n \\ \tilde{H}_{n,m}(s) = 0 & m \neq n \end{cases}, \tag{3.59}$$

or

$$\tilde{\mathbf{H}}(s) = \begin{bmatrix} \ddots & & & \\ & H(s-j\omega_0) & & 0 \\ & & H(s) & \\ & 0 & & H(s+j\omega_0) \\ & & & & \ddots \end{bmatrix} . \quad (3.60)$$

Hence, the HTM equivalent of an LTI system is a diagonal matrix with frequency-shifted copies of the transfer characteristic $H(s)$ on the diagonal. This diagonal structure makes sense since LTI systems cannot cause information to shift along the frequency axis.

3.4.1.2 Multiplication with a periodic signal $p(t)$

The kernel $h(t,\tau)$ corresponding to a memoryless multiplication $y(t) = p(t)u(t)$ of the input signal $u(t)$ with a periodic signal $p(t)$, where

$$p(t) = \sum_{k=-\infty}^{+\infty} P_k e^{jk\omega_0 t} , \quad (3.61)$$

is found by observing that

$$\begin{aligned} y(t) &= \int_0^{\infty} h(t,\tau)u(t-\tau)d\tau & (3.62) \\ &= p(t)u(t) & (3.63) \\ &= \int_0^{\infty} p(t)\delta(\tau)u(t-\tau)d\tau & (3.64) \\ &= \sum_{k=-\infty}^{+\infty} e^{jk\omega_0 t} \int_0^{\infty} P_k\delta(\tau)u(t-\tau)d\tau . & (3.65) \end{aligned}$$

We therefore find $h_k(\tau) = P_k\delta(\tau)$ with $\delta(\tau)$ being the Dirac delta function. Using (3.45), the elements of the HTM corresponding to the operation (3.63) are found to equal

$$\tilde{H}_{n,m} = P_{n-m} , \quad (3.66)$$

or

$$\tilde{\mathbf{H}}(s) = \begin{bmatrix} \ddots & \ddots & \ddots & & \\ & P_0 & P_{-1} & P_{-2} & \\ \ddots & P_1 & P_0 & P_{-1} & \ddots \\ & P_2 & P_1 & P_0 & \\ & & \ddots & \ddots & \ddots \end{bmatrix} . \quad (3.67)$$

Hence, the HTM corresponding to a memoryless multiplication with a periodic signal $p(t)$ is a frequency-independent Toeplitz matrix.

Example (Ideal mixer): According to (3.67), the HTM corresponding to the elementary mixing stage

$$y(t) = \cos(\omega_0 t)u(t) \tag{3.68}$$

is given by

$$\tilde{\mathbf{H}}_{mixer}(s) = \begin{bmatrix} \ddots & \ddots & & & & 0 \\ \ddots & 0 & 1/2 & & & \\ & 1/2 & 0 & 1/2 & & \\ & & 1/2 & 0 & \ddots & \\ 0 & & & \ddots & \ddots & \end{bmatrix} \tag{3.69}$$

which is a banded matrix.

As a test, let us apply this operator on a simple input signal: $u(t) = \sin(\omega_0 t)$. This is a periodic signal with its Fourier coefficients given by

$$\mathbf{U} = \begin{bmatrix} \cdots & 0 & -\frac{1}{2j} & 0 & \frac{1}{2j} & 0 & \cdots \end{bmatrix}^T . \tag{3.70}$$

In section 3.5.1, it will be shown that the Fourier coefficients of the equally periodic (steady-state) output signal $y(t)$ satisfy $\mathbf{Y} = \tilde{\mathbf{H}}_{mixer}(0) \cdot \mathbf{U}$. This evaluates as

$$\mathbf{Y} = \begin{bmatrix} \cdots & -\frac{1}{4j} & 0 & 0 & 0 & \frac{1}{4j} & \cdots \end{bmatrix}^T . \tag{3.71}$$

Hence, we find $y(t) = \left(e^{j2\omega_0 t} - e^{-j2\omega_0 t}\right)/4j = \sin(\omega_0 t)/2$. This result is readily verified by straightforward evaluation of $y(t) = \cos(\omega_0 t)\sin(\omega_0 t) = \sin(2\omega_0 t)/2$. ▲

3.4.2 HTMs of LPTV systems connected in parallel or in series

In constructing and manipulating linear systems, there are three types of basic connections: parallel, series and feedback connections. Together, they make up a complete set that can be used to realize any interconnect pattern. Within the HTM framework, it turns out to be particularly easy to compute the HTMs corresponding to these basic types of composed systems when given the HTMs of the composing systems. In what follows, we handle parallel en series connections. A separate section is devoted to feedback connections as these are harder to deal with.

Computing the HTM representation for a parallel and series connection, given the HTMs of the composing LPTV systems, is straightforward. With $\tilde{\mathbf{H}}_1(s)$ and $\tilde{\mathbf{H}}_2(s)$ the HTMs of two arbitrary LPTV systems, the HTM of their parallel respectively series connection, as illustrated in Fig. 3.12, is given by

$$\tilde{\mathbf{H}}_+(s) = \tilde{\mathbf{H}}_1(s) + \tilde{\mathbf{H}}_2(s) \tag{3.72}$$

$$\tilde{\mathbf{H}}_\times(s) = \tilde{\mathbf{H}}_2(s)\tilde{\mathbf{H}}_1(s) \tag{3.73}$$

It is important to note that the order of multiplication in (3.73) cannot be interchanged. Contrary to single-input single-output LTI systems, LPTV systems, in general, do not commute.

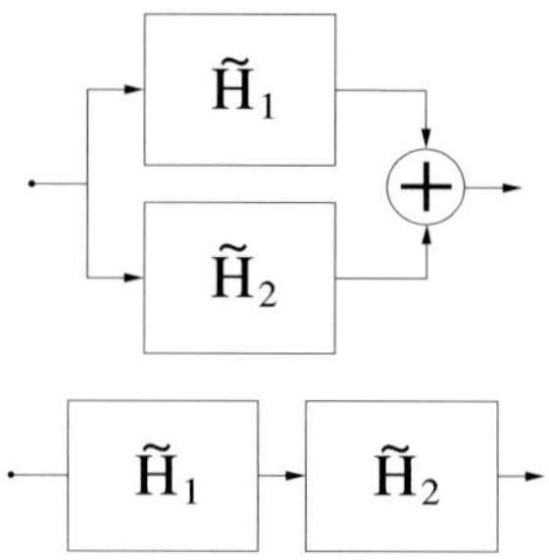

Figure 3.12: Parallel and series connection of two (LPTV) systems.

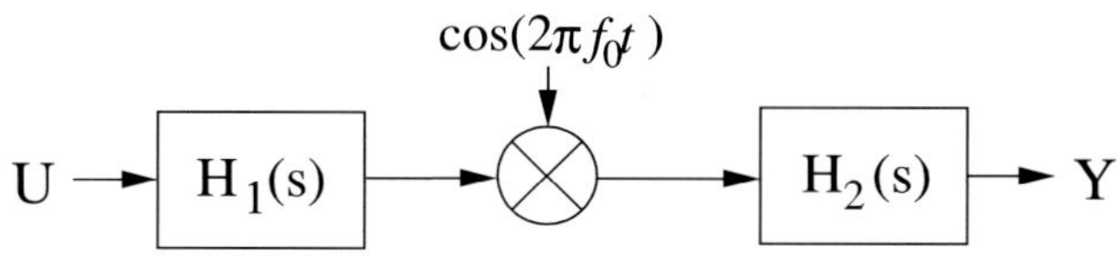

Figure 3.13: Block diagram of an elementary downconversion stage consisting of a (bandpass) pre-filter, an ideal mixing operation and a (lowpass) post-filter.

Example (Downconversion stage): Consider the block diagram of the elementary downconversion stage in Fig. 3.13, consisting of a (bandpass) pre-filter, an ideal mixing operation and a (lowpass) post-filter. Using (3.60), (3.69) and (3.73), it is readily shown that the elements of the input-output HTM $\tilde{\mathbf{H}}(s)$ characterizing the entire downconversion stage are given by

$$\tilde{H}_{n,m}(s) = \frac{1}{2} H_1(s + m\omega_0) H_2(s + n\omega_0) \delta(|n-m| - 1) \ . \tag{3.74}$$

Downconversion of signal content from the carrier frequency to baseband is therefore captured by the harmonic transfer function

$$\tilde{H}_{0,1}(s) = H_1(s + j\omega_0) H_2(s)/2 \ . \tag{3.75}$$

This transfer function relates the output equivalent baseband component at DC to the input equivalent baseband component at carrier frequency. ▲

3.4.3 Feedback systems and HTM inversions

LPTV feedback connections are more difficult to handle. For example, as the HTM of the feedback system in Fig. 3.14 is concerned, straightforward manipulation yields the closed-loop characteristic

$$\tilde{\mathbf{H}}_{cl}(s) = \left(\mathbf{I} + \tilde{\mathbf{G}}(s)\right)^{-1} \tilde{\mathbf{A}}(s) \ . \tag{3.76}$$

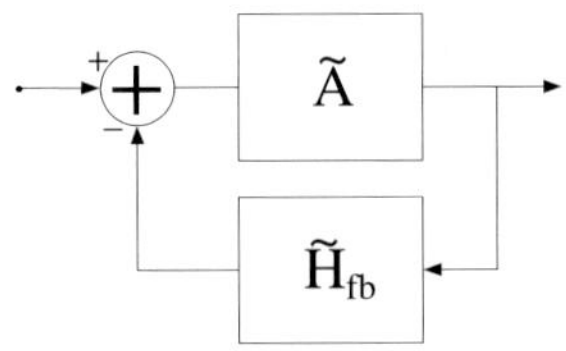

Figure 3.14: Feedback connection of two (LPTV) systems.

Here, $\tilde{\mathbf{G}}(s) = \tilde{\mathbf{A}}(s)\tilde{\mathbf{H}}_{fb}(s)$ is the open-loop gain HTM while $\mathbf{I}$ is the identity matrix. Hence, computing $\tilde{\mathbf{H}}_{cl}(s)$ involves a matrix inversion. Since, in principle, HTMs are ∞-dimensional matrices, LPTV feedback requires us to invert a very large matrix. This problem is equivalent to solving very large systems of equations.

Numerically, inverting very large linear systems is a hard problem to solve. Doing so symbolically even seems impossible. Dealing with these large matrix inversions in an efficient manner is one of the main challenges in rendering HTM-based analysis suited for system and circuit design practice. Since almost any practical application contains feedback, the inversions cannot be avoided. Therefore, it is very important to develop methods that handle the inversions efficiently. Methods for inverting HTMs are essential in bridging the gap between HTMs as an abstract theoretical framework and HTMs as a powerful tool for both numerical and symbolic computations.

In what follows, we discuss three different techniques to invert HTMs. The first one truncates the ∞-dimensional HTM and uses, for example, sparse-matrix techniques to handle the inversions. Although appropriate for numerical computations, this technique is very time consuming and ineffective for symbolic computations. Sections 3.4.3.2 and 3.4.3.3 introduce two methods that allow us to deal with LPTV feedback by assuming that the HTMs that are involved satisfy specific properties. The first method assumes the open-loop gain HTM $\tilde{\mathbf{G}}(s)$ to have a dominant diagonal (i.e. LTI) component. The second method is based on a rank-reduced approximation of $\tilde{\mathbf{G}}(s)$. As will be illustrated in the next chapter, most open-loop gain HTMs encountered in practice can be dealt with using one of both methods. Moreover, both approaches not only allow numerical inversion of HTMs, they also render symbolic computations possible.

3.4.3.1 Inversion by truncation

A straightforward approach tackles the inversion in (3.76) by truncating the HTM $\left(\mathbf{I}+\tilde{\mathbf{G}}(s)\right)$ to a finite N-dimensional square matrix. This amounts to neglecting all signal content beyond a particular frequency. The value of N depends on the number of frequency bands that are of importance in both the input and output signal. The matrix inverse can then be computed using Krylov subspace algorithms [Barr94, Saad96]. This approach is similar to harmonic balance algorithms [Kund86, Kund90]. It is adopted by most authors that use HTMs or a likewise representation in dealing with LPTV system behavior [Maas88, Moll00, Royc98, Royc99].

3.4.3.2 Power series approximation

In a lot of practical LPTV systems, a HTM $\tilde{\mathbf{H}}(s)$ that needs to be inverted ($\tilde{\mathbf{H}}(s) = \mathbf{I}+\tilde{\mathbf{G}}(s)$ in the case of the feedback example) tends to have a dominant diagonal (LTI) component. Hence, it is natural to decompose $\tilde{\mathbf{H}}(s)$ as

$$\begin{aligned} \tilde{\mathbf{H}}(s) &= \tilde{\mathbf{H}}_{LTI}(s) - \Delta\tilde{\mathbf{H}}(s) && (3.77) \\ &= \tilde{\mathbf{H}}_{LTI}(s)\left(\mathbf{I} - \tilde{\mathbf{H}}_{LTI}(s)^{-1}\Delta\tilde{\mathbf{H}}(s)\right) \; . && (3.78) \end{aligned}$$

Here, $\tilde{\mathbf{H}}_{LTI}(s)$ is a diagonal HTM corresponding to the diagonal (LTI subsection) of $\tilde{\mathbf{H}}(s)$ while $-\Delta\tilde{\mathbf{H}}(s)$ contains the off-diagonal elements. $\tilde{\mathbf{H}}(s)$ is dominated by $\tilde{\mathbf{H}}_{LTI}(s)$ if the major part of the output signal's energy is generated by this LTI component, or

$$\left\|\tilde{\mathbf{H}}_{LTI}(s)^{-1}\Delta\tilde{\mathbf{H}}(s)\right\| \ll 1 \; . \tag{3.79}$$

(See appendix A for a discussion on HTM norms.)

Assuming the presence of a dominant LTI component, i.e. with (3.79) being satisfied, the inverse of $\tilde{\mathbf{H}}(s)$ can be approximated by means of the series expansion

$$\tilde{\mathbf{H}}(s)^{-1} = \sum_{r=0}^{\infty} \left(\tilde{\mathbf{H}}_{LTI}(s)^{-1}\Delta\tilde{\mathbf{H}}(s)\right)^r \tilde{\mathbf{H}}_{LTI}(s)^{-1} \tag{3.80}$$

With the inverse of the diagonal HTM $\tilde{\mathbf{H}}_{LTI}(s)$ being straightforward to compute, this expansion requires only matrix multiplications and additions. Therefore, (3.80) can be evaluated using both numerical and symbolic techniques. For practical purposes, the expansion is truncated after a finite number of terms R. Note that, especially for symbolic computations, use of (3.80) is only practical if R can be kept low, i.e. if (3.80) converges sufficiently fast. Note that the theoretical justification of the expansion (3.80) builds on the Banach fixed-point theorem [Cor91, Huts80, Vana02c].

3.4.3.3 Rank-reduced approximation

Another strategy assumes that the HTM $\tilde{\mathbf{H}}(s)$ that is to be inverted can be approximated as

$$\tilde{\mathbf{H}}(s) \approx \tilde{\mathbf{Z}}(s) + \mathbf{U}(s)\mathbf{V}(s)^T \; , \tag{3.81}$$

i.e. the sum of an HTM $\tilde{\mathbf{Z}}(s)$ whose inverse can be determined at low computational cost and a low-rank update $\mathbf{U}(s)\mathbf{V}(s)^T$([4]). Here, $\mathbf{U}(s)$ and $\mathbf{V}(s)$ are column matrices with a finite number M of columns. Formally, $\mathbf{U}(s), \mathbf{V}(s) \in \mathbb{C}^{\infty \times M}$. For the feedback loop in Fig. 3.14, the approximation (3.81) is possible if, for example, the open-loop gain $\tilde{\mathbf{G}}(s) = \tilde{\mathbf{A}}(s)\tilde{\mathbf{H}}_{fb}(s) \approx \mathbf{U}(s)\mathbf{V}(s)^T$ is nearly rank-deficient. In such a case $\tilde{\mathbf{Z}}(s) = \mathbf{I}$. Often, the such an approximation only holds for s belonging to a bounded subset of the complex plane.

[4]For this approximation to be valid, the largest singular value of the approximation error matrix should be well below the smallest singular value of (3.81), i.e. $\sigma_{max}\left(\tilde{\mathbf{H}}(s) - \tilde{\mathbf{Z}}(s) + \mathbf{U}(s)\mathbf{V}(s)^T\right) \ll \sigma_{min}(\tilde{\mathbf{Z}}(s) + \mathbf{U}(s)\mathbf{V}(s)^T)$.

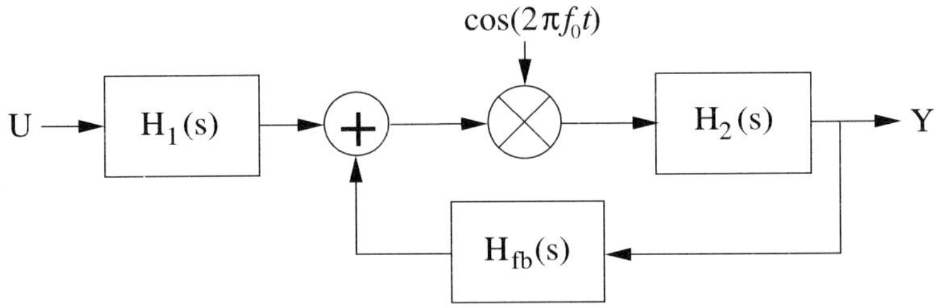

Figure 3.15: Block diagram of an elementary downconversion stage with parasitic feedback from the output to the mixer's input.

If (3.81) is satisfied, HTM inversions are efficiently computed by means of the Sherman-Morisson-Woodbury formula [Gol96][5]. This yields

$$\tilde{\mathbf{H}}(s)^{-1} \approx \tilde{\mathbf{Z}}(s)^{-1} - \tilde{\mathbf{Z}}(s)^{-1}\mathbf{U}(s)\left(\mathbf{I}+\mathbf{V}(s)^T\tilde{\mathbf{Z}}(s)^{-1}\mathbf{U}(s)\right)^{-1}\mathbf{V}(s)^T\tilde{\mathbf{Z}}(s)^{-1} \ . \tag{3.82}$$

Since it was assumed that to $\tilde{\mathbf{Z}}(s)$ is invertible at low computational cost, only the computation of $\left(\mathbf{I}+\mathbf{V}(s)^T\tilde{\mathbf{Z}}(s)^{-1}\mathbf{U}(s)\right)^{-1}$ might pose difficulties. However, as $\mathbf{U}(s), \mathbf{V}(s) \in \mathbb{C}^{\infty \times M}$, this involves inverting an $M \times M$ matrix. If M is sufficiently small, i.e. if $\mathbf{U}(s)\mathbf{V}(s)^T$ is of low rank, this is much less of a problem than inverting some large-dimensional matrix obtained by truncating the original ∞-dimensional HTM.

Example (Downconversion with feedback): Consider the system in Fig. 3.15. This is the same downconversion stage as in Fig. 3.13 but now we've added a time-invariant (parasitic) feedback connection from the output to the mixer's input. Hence, the overall input-output relation is given by

$$\tilde{\mathbf{H}}(s) = \left(\mathbf{I}-\tilde{\mathbf{G}}(s)\right)^{-1}\tilde{\mathbf{H}}_2(s)\tilde{\mathbf{H}}_{mixer}(s)\tilde{\mathbf{H}}_1(s) \ . \tag{3.83}$$

Here, the open-loop gain HTM equals

$$\tilde{\mathbf{G}}(s) = \tilde{\mathbf{H}}_2(s)\tilde{\mathbf{H}}_{mixer}(s)\tilde{\mathbf{H}}_{fb}(s) \ . \tag{3.84}$$

The HTMs $\tilde{\mathbf{H}}_1(s)$, $\tilde{\mathbf{H}}_2(s)$, $\tilde{\mathbf{H}}_{mixer}(s)$ and $\tilde{\mathbf{H}}_{fb}(s)$ respectively correspond to the filters $H_1(s)$ and $H_2(s)$, the multiplication with $\cos(2\pi f_0 t)$ and the feedback path $H_{fb}(s)$.

If the feedback loop is weak, i.e. if

$$\left\|\tilde{\mathbf{G}}(s)\right\| \ll 1 \ , \tag{3.85}$$

then, using a first-order power series approximation, we obtain

$$\tilde{\mathbf{H}}(s) \approx \left(\mathbf{I}+\tilde{\mathbf{G}}(s)\right)\tilde{\mathbf{H}}_2(s)\tilde{\mathbf{H}}_{mixer}(s)\tilde{\mathbf{H}}_1(s) \ . \tag{3.86}$$

Hence, the HTM elements $\tilde{H}_{0,1}(s)$, $\tilde{H}_{0,0}(s)$ and $\tilde{H}_{0,2}(s)$ —respectively characterizing the (wanted) transfers from carrier frequency to baseband and the (unwanted) transfers

[5] See also appendix B.

from baseband to baseband and twice the carrier frequency to baseband— we find (most important terms only)

$$\tilde{H}_{0,1}(s) \approx \frac{1}{2} H_2(s) H_1(s + j\omega_0) \tag{3.87}$$

and

$$\begin{aligned}\tilde{H}_{0,0}(s) &\approx \frac{1}{4} H_2(s) H_2(s - j\omega_0) H_{fb}(s - j\omega_0) H_1(s) \\ &\quad + \frac{1}{4} H_2(s) H_2(s + j\omega_0) H_{fb}(s + j\omega_0) H_1(s) &(3.88)\\ \tilde{H}_{0,2}(s) &\approx \frac{1}{4} H_2(s) H_2(s + j\omega_0) H_{fb}(s + j\omega_0) H_1(s + 2j\omega_0) &(3.89)\end{aligned}$$

It is observed that the presence of the feedback $H_{fb}(s)$ causes (parasitic) transfers from baseband and twice the carrier frequency to baseband. Often, these transfers are undesired and need to suppressed.

A second way to solve (3.83) builds on the observation that, with regard to the feedback loop, the lowpass filter $H_2(s)$ throws out most of the high-frequent information. If $H_2(s)$ rolls off sufficiently fast, we can use a rank-one approximation for $\tilde{\mathbf{G}}(s)$ (valid in the frequency interval $[-\omega_0/2, \omega_0/2]$), i.e.

$$\mathbf{I} + \tilde{\mathbf{G}}(s) \approx \mathbf{I} + \begin{bmatrix} \vdots \\ 0 \\ 1 \\ 0 \\ \vdots \end{bmatrix} \tilde{\mathbf{G}}_{0,\cdot}(s) \ . \tag{3.90}$$

Here,

$$\tilde{\mathbf{G}}_{0,\cdot}(s) = \begin{bmatrix} \cdots & 0 & H_2(s) H_{fb}(s - j\omega_0) & 0 & H_2(s) H_{fb}(s + j\omega_0) & 0 & \cdots \end{bmatrix} \tag{3.91}$$

is the 0-th row of $\tilde{\mathbf{G}}(s)$. Using (3.82), it is then straightforward to obtain the same result as in (3.87)-(3.89). This illustrates that, often, both power series approximation and a rank-reduced approximations can be used to obtain (approximate) expressions for the input-output HTM elements of feedback systems[6]. Accuracy of these approximations will be verified for some practical cases in the next chapter. ▲

3.4.4 Relating HTMs to state-space representations

A final topic related to LPTV system manipulation using HTMs concerns the link between an LPTV system's HTM representation and its state-space equations. The latter are, for example, obtained by linearizing a set of circuit equations in the neighborhood

[6]This, however, does not hold true for all cases. For analyzing PLLs with sampling phase-frequency detectors (see the next chapter), only the rank-reduced approximation turns out to be successful.

of a time-varying operating point. This typically results in a set of linear differential-algebraic equations (DAEs) that are structured as

$$\mathbf{E}(t)\frac{\mathrm{d}\mathbf{x}}{\mathrm{d}t} = \mathbf{A}(t)\mathbf{x}+\mathbf{b}(t)u \tag{3.92}$$

$$y = \mathbf{c}(t)\mathbf{x}+d(t)u \ . \tag{3.93}$$

Here, $\mathbf{E}(t),\mathbf{A}(t) \in \mathbb{R}^{N\times N}$, $\mathbf{b}(t) \in \mathbb{R}^{N\times 1}$, $\mathbf{c}(t) \in \mathbb{R}^{N\times 1}$ and $d(t) \in \mathbb{R}$ are T-periodic matrices (with $T = 2\pi/\omega_0$). Using the expressions for constructing elementary HTMs and the rules for addition and concatenation, it is straightforward to obtain

$$\tilde{\mathbf{H}}(s) = \tilde{\mathbf{H}}_{\mathbf{c}}\left(\tilde{\mathbf{H}}_{\mathbf{E}}\cdot\tilde{\mathbf{H}}_{diff}(s)-\tilde{\mathbf{H}}_{\mathbf{A}}\right)^{-1}\tilde{\mathbf{H}}_{\mathbf{b}}+\tilde{\mathbf{H}}_{d} \tag{3.94}$$

as the HTM relating $y(t)$ to $u(t)$. Here, $\tilde{\mathbf{H}}_{\mathbf{E}}$ represents a block Toeplitz matrix structured like the one in (3.67)[7]. The elements of $\tilde{\mathbf{H}}_{\mathbf{E}}$ equal the matrix-valued Fourier coefficients of $\mathbf{E}(t) = \sum_k \mathbf{E}_k e^{jk\omega_0 t}$. $\tilde{\mathbf{H}}_{\mathbf{A}}$, $\tilde{\mathbf{H}}_{\mathbf{b}}$, $\tilde{\mathbf{H}}_{\mathbf{c}}$ and $\tilde{\mathbf{H}}_{d}$ are likewise obtained from $\mathbf{A}(t)$, $\mathbf{b}(t)$, $\mathbf{c}(t)$ and $d(t)$. Furthermore,

$$\tilde{\mathbf{H}}_{diff}(s) = \begin{bmatrix} \ddots & & & \\ & (s-j\omega_0)\mathbf{I}_N & & 0 \\ & & s\mathbf{I}_N & \\ & 0 & & (s+j\omega_0)\mathbf{I}_N \\ & & & & \ddots \end{bmatrix} \tag{3.95}$$

is the (block) HTM representation for the differentiation operator $\mathrm{d}/\mathrm{d}t[\cdot]$. Here $\mathbf{I}_N \in \mathbb{R}^{N\times N}$ is the identity matrix. The numerical extraction of HTMs or related quantities can now be accomplished by means of the solution of (3.94) whereby all matrices are properly truncated [Phil98, Royc98]. Typically, one needs Krylov subspace solvers [Barr94, Saad96] to handle the large-scale matrix inversions that arise.

Example (Downconversion stage): Let us consider the downconversion circuit in Fig. 3.16. This circuit implements a mixing stage followed by a lowpass filtering operation. By linearizing the circuit equations in the neighborhood of the periodic operating point, we obtain the linear periodically time-varying system of DAEs

$$C\frac{\mathrm{d}x}{\mathrm{d}t} = -\frac{x}{R}-\beta\cos(\omega_0 t)u(t) \tag{3.96}$$

$$y(t) = x(t) \ . \tag{3.97}$$

Here, we assumed the opamp to be ideal and we used the results of the example in section 3.1.3 (see equation (3.10)) to model the transistor's time-varying small-signal behavior. If we map each of the terms in (3.96)-(3.97) on their HTM equivalent, then

[7] Since $\mathbf{E}(t)$ is a matrix, multiplication with $\mathbf{E}(t)$ represents a multi-input multi-output (MIMO) LPTV system. The HTM corresponding with such a MIMO system has the same structure as a HTM corresponding to a single-input single-output (SISO) system. The only difference is that the HTM elements are now matrix valued instead of scalar valued.

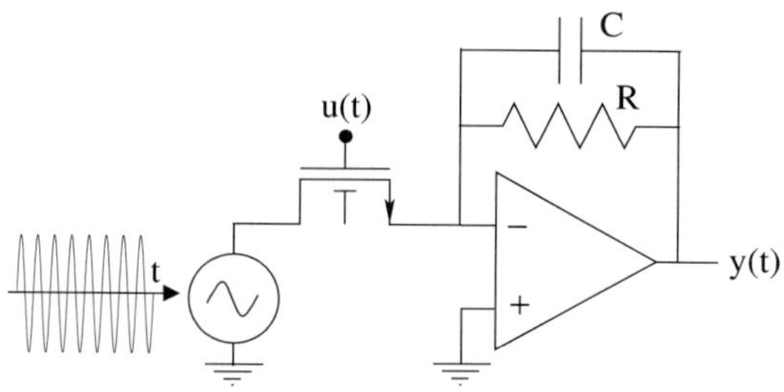

Figure 3.16: Downconversion circuit that implements a mixing stage followed by a lowpass filtering operation.

we obtain that $\tilde{\mathbf{H}}_{\mathbf{E}} = C \cdot \mathbf{I}$, with $\mathbf{I}$ the unity matrix, $\tilde{\mathbf{H}}_{\mathbf{A}} = -(1/R) \cdot \mathbf{I}$, $\tilde{\mathbf{H}}_{\mathbf{c}} = \mathbf{I}$, $\tilde{\mathbf{H}}_d = 0$ and

$$\tilde{\mathbf{H}}_{\mathbf{b}} = \begin{bmatrix} \ddots & \ddots & & & 0 \\ \ddots & 0 & -\frac{\beta}{2} & & \\ & -\frac{\beta}{2} & 0 & -\frac{\beta}{2} & \\ & & -\frac{\beta}{2} & 0 & \ddots \\ 0 & & & \ddots & \ddots \end{bmatrix} . \tag{3.98}$$

Substituting this into (3.94) yields the input-output relation

$$\mathbf{Y}(s) = \begin{bmatrix} \ddots & \ddots & & & 0 \\ \ddots & 0 & \frac{-\beta R}{2(1+(s-j\omega_0)RC)} & & \\ & \frac{-\beta R}{2(1+sRC)} & 0 & \frac{-\beta R}{2(1+sRC)} & \\ & & \frac{-\beta R}{2(1+(s+j\omega_0)RC)} & 0 & \ddots \\ 0 & & & \ddots & \ddots \end{bmatrix} \mathbf{U}(s) . \tag{3.99}$$

The matrix in the equation above is the HTM associated with the state-space description (3.96)-(3.97). If we take a look at the HTM element that models the transfer of signal content from the carrier frequency to baseband, i.e $\tilde{H}_{0,1}(s) = -\beta R/2\,(1+sRC)$, then, as expected, we clearly observe a lowpass filtering action. ▲

3.5 LPTV system analysis using HTMs

Having discussed HTMs, their basic properties and the ways in which we can manipulate them, we now turn to their use in analyzing various aspects of LPTV system behavior. In what follows, it is shown how HTMs can be used to perform multi-tone (section 3.5.1), stability (section 3.5.2) and noise (section 3.5.3) analysis of LPTV systems.

3.5.1 Multi-tone analysis

Let us be given an LPTV system characterized by the HTM $\tilde{\mathbf{H}}(s)$. To this system, we apply an input signal $u(t)$ with equivalent baseband components (envelopes) that equal

$$u_m(t) = U_{r,m} e^{j\Delta\omega t} \ , \tag{3.100}$$

where $U_m \in \mathbb{C}$. This corresponds to injecting extra (small-signal) tones into the system. These tones are located at $\Delta\omega + m\omega_0$. The Laplace transforms of the envelopes in (3.100) equal

$$U_m(s) = \frac{U_{r,m}}{s - j\Delta\omega} \ . \tag{3.101}$$

They all contain a pole on the imaginary axis with $U_{r,m}$ as the associated residue.

With an LPTV system's input-output behavior captured by (3.43), the output signal's equivalent baseband components are obtained from

$$\mathbf{Y}(s) = \tilde{\mathbf{H}}(s)\mathbf{U}(s) = \frac{1}{s - j\Delta\omega} \tilde{\mathbf{H}}(s)\mathbf{U}_r \tag{3.102}$$

where $\mathbf{U}_r = \begin{bmatrix} \cdots & U_{r,-1} & U_{r,0} & U_{r,1} & \cdots \end{bmatrix}^T$. Just as it is the case for the input signal components, the Laplace transforms of the output signal components also contain a pole on the imaginary axis at $s = j\Delta\omega$. The associated vector of residues equals

$$\mathbf{Y}_r = \lim_{s \to j\Delta\omega} (s - j\Delta\omega)\tilde{\mathbf{H}}(s)\mathbf{U}(s) = \tilde{\mathbf{H}}(j\Delta\omega)\mathbf{U}_r \tag{3.103}$$

with $\mathbf{Y}_r = \begin{bmatrix} \cdots & Y_{r,-1} & Y_{r,0} & Y_{r,1} & \cdots \end{bmatrix}^T$. In steady state, the output signal's equivalent baseband components hence become

$$y_n(t) = Y_{r,n} e^{j\Delta\omega t} \tag{3.104}$$

Equation (3.103) relates the tones in the output spectrum to the tones in the input spectrum and allows efficient evaluation of the LPTV system's steady-state response to multi-tone input signals.

3.5.2 Stability analysis

The similarity between the input-output relation (3.76), characterizing the closed-loop behavior of a time-varying feedback system, and its counterpart in traditional LTI feedback analysis raises the question whether powerful techniques for frequency-domain stability analysis, like Nyquist and Bode [Bode45], can be extended to the time-varying case. The answer turns out to be affirmative. Results that follow are similar to the ones in [Were91a, Were91b]. They are slightly simplified by exploiting the repetitive structure of both HTM poles and eigenvalues.

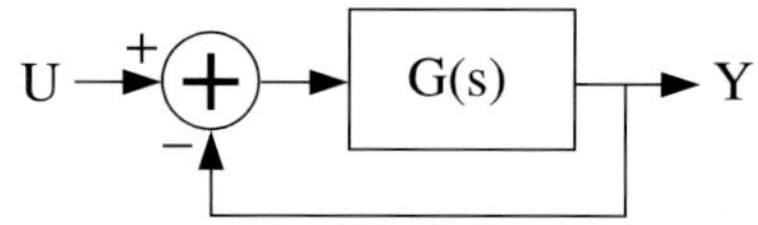

Figure 3.17: Elementary feedback loop.

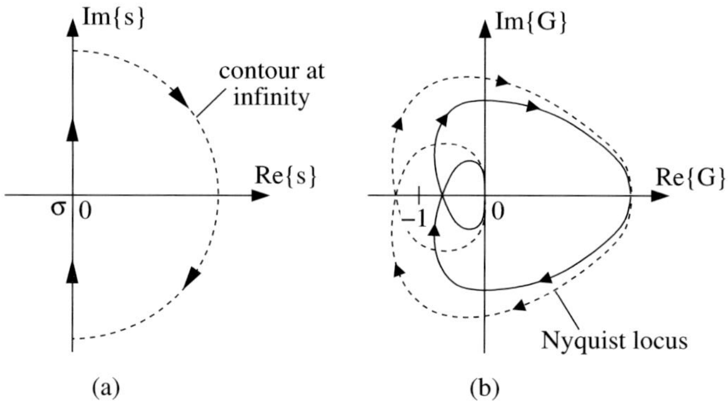

Figure 3.18: Nyquist stability is based on the study of Nyquist loci. A Nyquist locus is the graph traced by $G(s)$ along (a) a contour that encircles the entire right-hand plane . (b) If $G(s)$ contains no poles in the encircled area, the solid line indicates that the closed loop $H(s)$ has no unstable poles as it does not encircle $-1+j0$. On the other hand, the dashed line, that does encircle $-1+j0$ a single time, denotes the presence of an unstable pole in the closed-loop behavior.

3.5.2.1 Recapitulating stability analysis of SISO LTI systems

We begin our treatment with a brief review of traditional frequency-domain stability theory for single-input single-output (SISO) LTI systems [Fran94, Mac89]. Here, the starting point is the work of Nyquist[8]. Given the feedback loop in Fig. 3.17, with

$$H(s) = \frac{Y(s)}{U(s)} = \frac{G(s)}{1+G(s)} , \tag{3.105}$$

Nyquist studied the stability of the closed-loop system $H(s)$ in terms of the characteristics of the open-loop gain $G(s)$.

Nyquist stability analysis is based on the study of a *Nyquist locus*. As illustrated in Fig. 3.18, a Nyquist locus is the graph traced by $G(s)$ as s traces a *Nyquist contour*, i.e. a contour that encircles the entire right-hand plane. If $\lim_{s\to\infty} G(s)$ exists, then this

[8]As a historical note aside: Nyquist's stability criterion essentially is an application of the *argument principle* which in turn builds on Cauchy's work on complex functions [Arf85]. Hence, the mathematical apparatus needed for the stability analysis of linear systems was already available by 1830.

graph is entirely determined by the values $G(j\omega)$ for $-\infty < \omega < \infty$. Closing the loop using a contour at infinity is only relevant for theoretical purposes.

The link between the properties of the closed-loop transfer function $H(s)$ and the properties of the open-loop gain $G(s)$ is established by Nyquist's theorem [Bode45, Fran94, Mac89]:

Theorem (Nyquist): *A closed-loop system $H(s) = G(s)/(1+G(s))$ contains no unstable poles if the Nyquist locus of the open-loop gain $G(s)$ encircles $-1+j0$ as many times anti-clockwise as $G(s)$ has poles located in the right-hand plane.*

For example, if $G(s)$ contains no poles in the right-hand plane, the Nyquist locus indicated by the solid line in Fig. 3.18(b) proves that the closed-loop transfer function $H(s)$ is stable as it does not encircle $-1+j0$. The dashed line, encircling $-1+j0$ a single time, indicates the presence of a single unstable closed-loop pole.

Note that by tracing curves other than the one in Fig. 3.18(a), similar results are obtained that provide more detailed information on the position of the closed-loop poles. For example, plotting the graph of $G(s)$ for s varying along a σ-shifted frequency axis, i.e. $s \in \{\sigma + j\omega | -\infty < \omega < \infty\}$, helps us to determine whether the closed-loop system contains poles with a real part that is greater than σ. This is useful as, often, designers are not interested in shear stability. They would also like the closed-loop system to settle sufficiently fast, i.e. all poles must have a real part that is smaller than a certain negative real σ. This implies that the closed loop should not contain any poles located in the plane to the right of the σ-shifted frequency axis. For a more detailed discussion, see [Mac89].

3.5.2.2 Recapitulating stability analysis of MIMO LTI systems

Since HTMs can be considered as the limit of a sequence of finite-dimensional, multi-input multi-output (MIMO) LTI systems, the question rises whether Nyquist can be generalized to handle this kind of systems. A generalization like this is discussed in [Deso80, Mac89].

For MIMO LTI systems, the open-loop gain in Fig. 3.17 is no longer scalar- but matrix-valued, i.e.

$$\mathbf{G}(s) = \begin{bmatrix} G_{1,1}(s) & \cdots & G_{1,N}(s) \\ \vdots & \ddots & \vdots \\ G_{N,1}(s) & \cdots & G_{N,N}(s) \end{bmatrix} . \tag{3.106}$$

For the moment, $\mathbf{G}(s) \in \mathbb{C}^{N\times N}$ is considered an arbitrary matrix-valued transfer function (no structural constraints). As for LTI transfer functions, if

$$\lim_{s\to p} \|\mathbf{G}(s)\| = \infty \ , \tag{3.107}$$

then $p \in \mathbb{C}$ is said to be a pole of $\mathbf{G}(s)$.

For the feedback system in Fig. 3.17, the closed-loop transfer matrix $\mathbf{H}(s)$ is found to equal

$$\mathbf{H}(s) = (\mathbf{I}+\mathbf{G}(s))^{-1}\mathbf{G}(s) \tag{3.108}$$

with $\mathbf{I} \in \mathbb{C}^{N\times N}$ the N-dimensional unity matrix. In analyzing the stability or settling behavior of the closed-loop system, i.e. the absence of poles of $\mathbf{H}(s)$ in a particular region of the complex plane, we compute the eigenvalues $\lambda_k(s)$ of the open-loop gain $\mathbf{G}(s)$, i.e.

$$\exists \mathbf{U}_k(s) \in \mathbb{C}^N : \mathbf{G}(s)\mathbf{U}_k(s) = \lambda_k(s)\mathbf{U}_k(s) \tag{3.109}$$

for $k = 1,\ldots,N$. The graphs of the $\lambda_k(s)$ as s traces a Nyquist contour are called the characteristic loci of $\mathbf{G}(s)$. The properties of the characteristic loci of $\mathbf{G}(s)$ are linked to the properties of the closed-loop system $\mathbf{H}(s)$ by the following theorem [Deso80, Mac89]:

Theorem (Generalized Nyquist): *A closed-loop MIMO system* $\mathbf{H}(s) = (\mathbf{I}+\mathbf{G}(s))^{-1} \times \mathbf{G}(s)$ *contains no unstable poles if the characteristic loci of the open-loop gain* $\mathbf{G}(s)$*, taken together, encircle* $-1+j0$ *as many times anti-clockwise as* $\mathbf{G}(s)$ *has poles located in the right-hand plane.*

Example (MIMO stability): Consider the system with open-loop gain

$$\mathbf{G}(s) = \begin{bmatrix} 0 & 1 \\ 10\frac{s-1}{s+1} & 0 \end{bmatrix} . \tag{3.110}$$

The eigenvalues of this system are

$$\lambda_{1,2}(s) = \pm\sqrt{10\frac{s-1}{s+1}} . \tag{3.111}$$

Fig. 3.19 plots the characteristic loci for both $\lambda_1(s)$ and $\lambda_2(s)$. Since $\mathbf{G}(s)$ has no unstable poles and since the joint characteristic loci encircle $-1+j0$ a single time, it is concluded that the closed-loop system $\mathbf{H}(s)$ determined by (3.108) will be unstable. Note also the importance of drawing *both* characteristic loci. If only one of them is drawn, the overall locus is not closed and the generalized Nyquist criterion has therefore no meaning. ▲

3.5.2.3 Stability analysis of LPTV systems

The correspondence between the HTM representation of an LPTV system and the transfer matrix of a MIMO LTI system suggests to apply the results above to establish a frequency-domain Nyquist criterion that can handle LPTV system stability. However, the ∞-dimensional structure of the HTMs causes trouble. It implies both an infinite number of poles and eigenvalues. Straightforward application of the generalized Nyquist criterion for MIMO LTI systems would therefore result in a great waste of efforts.

Fortunately, the structure inherent to HTMs makes simplification possible. As mentioned in section 3.3.4, both the poles and eigenvalues of a HTM $\tilde{\mathbf{G}}(s)$ exhibit a repetitive structure. If p is a pole of $\tilde{\mathbf{G}}(s)$, so is $p+j\omega_0$. Furthermore, the set $\{\lambda_k(s)\}$ of

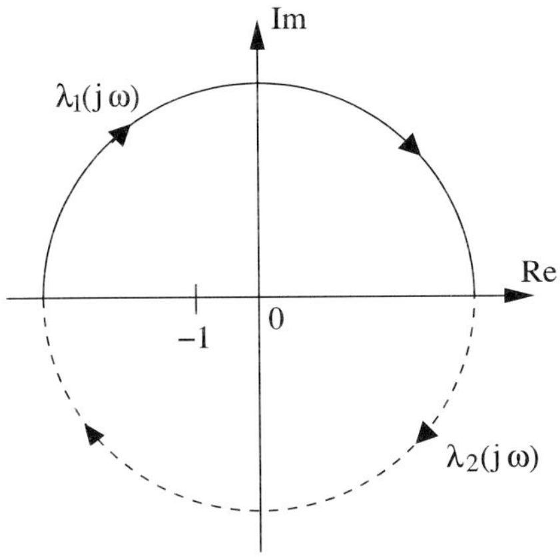

Figure 3.19: The two characteristic loci for the MIMO system defined by equation (3.110).

eigenvalues of $\tilde{\mathbf{G}}(s)$ is periodic in a sense that

$$\lambda_k(s) \in \text{eig}\left\{\tilde{\mathbf{G}}(s)\right\} \Rightarrow \lambda_k(s) \in \text{eig}\left\{\tilde{\mathbf{G}}(s+j\omega_0)\right\} \quad . \tag{3.112}$$

Both observations taken together imply that it is sufficient to trace s over the frequency interval $[-\omega_0/2, \omega_0/2]$ in drawing the characteristic loci of $\tilde{\mathbf{G}}(s)$ and in determining its poles. Due to the repetitive nature of the poles and eigenvalues of $\tilde{\mathbf{G}}(s)$ in the direction of the imaginary axis, no new information will be revealed as we trace s beyond this interval. This brings us to the Nyquist criterion for LPTV systems [Were91a, Moll00][9]:

Theorem (Nyquist LPTV): *A closed-loop LPTV system that is represented by the HTM* $\tilde{\mathbf{H}}(s) = \left(\mathbf{I}+\tilde{\mathbf{G}}(s)\right)^{-1}\tilde{\mathbf{G}}(s)$ *contains no unstable poles if the characteristic loci of the open-loop gain* $\tilde{\mathbf{G}}(s)$*, taken together and drawn for* $s = j\omega$ *for* $-\omega_0/2 < \omega < \omega_0/2$*, encircle* $-1+j0$ *as many times anti-clockwise as* $\tilde{\mathbf{G}}(s)$ *has poles located in the right-hand plane with an imaginary part that lies in the interval* $[-\omega_0/2, \omega_0/2]$.

Although satisfying from a theoretical point of view, questions rise as to the practical applicability of this Nyquist criterion for LPTV systems. For a particular value $s = j\omega$, $\tilde{\mathbf{G}}(s)$ is still an ∞-dimensional matrix. Hence, it has an infinite number of eigenvalues. Fortunately, for many practical systems, most of the eigenvalues have a magnitude that is well below 1. Since their locus cannot help to encircle $-1+j0$, they are irrelevant as far as stability is concerned.

[9] As presented here, the criterion slightly deviates from the one presented in [Were91a, Moll00]. Contrary to the procedure suggested there, the periodic structure of the set of eigenvalues $\{\lambda_k(s)\}$ renders it unnecessary to close the curve by tracing s to $+\infty$ and back again along the curves $\{\sigma \pm j\omega_0/2 \,|\, 0 < \sigma < \infty\}$. There is no error in doing so, but, due to periodicity of the eigenvalues, the net number of encirclements induced by tracing these trajectories will always be zero.

3.5.2.4 Computing the dominant HTM eigenvalues

As was the case for HTM inversions, computing the eigenvalues of an ∞-dimensional HTM is a hard problem to solve. However, keeping in mind that we are only interested in eigenvalues that are sufficiently large, the problem can be simplified by exploiting the characteristics of open-loop gain HTMs as they occur in practice. In what follows, we discuss three methods for HTM eigenvalue computations. The methods parallel the ones introduced in sections 3.4.3.1-3.4.3.3 to compute the inverse of a HTM.

Truncation-based eigenvalue computations

Again, the most straightforward strategy is to truncate the open-loop gain HTM $\tilde{\mathbf{G}}(s)$, i.e. reduce it to a finite-dimensional matrix. Eigenvalues can then be computed using standard textbook algorithms [Gol96] . If the truncated matrix is very large, it is better to use subspace methods [Saad92, Sore00] to find the largest eigenvalues.

Loops with a dominant LTI component

A second approach towards estimating eigenvalue locations makes use of Gershgorin's theorem [Gol96, Mac89]:

Theorem (Gershgorin's theorem): *Let $\mathbf{Z} = [z_{k,l}]$ be a complex matrix of dimensions $M \times M$. Then the eigenvalues of $\mathbf{Z}$ lie in the union of the M (Gershgorin) circles, each with center $z_{k,k}$ and radius*

$$R_k = \sum_{\substack{l=1 \\ l \neq k}}^{m} |z_{k,l}| \ . \tag{3.113}$$

As $\tilde{\mathbf{G}}(s)$ can be considered as the limit of a sequence of finite-dimensional matrices, we obtain that for each $s \in \mathbb{C}$ and for each eigenvalue $\lambda(s)$ of $\tilde{\mathbf{G}}(s)$:

$$\exists k : \left|\lambda(s) - \tilde{G}_{k,k}(s)\right| \leq R_k(s) = \sum_{l \neq k} \left|\tilde{G}_{l,k}(s)\right| \ . \tag{3.114}$$

All eigenvalues $\lambda(s)$ of $\tilde{\mathbf{G}}(s)$ are hence contained within the Gershgorin circles with center points $\tilde{G}_{k,k}(s)$ and radii $R_k(s) = \sum_{l \neq k} \left|\tilde{G}_{l,k}(s)\right|$. This theorem is especially useful when $\tilde{\mathbf{G}}(s)$ is a near-diagonal matrix, i.e. if the time-varying feedback loop contains a dominant LTI component. In that case, it roughly holds that

$$\lambda_k(s) \approx \tilde{G}_{k,k}(s) \tag{3.115}$$

whereby the radius $R_k(s)$ of the corresponding Gershgorin circle provides a worst-case upperbound for the approximation error. Gershgorin circles can therefore be used to estimate how badly system stability is (potentially) deteriorated by the presence of time-varying components in the feedback path.

Rank-reduced approximations

Suppose that there exists a pair of finite-rank matrices $\mathbf{U}(s), \mathbf{V}(s) \in \mathbb{C}^{\infty \times M}$ such that

$$\tilde{\mathbf{G}}(s) \approx \mathbf{U}(s)\mathbf{V}(s)^T \tag{3.116}$$

constitutes a valid approximation, i.e. $\left\|\tilde{\mathbf{G}}(s) - \mathbf{U}(s)\mathbf{V}(s)^T\right\| \ll 1$. In that case, the eigenvalues of the ∞-dimensional HTM $\tilde{\mathbf{G}}(s)$ that can have a magnitude larger than 1 are roughly equal to the eigenvalues of the finite-dimensional matrix

$$\mathbf{G}_{reduced}(s) = \mathbf{V}(s)^T\mathbf{U}(s) \ . \tag{3.117}$$

Here, $\mathbf{G}_{reduced}(s) \in \mathbb{C}^{M \times M}$. Basically, this approach amounts to a subspace projection method whereby $\mathbf{U}(s)$ spans the M-dimensional subspace containing the eigenvectors associated with the largest eigenvalues of $\tilde{\mathbf{G}}(s)$.

Example (Downconversion with feedback): Consider again the downconversion stage with (parasitic) feedback in Fig. 3.15. The closed-loop input-output behavior is characterized by (3.83) with the open-loop gain HTM $\tilde{\mathbf{G}}(s)$ given by (3.84). Based on the Gershgorin theorem, we find the magnitude of the eigenvalues $\lambda_k(s)$ of $\tilde{\mathbf{G}}(s)$ to be bounded by

$$|\lambda_k(s)| \leq \frac{1}{2} \max_{k \in \mathbb{Z}} \left|H_2(s+jk\omega_0)\left(H_{fb}(s+j(k-1)\omega_0) + H_{fb}(s+j(k+1)\omega_0)\right)\right| \ . \tag{3.118}$$

The hatched region in Fig. 3.20(a) marks the region bounded by (3.118) for the filters $H_2(s)$ and $H_{fb}(s)$ in Fig. 3.15 respectively a second-order lowpass filter and a first-order highpass filter. As is observed, this region does not include $-1+j0$. As a consequence, it is guaranteed that the time-varying feedback path does not cause instabilities to occur.

More precise results are obtained using rank-reduced approximations. According to the rank-one approximation (3.90)-(3.91), $\mathbf{G}_{reduced}(s) = 0$. Therefore, as long as the rank-one approximation is valid, the feedback loop only has marginal impact on the system's dynamic behavior.

However, when the rank-one approximation starts to break down, e.g. due to a growing bandwidth of $H_2(s)$, we need to increase the number of rows of $\tilde{\mathbf{G}}(s)$ that we take into account. Using, for example, the rank-three approximation

$$\tilde{\mathbf{G}}(s) \approx \begin{bmatrix} \vdots & \vdots & \vdots \\ 0 & 0 & 0 \\ 1 & 0 & 0 \\ 0 & 1 & 0 \\ 0 & 0 & 1 \\ 0 & 0 & 0 \\ \vdots & \vdots & \vdots \end{bmatrix} \tilde{\mathbf{G}}_{[-1,1],\cdot}(s) \tag{3.119}$$

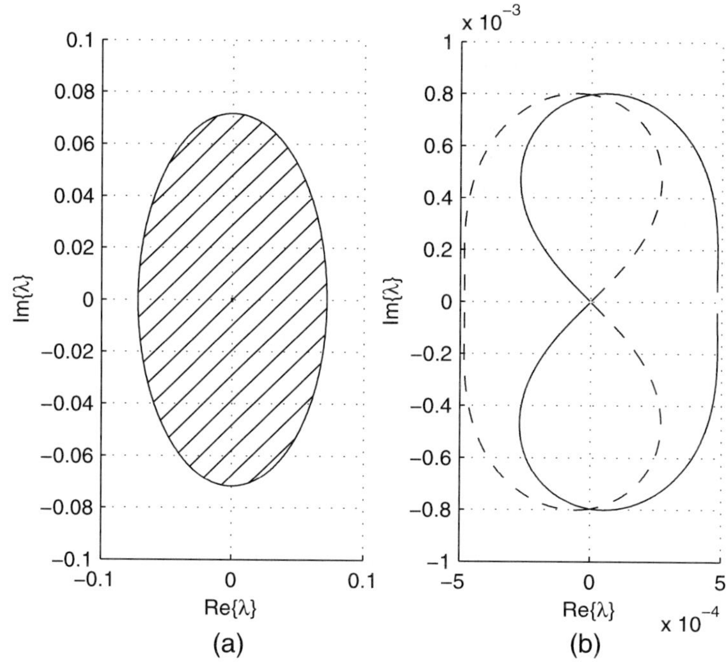

Figure 3.20: Estimating the position of the eigenvalues of $\tilde{\mathbf{G}}(s)$ using (a) the Gershgorin theorem and (b) the eigenvalues obtained from a rank-reduced approximation.

with $\tilde{\mathbf{G}}_{[-1,1],\cdot}(s)$ containing the rows -1, 0 and 1 of $\tilde{\mathbf{G}}(s)$, we find stability to be determined by the eigenvalues of

$$\mathbf{G}_{reduced} = \begin{bmatrix} 0 & \frac{H_2(s-j\omega_0)H_{fb}(s)}{2} & 0 \\ \frac{H_2(s)H_{fb}(s-j\omega_0)}{2} & 0 & \frac{H_2(s)H_{fb}(s+j\omega_0)}{2} \\ 0 & \frac{H_2(s-j\omega_0)H_{fb}(s)}{2} & 0 \end{bmatrix}. \quad (3.120)$$

These eigenvalues are given by

$$\left\{0, \pm\frac{1}{2}\sqrt{H_{fb}(s)H_2(s)\left(H_2(s-j\omega_0)H_{fb}(s-j\omega_0)+H_2(s+j\omega_0)H_{fb}(s+j\omega_0)\right)}\right\}. \quad (3.121)$$

Fig. 3.20(b) plots the non-zero eigenvalues in (3.121) for $s = j\omega$ with ω ranging from $-\omega_0/2$ to $\omega_0/2$. The eigenvalues turn out to be over an order of magnitude smaller than the worst-case upperbound (3.118) predicted by the Gershgorin theorem. Observing (3.121), it is seen that the eigenvalues can only be large in magnitude when the product $H_2(s)H_{fb}(s)$ remains large for frequencies up to ω_0. Only then, the stability of the time-varying feedback loop becomes an issue. ▲

3.5.3 Noise analysis

The remainder of this discussion on LPTV system analysis using HTMs is devoted to noise analysis. Just as LPTV system behavior transfers deterministic signal components from one frequency band to another, so does it transfer random (stochastic) signal components. A complex-valued stochastic process $n(t)$ is characterized by a probability density function (PDF) $P_n(s(t))$. This PDF specifies the probability for $n(t)$ to be near a particular sample $s(t)$. For a rigorous introduction to stochastic processes, see [Mid87].

Given a particular system and a random input signal's PDF, the task of finding the PDF that characterizes the output signal is often a very difficult one. However, in many instances, one is not interested in the entire stochastic characterization of $n(t)$. Often, first- and second-order stochastic behavior, i.e. the mean and autocorrelation function

$$m_n(t) = E\{n(t)\} \tag{3.122}$$

$$R_{nn}(t,\tau) = E\{n(t)n(t-\tau)^*\} \tag{3.123}$$

contain all the information that is needed on $n(t)$. Here, $E\{\cdot\}$ denotes the expected value operator while $n(t)^*$ is the adjoint of $n(t)$. In case of Gaussian stochastic processes, the quantities (3.122)-(3.123) even fix the entire PDF [Jayn03, Mid87].

This section develops the formalism necessary to deal with noise injected in LPTV systems. In doing so, it is mainly concerned with a second-order noise characterization (mean and autocorrelation function). It will be seen that traditional stationary noise models —and their power spectral densities (PSDs)— are no longer sufficient to characterize the kind of noise produced by LPTV systems. An extended class of noise processes is needed: cyclostationary noise processes.

The treatment that follows is to some extent similar to the one in [Royc98]. However, it differs by allowing a non-unique characterization for cyclostationary noise processes. These extra degrees of freedom can, for example, be used to simplify the problem setup. Furthermore, we always assume all noise sources to be zero-mean, or

$$m_n(t) = 0 \ .$$

Since a non zero-mean noise source can be modeled as the sum of a deterministic signal and a zero-mean noise source, there is no loss in generality.

3.5.3.1 Cyclostationarity: does it matter for design practice?

Before introducing a whole set of new concepts for dealing with noise in LPTV systems, we pause for a moment and ask the question: is it really needed? What is the problem with traditional noise analysis based on stationary noise models and their power spectral density (PSD) representations? The answer lies in the fact that a stationary noise model assumes the noise content in different, non-overlapping frequency bands to be uncorrelated.

Since LPTV systems cause up- or downconversions, they destroy stationarity by introducing cross-correlations between the noise content in different, non-overlapping frequency bands. Consider, for example, the leaky upconversion mixer shown in Fig. 3.21.

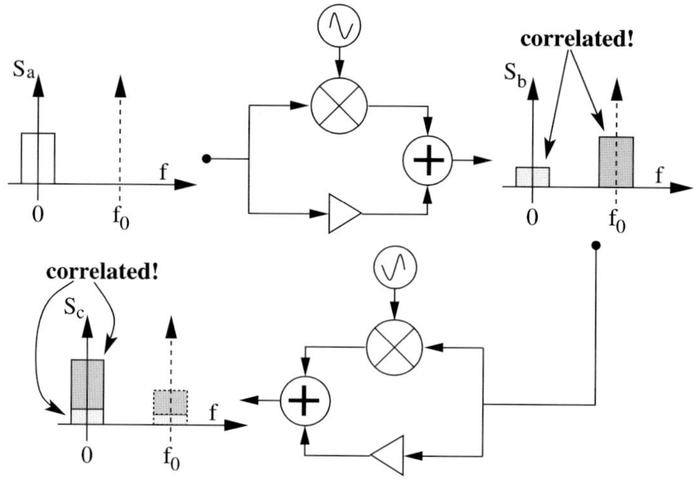

Figure 3.21: Any LPTV system involving up- or downconversion to different frequency bands destroys the stationarity of a noise source. As shown in the upper part of the figure, a "leaky" upconversion mixer transfers input noise to both baseband and the carrier frequency, causing the output noise in these frequency bands to be correlated. If noise components are correlated, simply adding their PSDs, e.g. to model the result of a leaky downconversion, is incorrect as it may lead to underestimating noise strength.

This mixer transfers input noise to both baseband and the carrier frequency. Clearly, the noise components in these bands is correlated as they originate from a common source. Hence, the resulting noise signal is no longer stationary.

Downconverting this signal using a second leaky mixer results in baseband output noise that combines input noise components coming from both baseband and the carrier frequency. As indicated in Fig. 3.21, these components are correlated. Therefore, it is no longer allowed to characterize the output noise PSD by adding the PSDs of its constituting components. Such a procedure may lead to over-optimistic noise analysis in a sense that the output noise strength is underestimated (by neglecting what can be seen as stochastic positive interference). As always, it is not a healthy design practice to underestimate potential trouble. It is therefore important to keep track of the cross-correlations that exist between the noise components in the different (non-overlapping) frequency bands.

In summary, the basic problem with stationary noise models lies in the fact that they are unable to capture cross-correlations between the noise components in different frequency bands. As illustrated by the example above, many common LPTV systems bring such cross-correlations about. Moreover, knowing the cross-correlations is important since neglecting them may lead to the underestimation of the output noise strength. Hence, we need a noise model that is able to capture cross-correlations between noise components in different frequency bands. This is precisely what a cyclostationary noise model has to offer. However, before elaborating cyclostationary processes in depth, we first give a brief review on traditional stationary noise theory as it contains many of its results will be needed further on.

3.5.3.2 A brief review on stationary noise theory

Stationary noise processes [Mid87] are the type of stochastic processes most common to designers. Mathematically, a zero-mean stochastic process $n(t)$ is (wide-sense) stationary if its autocorrelation function does not depend on the time t, or

$$E\{n(t)n(t-\tau)^*\} = R_{nn}(\tau) \ . \tag{3.124}$$

The way in which the behavior of the noise process at time t interrelates with that at a past or future time instance $t-\tau$, only depends on the time difference τ. As was already explained, this comes down to the absence of cross-correlations between noise components in different (non-overlapping) frequency bands. Many noise sources that occur in practice are stationary, e.g. the ones associated with a transistor[10]. A special case of stationary noise is *white noise*. For a zero-mean white noise process $n_w(t)$, it holds that

$$E\{n_w(t)n_w(t-\tau)^*\} = N\delta(\tau) \tag{3.125}$$

where $\delta(\tau)$ is the Dirac delta function. This implies that there is no correlation whatsoever between samples of a white noise process taken at different instances in time. Non-white noise processes are also said to be *colored*.

Stationarity turns out to be a most convenient property in dealing with a noise process $n(t)$. Frequency-domain manipulation becomes possible by introducing the noise process *power spectral density* (PSD) function $S_{nn}(j\omega)$. This PSD function is the time-averaged energy content in a narrow frequency band around ω, or

$$\begin{aligned} S_{nn}(j\omega) &= \lim_{T\to\infty} \frac{1}{T} E\left\{ \left| \int_{-T/2}^{+T/2} n(t)e^{-j\omega t}\,dt \right|^2 \right\} \\ &= \int_{-\infty}^{+\infty} R_{nn}(\tau)e^{-j\omega\tau}\,d\tau \ . \end{aligned} \tag{3.126}$$

From the definition it is immediately clear that $S_{nn}(j\omega)$ is real and positive $\forall \omega \in \mathbb{R}$. The lower equality follows from the Wiener-Khinchin theorem [Mid87]. Note that for a white noise source $S_{nn}(j\omega) = N, \forall \omega$.

Use of the PSD function $S_{nn}(j\omega)$ makes it very easy to compute changes in noise characteristics as noise passes through time-invariant systems, e.g. filters. Given a stationary input noise process $n(t)$ and the impulse response $h(\tau)$ of an LTI system, then the PSD function of the filtered noise process

$$y(t) = H[n(t)] = \int_0^{\infty} h(\tau)n(t-\tau)\,d\tau \tag{3.127}$$

[10] This is only approximately true for $1/f$ noise. In [Kesh82], it is argued that $1/f$ noise is time-varying in nature but behaves in a time-invariant manner when observed over a sufficiently small window in time.

equals

$$S_{yy}(j\omega) = H(j\omega)S_{nn}(j\omega)H(j\omega)^* \ . \tag{3.128}$$

Here, $H(s) = \mathcal{L}\{h(t)\}$. By using (3.128), it is possible to show that any colored noise source $n(t)$ can be obtained by passing white noise through an appropriate causal LTI filter.

Equation (3.128) is a special case of an important result that will be used later on. Let us be given two noise sources $n_1(t)$, $n_2(t)$ with a stationary cross-correlation function

$$R_{n_1n_2}(\tau) = E\{n_1(t)n_2(t-\tau)^*\} \ . \tag{3.129}$$

Furthermore, let us be given two time-invariant filters with impulse responses $h_1(\tau)$ and $h_2(\tau)$. The cross-correlation function of the filtered processes

$$y_1(t) = \int_0^\infty h_1(\tau)n_1(t-\tau)d\tau \tag{3.130}$$

$$y_2(t) = \int_0^\infty h_2(\tau)n_2(t-\tau)d\tau \ , \tag{3.131}$$

the satisfies

$$\mathcal{F}\{R_{y_1y_2}(\tau)\} = S_{y_1y_2}(j\omega) = H_1(j\omega)S_{y_1y_1}(j\omega)H_2(j\omega)^* \ . \tag{3.132}$$

Here $S_{n_1n_2}(j\omega) = \mathcal{F}\{R_{n_1n_2}(\tau)\}$ while $H_1(s)$ and $H_2(s)$ respectively equal the Laplace transforms of the impulse responses $h_1(t)$ and $h_2(t)$.

3.5.3.3 Cyclostationary noise

As observed from the above, stationarity makes it easy to perform noise computations. Unfortunately, as was discussed above, LPTV system behavior destroys stationarity. From a mathematical point of view, the autocorrelation function no longer solely depends upon the time difference τ. This is readily seen by considering the elementary mixing operation

$$y(t) = \cos(\omega_0 t)n(t) \ . \tag{3.133}$$

With $n(t)$ stationary and characterized by (3.124), the autocorrelation function of the output process $y(t)$ equals

$$R_{yy}(t,\tau) = E\{y(t)y(t-\tau)^*\} = \frac{1}{2}\left(\cos(2t-\tau)+\cos(\tau)\right)R_{nn}(\tau) \ . \tag{3.134}$$

This is a function of both τ and t. In characterizing noisy input-output behavior of LPTV systems, we therefore need to extend stochastic processes beyond stationary ones.

The direction in which to extend the class of relevant noise processes is suggested by a result that was already obtained for stationary colored noise: any stationary colored noise process can be obtained by passing a white noise signal through an appropriate LTI filter. Such a filter is also called an LTI *operator*. Hence, the class of stationary colored noise processes is generated by the algebra of LTI operators. Here, the word "algebra" simply denotes the fact that connecting two LTI operators in series or in parallel again yields an LTI operator.

The result above suggests us to define a class of noise processes that is generated by LPTV operators:

Definition (Cyclostationary noise): *A noise process $n(t)$ is called cyclostationary if there exists an LPTV operator $H[\cdot]$ with kernel $h(t,\tau)$ such that*

$$n(t) = H[n_w(t)] = \int_0^\infty h(t,\tau)n_w(\tau)d\tau \tag{3.135}$$

where $n_w(t)$ is a white noise process. In words, cyclostationary noise is obtained by passing white noise through an LPTV system.

Since the set of all LPTV systems makes up an operator algebra, LPTV system behavior will, by definition, preserve cyclostationarity. Cyclostationary noise processes are therefore associated with LPTV systems just as stationary noise processes are associated with LTI systems.

It should be noted that, traditionally, cyclostationarity is defined by means of the periodicity of the autocorrelation function $R_{nn}(t,\tau)$ with respect to t [Royc98, Strom77]. However, the definition above gives a better insight into the nature of cyclostationary noise: it relates the structure of a noise process to that of a "generating" operator. Furthermore, this definition is readily generalized: for each algebra of LTV operators, there exists a corresponding class of noise processes.

3.5.3.4 Characterizing cyclostationary noise

Consistent with the ideas underlying the HTM framework, we choose to represent a cyclostationary noise signal $n(t)$ as a vector

$$\mathbf{n}(t) = \begin{bmatrix} \vdots \\ n_{-1}(t) \\ n_0(t) \\ n_1(t) \\ \vdots \end{bmatrix} \tag{3.136}$$

of stochastic equivalent baseband components. The overall noise process $n(t)$ then equals

$$n(t) = \begin{bmatrix} \cdots & e^{-j\omega_0 t} & 1 & e^{j\omega_0 t} & \cdots \end{bmatrix} \times \mathbf{n}(t) = \sum_{k=-\infty}^{+\infty} e^{jk\omega_0 t} n_k(t) \ . \tag{3.137}$$

This equality holds in a mean square sense. The processes $n_k(t)$ can be interpreted as the noisy signal content in the k-th frequency band[11]. Up to second order, the vector

[11] As discussed in section 3.3.2, this concerns an interpretation and not a requirement. If, for example, $n_0(t)$ is chosen to be a white noise process, then it contributes to signal content in all frequency bands. Selecting the $n_k(t)$ to be largely constrained to baseband, however, eases interpreting results.

$\mathbf{n}(t)$ of noise processes is completely characterized by the correlation matrix

$$E\{\mathbf{n}(t)\mathbf{n}(t-\tau)^*\} = [E\{n_k(t)n_l(t-\tau)^*\}] \tag{3.138}$$
$$= \left[R_{n_k n_l}(\tau)\right] \tag{3.139}$$
$$= \mathbf{R}_{\mathbf{nn}}(\tau) \; . \tag{3.140}$$

Note that the cross-correlation function $R_{n_k n_l}(\tau)$ between any two baseband components $n_k(t)$ and $n_l(t)$ is assumed to be stationary. The matrix $\mathbf{R}_{\mathbf{nn}}(\tau)$ is termed a *harmonic correlation matrix.*

Question of course is whether, for an arbitrary cyclostationary noise signal $n(t)$, it is always possible to find a vector $\mathbf{n}(t)$ that satisfies both (3.137) and (3.140). A first part of the answer lies in the fact that for a white noise process $n_w(t)$, finding such a vector $\mathbf{n}_w(t)$ is trivial. One could, for instance, simply select

$$n_{w,0}(t) = n_w(t) \tag{3.141}$$
$$n_{w,k}(t) = 0, \forall k \neq 0 \; . \tag{3.142}$$

A choice that is better in line with the interpretation of the $n_k(t)$ as equivalent baseband components is given by

$$n_{w,k}(t) = \int_{-\infty}^{+\infty} \frac{\sin(\omega_0 \tau)}{\tau} e^{-jk\omega_0(t-\tau)} n_w(t-\tau) d\tau \; . \tag{3.143}$$

In this case, all $n_{w,k}(t)$ are bandlimited within $[-\omega_0/2, \omega_0/2]$ and uncorrelated. The second part of the noise representation problem is solved in section 3.5.3.7. There, it will be shown that LPTV input-output signal processing preserves the existence of a vector representation that satisfies both (3.137) and (3.140). If such a vector exists for the noise process at the input, it is straightforward to derive one that represents the noise process at the output. If we combine both facts above with the definition of cyclostationary noise processes, then it is readily proven that it is always possible to represent a cyclostationary noise process by means of a vector $\mathbf{n}(t)$ —of stochastic equivalent baseband components— that satisfies both (3.137) and (3.140).

For the stochastic characterization of the vector $\mathbf{n}(t)$, we rarely use harmonic correlation matrix $\mathbf{R}_{\mathbf{nn}}(\tau)$ itself. Instead, it is more convenient to perform computations based on

$$\mathbf{S}_{\mathbf{nn}}(j\omega) = \mathcal{F}\{\mathbf{R}_{\mathbf{nn}}(\tau)\} \; . \tag{3.144}$$

The matrix $\mathbf{S}_{\mathbf{nn}}(j\omega)$ is termed a *harmonic power spectral density* (HPSD) matrix. A HPSD matrix contains all the information needed for the second-order, stochastic characterization of a given cyclostationary noise process $n(t)$.

3.5.3.5 On the non-uniqueness of a HPSD matrix

As presented in this treatment, *a HPSD matrix is not unique*. Different sets of baseband components $\{n_k(t)\}$ can be chosen to represent the same overall process $n(t)$. It is only required that (3.137) and (3.140) are satisfied. As was discussed in section 3.3.2, this non-uniqueness is inherent to the degrees of freedom available in the linear periodically modulated signal model (3.137). We do not fix these degrees of freedom in advance. In some cases, especially for symbolic computations (pencil and paper analysis), a proper choice may help to simplify the problem setup.

However, within a CAD framework, non-uniqueness hampers automating procedures. In this case, uniqueness can be enforced by requiring all stochastic baseband components $n_k(t)$ to be strictly bandlimited to $[-\omega_0/2, \omega_0/2]$ (note that this is imposed implicitly in [Royc98]). As an advantage, it then becomes sufficient to compute $\mathbf{S_{nn}}(j\omega)$ for $\omega \in [-\omega_0/2, \omega_0/2]$; outside this interval, $\mathbf{S_{nn}}(j\omega)$ is zero (or, as in [Royc98], it repeats itself in the sense of (3.53)). Furthermore, bandlimiting the $n_k(t)$ gives a clear-cut interpretation to the elements of $\mathbf{S_{nn}}(j\omega)$: the diagonal element $S_{n_k n_k}(j\omega)$ captures the autocorrelation of the noise content in the k-th frequency band; the off-diagonal element $S_{n_k n_l}(j\omega)$ (for $k \neq l$) captures the cross-correlation between the signal content in two different (non-overlapping) frequency bands.

3.5.3.6 Properties of a HPSD matrix

Although the choice of the stochastic baseband components $n_k(t)$ is to some extent arbitrary, different HPSD matrices $\mathbf{S_{nn}}(j\omega)$ that characterize the same cyclostationary noise process $n(t)$ do have a number of properties in common.

Semi-positive definiteness

A first property of a HPSD matrix $\mathbf{S_{nn}}(j\omega)$ is its semi-positive definiteness for any $\omega \in \mathbb{R}$. This follows directly from the properties of PSDs of stationary noise sources. Given a complex-valued vector $\mathbf{x} = \begin{bmatrix} \cdots & x_{-1} & x_0 & x_1 & \cdots \end{bmatrix}^T$, it is easily verified that the process

$$n_x(t) = \mathbf{x}^* \mathbf{n}(t) \tag{3.145}$$

is stationary with its PSD satisfying

$$S_{xx}(j\omega) = \mathbf{x}^* \mathbf{S_{nn}}(j\omega) \mathbf{x} \geq 0, \forall \omega \ . \tag{3.146}$$

The latter inequality follows from the properties of stationary noise processes. Since $\mathbf{x}$ is arbitrary, (3.146) implies that the HPSD matrix $\mathbf{S_{nn}}(j\omega)$ is a semi-positive definite matrix for any $\omega \in \mathbb{R}$.

Structural constraints

As was already mentioned, it is possible to characterize the same overall cyclostationary noise process $n(t)$ by means of different HPSD matrices $\mathbf{S_{nn}}(j\omega)$. This corresponds to different choices for the baseband components $n_k(t)$. However, all HPSD matrices associated with a same noise process $n(t)$ will have a number of quantities in common.

These quantities are revealed by expressing the autocorrelation function $R_{nn}(t,\tau)$ of the overall cyclostationary process $n(t)$ in terms of the elements of a possible HPSD matrix $\mathbf{S_{nn}}(j\omega)$. Using (3.137) and some algebra, we obtain

$$\begin{aligned} R_{nn}(t,\tau) &= E\{n(t)n(t-\tau)^*\} \\ &= \sum_{k=-\infty}^{\infty} R_{nn,k}(\tau) e^{jk\omega_0 t} \end{aligned} \tag{3.147}$$

with coefficients $R_{nn,k}(\tau)$ that equal

$$R_{nn,k}(\tau) = \sum_{l=-\infty}^{+\infty} R_{n_{k+l}n_l}(\tau) e^{jl\omega_0\tau} \ . \tag{3.148}$$

Here, the $R_{n_{k+l}n_l}(\tau)$ are the elements of the harmonic correlation matrix $\mathbf{R_{nn}}(\tau)$ defined in (3.139). The PSDs associated with the $R_{nn,k}(\tau)$ are then given by

$$S_{nn,k}(j\omega) = \mathcal{F}\{R_{nn,k}(\tau)\} = \sum_{l=-\infty}^{+\infty} S_{n_{k+l}n_l}(j\omega - jl\omega_0) \ . \tag{3.149}$$

Observing (3.147), it is seen that $R_{nn}(t,\tau)$ is periodic in t since it can be expanded in a Fourier series. Furthermore, (3.149) shows us how to interpret the PSD $S_{nn,k}(j\omega)$ associated with the k-th Fourier coefficient $R_{nn,k}(\tau)$: the amount of energy in $S_{nn,k}(j\omega)$ in the band around $l\omega_0$ is a measure for the strength of the cross-correlation between the noise components in the frequency bands around respectively $l\omega_0$ and $(k+l)\,\omega_0$.

Since the quantities $S_{nn,k}(j\omega)$ are unique, the relation (3.149) represents a number of structural constraints that must be satisfied by all different HPSD characterizations of a same cyclostationary noise process $n(t)$. Whatever characterization is used, the sum on the right-hand side of (3.149) does not vary from one characterization to another.

As a final note, we mention that (3.149) allows us to construct a uniquely determined HPSD matrix. If the baseband components $n_k(t)$ are constrained to to $[-\omega_0/2, \omega_0/2]$, then the same holds for all corresponding HPSD matrix elements $S_{n_kn_l}(j\omega)$. Using (3.149), we then find that

$$S_{n_kn_l}(j\omega) = S_{nn,k-l}(j\omega + jl\omega_0) \ , \tag{3.150}$$

for $\omega \in [-\omega_0/2, \omega_0/2]$ while equal to zero elsewhere. Hence, imposing bandwidth restrictions on the $n_k(t)$ hence results in a uniquely determined HPSD matrix.

3.5.3.7 Noise computations using HTMs and HPSD matrices

Next, we address the characterization for the output $y(t)$ of an LPTV system to which a cyclostationary input signal $n(t)$ is applied. Hereby, we keep in mind that LPTV operators preserve cyclostationarity. Hence, we look for a HPSD matrix that characterizes

$$y(t) = \int_0^{\infty} h(t,\tau) n(t-\tau) \ . \tag{3.151}$$

Here, $h(t,\tau)$ is an LPTV system kernel.

With $\tilde{\mathbf{H}}(s)$ the HTM associated with $h(t,\tau)$, it follows from (3.137) and the HTM-based input-output relation (3.43) that

$$y(t) = \sum_{l=-\infty}^{+\infty} y_l(t) e^{jl\omega_0 t} \ . \tag{3.152}$$

Here,

$$y_l(t) = \sum_{k=-\infty}^{+\infty} \int_0^{\infty} \tilde{h}_{l,k}(\tau) n_k(t-\tau) d\tau \tag{3.153}$$

with $\tilde{h}_{l,k}(\tau) = \mathcal{L}^{-1}\{\tilde{H}_{l,k}(s)\}$. Using (3.132) it is then readily obtained that the matrix

$$\mathbf{S_{yy}}(j\omega) = \tilde{\mathbf{H}}(j\omega)\mathbf{S_{nn}}(j\omega)\tilde{\mathbf{H}}(j\omega)^* \tag{3.154}$$

represents a valid HPSD matrix that characterizes the correlations between the baseband components $y_l(t)$. Note that all correlations between the different components $y_l(t)$ remain stationary. The relation (3.154) is a straightforward extension of (3.128). Here, the former captures the impact of an LPTV system on a cyclostationary noise process while the latter models the impact of an LTI system on a stationary noise processes.

Example (Mixing noise): Consider the elementary downconversion stage in Fig. 3.13. The elements of the HTM corresponding to this stage are determined by (3.74). As an input signal, we apply

$$u(t) = \cos(\omega_0 t)n(t) \tag{3.155}$$

where $n(t)$ is a stationary noise process with its PSD $S_{nn}(j\omega)$ largely constrained to $[-\omega_0/2, \omega_0, 2]$. A possible HPSD matrix for the cyclostationary input process $u(t)$ is given by

$$\mathbf{S_{uu}}(j\omega) = \begin{bmatrix} & \vdots & \vdots & \vdots & \\ \cdots & \frac{1}{4}S_{nn}(j\omega) & 0 & \frac{1}{4}S_{nn}(j\omega) & \cdots \\ \cdots & 0 & 0 & 0 & \cdots \\ \cdots & \frac{1}{4}S_{nn}(j\omega) & 0 & \frac{1}{4}S_{nn}(j\omega) & \cdots \\ & \vdots & \vdots & \vdots & \end{bmatrix}. \tag{3.156}$$

Note the presence of the off-diagonal elements. This denotes that the noise components in different frequency bands are correlated. For the baseband component $y_0(t)$ of the noise at the output of the downconversion stage, we find, by means of (3.154) and (3.74), that

$$\begin{aligned} S_{y_0y_0}(j\omega) &= \frac{1}{16}\left|H_1(j\omega - j\omega_0) + H_1(j\omega + j\omega_0)\right|^2 \left|H_2(j\omega)\right|^2 S_{nn}(j\omega) \\ &\approx \frac{1}{4}\mathrm{Re}\left\{H_1(j\omega_0)\right\}^2 \left|H_2(j\omega)\right|^2 S_{nn}(j\omega) . \end{aligned} \tag{3.157}$$

The latter approximation holds for $\omega \ll \omega_0$. When neglecting the cross-correlations between the noise components in the different input frequency bands, i.e. when zeroing the off-diagonal elements in (3.156), we obtain

$$\begin{aligned} [\mathbf{S_{yy}}(j\omega)]_{0,0} &= \frac{1}{16}\left(\left|H_1(j\omega - j\omega_0)\right|^2 + \left|H_1(j\omega - j\omega_0)\right|^2\right) \left|H_2(j\omega)\right|^2 S_{nn}(j\omega) \\ &\approx \frac{1}{8}\left|H_1(j\omega_0)\right|^2 \left|H_2(j\omega)\right|^2 S_{nn}(j\omega) . \end{aligned} \tag{3.158}$$

For a typical design, (3.158) predicts a noise strength that is a factor two below the correct solution in (3.157). Hence, a stationary noise approximation may be over optimistic in predicting output noise levels. ▲

This concludes our treatment on LPTV system analysis using HTMs. The theoretical results above demonstrate that use of HTMs allows us to extend traditional LTI frequency-domain techniques to the realm of LPTV system analysis. As will be illustrated by the practical applications in the next chapter, this is of great use in system and circuit design practice.

3.6 Conclusions and directions for further research

By considering the signal content stored on different carrier waves, i.e. in different frequency bands, as independent input signals, a SISO LPTV system can be described as a limiting case of a MIMO LTI system. As a result, the behavior of an LPTV system can be captured by a matrix of transfer functions, called a HTM. Techniques for multi-variable LTI system analysis and design can therefore be applied to analyze the behavior of LPTV systems. These techniques are made efficient by exploiting both the structure inherent to HTMs and the properties of HTMs as they occur in practical applications. This results in a consistent framework for handling LPTV system analysis.

Furthermore, the theory has been presented in a manner that is straightforward generalize towards other classes of LTV systems. These classes are then associated with sets of carrier waves other than the harmonic functions $\left\{\psi_m(t) = e^{jm\omega_0 t}\right\}$ used in this chapter. Suggested directions for further research involve, for instance, linear quasi-periodic time-varying system analysis and the use of principal component analysis for the autodetection of relevant sets of carrier waves. The former is useful to analyze the behavior of heterodyne transceiver stages, while the latter can, for example, be applied to capture the noise behavior of nonlinear systems.

Chapter 4

Applications of LPTV system analysis using harmonic transfer matrices

Whoever in the pursuit of science, seeks after immediate practical utility may rest assured that he seeks in vain. —*Hermann von Helmholtz*

Harmonic transfer matrices (HTM) constitute a natural framework for frequency-domain analysis of linear periodically time-varying (LPTV) system behavior. They directly describe how an LPTV system transfers signal content from one frequency band to another. Furthermore, HTMs not only satisfy from a theoretical point of view, they also have numerous practical applications. This chapter illustrates the practical use of HTMs for analyzing phase-locked loop (PLL) and mixer behavior. Hereby, special focus is given to symbolic results and techniques to obtain them. These symbolic results are most useful for guiding design decisions. It demonstrates the power of HTM-based analysis as a practical means to capture a circuit's time-varying small-signal behavior.

After a brief reminder on HTM basics in section 4.1, section 4.2 addresses frequency-domain modeling and analysis of PLL small-signal behavior, including the time-varying aspects. Using the HTM formalism, the well-known continuous-time, linear time-invariant (LTI) approximation is extended to take the impact of a PLL's time-varying behavior into account. The main source of this time-varying behavior lies in the nature of the sampling/mixing phase-frequency detector (PFD). Especially for PLLs with a fast feedback loop, this time-varying behavior has severe impact on, for example, loop stability. Hence, it cannot be neglected. Contrary to traditional LTI analysis, HTM-based analysis is able to predict and quantify these effects.

As holds for all pencil and paper analysis, HTM-based computations, especially symbolic ones, can be tedious and error prone. Section 4.3 therefore introduces an algorithm for automated HTM-based symbolic analysis of LPTV system behavior. It demonstrates that traditional techniques for automated symbolic analysis of LTI systems [Fern98, Gie91, Lin91] can be extended to handle LPTV behavior as well. The algorithm is applied to a number of different downconversion stages.

4.1 HTMs in a nutshell

The previous chapter elaborated the HTM framework in full mathematical detail. This may have left the impression that it requires an enormous amount of knowledge to apply HTMs in practice. This is not the case! Anyone mastering frequency-domain LTI system analysis can, with limited effort, get sufficient grasp on HTMs to put them to use in practice. This section summarizes the basic knowledge that is needed to get started. Of course, the better one masters the complete theory on HTMs, the easier it becomes to handle more complex systems. Note that this section can be skipped by readers who feel sufficiently familiar with basic HTM theory.

A HTM-based representation of LPTV system behavior relies on the observation that LPTV systems transfer signal content from one frequency band to another. In a first step, this requires the formalization of the concept of "signal content stored in a particular frequency band". This is accomplished by means of a *linear periodically modulated* signal model

$$u(t) = \sum_{m=-\infty}^{+\infty} u_m(t)e^{jm\omega_0 t} \quad . \tag{4.1}$$

Here, the $u_m(t)$ are termed *envelopes* or *equivalent baseband components*. They model the signal content of $u(t)$ in the frequency band around $m\omega_0$. When dealing with LPTV system behavior, it turns out to be more efficient to represent a signal $u(t)$ using the vector of spectra

$$\mathbf{U}(s) = \begin{bmatrix} \cdots & U_{-2}(s) & U_{-1}(s) & U_0(s) & U_1(s) & U_2(s) & \cdots \end{bmatrix}^T \tag{4.2}$$

rather than the single scalar spectrum $U(s) = \mathcal{L}\{u(t)\}$. Here, $U_m(s) = \mathcal{L}\{u_m(t)\}$ is the Laplace transform of the equivalent baseband component $u_m(t)$[1].

Let us be given an input signal $u(t)$ and an LPTV system

$$y(t) = H[u(t)] = \int_0^\infty h(t,\tau)u(t-\tau)d\tau \quad . \tag{4.3}$$

Using the signal representation (4.2) for both $u(t)$ and $y(t)$, it can then be shown that the operation of this LPTV system is described by frequency-domain matrix-vector multiplication, or

$$\mathbf{Y}(s) = \tilde{\mathbf{H}}(s)\mathbf{U}(s) \quad . \tag{4.4}$$

The matrix $\tilde{\mathbf{H}}(s)$ is called the harmonic transfer matrix (HTM) [Moll00, Vana02c, Were91a] corresponding to the LPTV system $H[\cdot]$. When truncated and evaluated at $s = j\omega$, it corresponds to the harmonic conversion matrix introduced in [Maas88].

The meaning of the HTM $\tilde{\mathbf{H}}(s)$ and its elements $\tilde{H}_{n,m}(j\omega)$ becomes clear when we consider a single row (with index n) in (4.4), i.e.

$$Y_n(s) = \sum_{m=-\infty}^{+\infty} \tilde{H}_{n,m}(s)U_m(s) \quad . \tag{4.5}$$

The output signal content in the frequency band around $n\omega_0$ is seen to be a linear combination of contributions coming from all possible input frequency bands. The

[1] It often makes good sense to require the equivalent baseband spectra $U_m(s)$ to be bandlimited to $[-\omega_0/2, \omega_0/2]$. However, as was discussed in section 3.3.2 in the previous chapter, this is not strictly necessary.

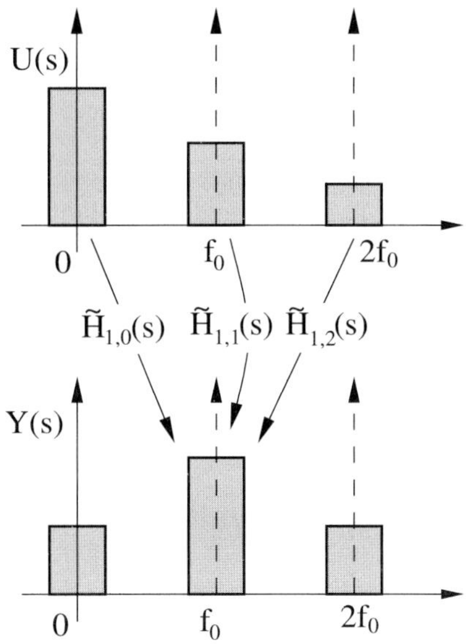

Figure 4.1: The HTM elements $\tilde{H}_{1,0}(s)$, $\tilde{H}_{1,1}(s)$ and $\tilde{H}_{1,2}(s)$ characterize the transfer of signal content from different input frequency bands to the output frequency band around the carrier frequency.

HTM element $\tilde{H}_{n,m}(s)$ hereby characterizes the transfer of the signal content from the input signal frequency band around $m\omega_0$ to the output signal frequency band around $n\omega_0$. In short, the HTM $\tilde{\mathbf{H}}(s)$ characterizes how information moves from one frequency band to another. This process is illustrated in Fig. 4.1.

Besides offering a compact frequency-domain characterization of LPTV system behavior, HTMs allow efficient manipulation of LPTV systems and straightforward computation of the overall input-output behavior, given a system's building-block HTMs. For example, the HTMs corresponding to the parallel connection $y = H_1[u] + H_2[u]$ and the series connection $y = H_2[H_1[u]]$ are respectively given by

$$\tilde{\mathbf{H}}_{+}(s) = \tilde{\mathbf{H}}_1(s) + \tilde{\mathbf{H}}_2(s) \tag{4.6}$$
$$\tilde{\mathbf{H}}_{\times}(s) = \tilde{\mathbf{H}}_2(s)\tilde{\mathbf{H}}_1(s) \ . \tag{4.7}$$

Note the resemblance with results from LTI system analysis. Furthermore, HTMs corresponding to basic building-block behavior are straightforward to construct. An LTI system with (scalar) transfer function $H(s)$ corresponds to a diagonal HTM whose elements are given by

$$\begin{cases} \tilde{H}_{n,m}(s) = H(s + jm\omega_0) & m = n \\ \tilde{H}_{n,m}(s) = 0 & m \neq n \end{cases} . \tag{4.8}$$

The elements of a HTM corresponding to a memoryless multiplication $y(t) = p(t)u(t)$, with $p(t) = \sum_{k=-\infty}^{+\infty} P_k e^{jk\omega_0(t)}$, are specified by

$$\tilde{H}_{n,m}(s) = P_{n-m} \quad . \tag{4.9}$$

Using (4.6)-(4.9), HTMs corresponding to more complex systems can be derived from the system's topology and the HTMs of its underlying building blocks.

In summary, HTMs allow us to handle LPTV system behavior in a way that largely resembles LTI system analysis. Hereby, the traditional scalar transfer function is replaced by a matrix of transfer functions: the harmonic transfer matrix. Understanding this basic notion is sufficient to make a beginning with practical HTM-based analysis of LPTV system behavior as occurring in PLLs and mixers.

4.2 Phase-Locked Loop analysis

Phase-locked loops (PLLs) are used in both analog and digital systems for generating signals that track the phase of a given input signal. They can be used to reduce oscillator phase noise by locking it to a high-quality reference, to synthesize frequencies which are multiples of the input frequency or, in digital applications, to buffer and deskew clock signals. A rigorous introduction to PLLs and their applications can be found in [Gard79]. In a lot of applications, PLLs are among those blocks whose performance is crucial in meeting system-level specifications. Hence, adequate and correct analysis of their behavior is of great importance.

Traditionally, PLLs are handled as LTI feedback loops [Gard79]. However, the behavior of many PLL building blocks exhibits time-varying components. In an LTI model, these components are simply neglected. However, as the PLL feedback loop grows faster, time-varying behavior becomes important as it causes, for example, stability degradation and noise folding. Analyzing these phenomena therefore requires theory and methods to capture a PLL's time-varying behavior.

This chapter analyzes PLL time-varying behavior using HTM-based methods. By means HTMs, we are able to predict and quantify the aspects of a PLL's behavior that are due to the time-varying nature of its building blocks. The analysis establishes rigorous mathematical underpinnings for PLL small-signal analysis. In doing so, it demarcates the region of validity of traditional PLL models build on LTI theory.

The discussion on time-varying PLL behavior is organized as follows. Section 4.2.1 introduces a PLL's basic architecture together with a short discussion on its building blocks. Next, section 4.2.2 reviews prior art on modeling and analysis of PLLs. In section 4.2.3, we provide a brief discussion on the construction of signal models appropriate for analyzing PLL behavior. Section 4.2.4 continues with models for the behavior of the PLL building blocks. These models are expressed in terms of HTMs. In section 4.2.5, the building block models are connected in order to obtain the input-output HTM of the overall PLL. Sections 4.2.6 and 4.2.7 show practical results by applying the method to PLLs implemented respectively using a sampling and a mixing PFD. Finally, in section 4.2.8 we come to the conclusions.

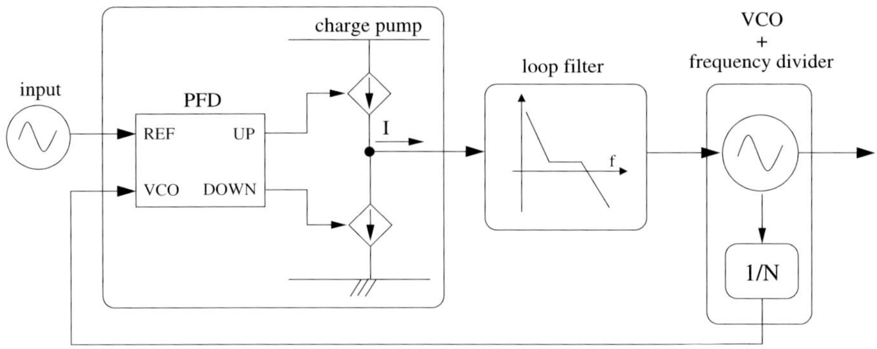

Figure 4.2: Typical PLL architecture consisting of a phase-frequency detector (PFD), a loop filter and a voltage-controlled oscillator (VCO) – frequency divider combination. In the architecture above, the PFD is implemented digitally, driving a charge pump.

4.2.1 PLL architectures and PLL building blocks

Fig. 4.2 illustrates a typical PLL architecture. It consists of a phase-frequency detector (PFD), a loop filter $H_{LF}(s)$ and a voltage-controlled oscillator (VCO) – frequency divider combination. The PFD measures the phase difference between the input signal that drives the PLL and the frequency-divided VCO signal. Its output is passed through the loop filter. This filtered error signal is then used to steer the VCO's oscillation frequency. Due to the correcting actions of the PLL feedback loop, the phase (and therefore also the frequency) of the frequency-divided VCO signal tracks that of the input[2]. This can be used for frequency synthesis, VCO phase noise suppression or frequency/phase modulation.

If the input signal is periodic and if the PLL is stable, then there exists a periodic large-signal operating point. As outlined in section 3.1.3 of the previous chapter, linearizing the PLL's behavior in the neighborhood of this operating point results in an LPTV small-signal model. So, in analyzing PLL small-signal behavior, we are essentially dealing with a time-varying system. Especially for PLLs with a fast feedback loop, this time-varying behavior has a severe impact on, for instance, loop stability and cannot be neglected.

The building block that contributes the most to a PLL's time-varying nature is the PFD. because of their superior acquisition of phase-lock, one nowadays most commonly uses digital PFDs that steer a charge-pump [Lee98]. These PFDs measure the phase error as the distance between the zero-crossings of the reference signal and the (frequency-divided) VCO signal. Hence, they only compute the phase error once per period of the reference signal, i.e. they sample the phase error. For this reason, we call them

[2]For this to hold, the frequency of the input signal must be within a sufficiently small range (the capture range) of that of the divided free-running VCO signal.

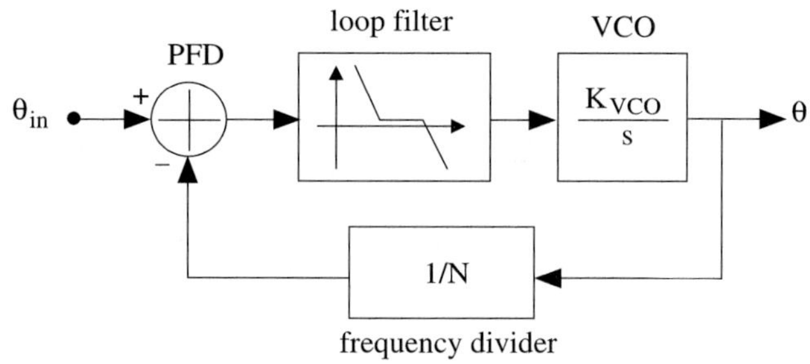

Figure 4.3: Continous-time linear time-invariant PLL model. This model neglects all time-varying behavior. Its validity is limited to slow feedback loops.

sampling PFDs. A second PFD implementation builds on the multiplication of the frequency-divided VCO signal and the reference signal by means of a mixer . The phase error is then computed as the DC component of the mixing product. This type of PFD is called a *mixing (or multiplying) PFD* and is currently still often used in microwave applications.

4.2.2 Prior art

In an attempt to capture a PLL's behavior, different models have been proposed in the past. The one best known from textbook analyses is the continuous-time linear time-invariant (CT-LTI) model [Cran98, Gard79, Lee98]. For PLLs with a sampling PFD, a discrete-time linear time-invariant (DT-LTI) model has been developed [Gard80, Hein88]. Other existing techniques for analyzing PLL behavior are simulation-based [Perr02].

4.2.2.1 Continuous-time, linear time-invariant model

Classical textbook treatments typically capture a PLL's small-signal behavior using CT-LTI feedback theory [Gard79, Lee98]. Analysis builds on the loop model illustrated in Fig. 4.3. In the frequency domain, input-output behavior is governed by the relation

$$H_{PLL}(s) = \frac{\theta(s)}{\theta_{in}(s)} = N\frac{G(s)}{1+G(s)} \quad . \tag{4.10}$$

Here,

$$G(s) = \frac{1}{N} H_{LF}(s) \frac{K_{VCO}}{s} \tag{4.11}$$

is the PLL's open-loop gain. The quantities θ_{in} and θ respectively represent the phase of the input and the VCO signal. $H_{LF}(s)$ is the loop filter's transfer function. Furthermore, the VCO behavior is modeled by means of an integrator. This stems from

$$\theta(t) = \int_0^t \omega_{VCO}(\tau)d\tau = \int_0^t K_{VCO} v_{control}(\tau)d\tau \tag{4.12}$$

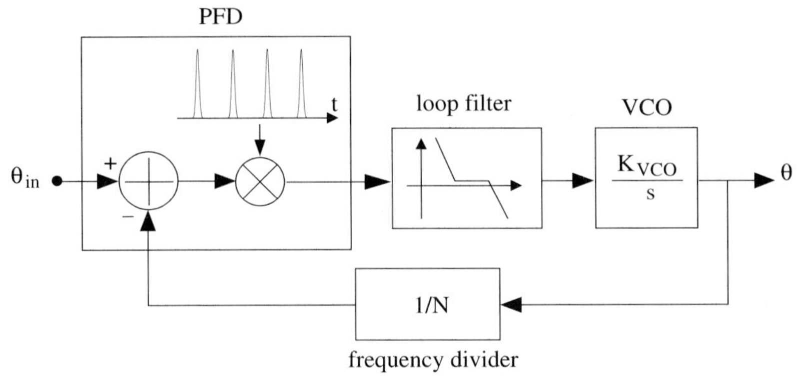

Figure 4.4: PLL implemented using a sampling PFD. Here, the PFD samples the instantaneous phase error once (or twice) per period of the input signal. Sampling is modeled as a multiplication of the phase error with a train of Dirac impulses.

where $\omega_{VCO}(\tau)$ is the instantaneous oscillation frequency of the VCO. This frequency is assumed to be proportional to the control voltage that steers the VCO.

As long as the models in Fig. 4.3 are valid, PLL design amounts to the design of a CT-LTI feedback loop. This problem is handled using techniques for traditional time-invariant feedback theory [Fran94]. However, as will soon be demonstrated, this CT-LTI PLL model only works fine as long as the unity-gain frequency of the open-loop gain $G(s)$ is well below the frequency of the input signal. If this condition does not hold, the CT-LTI model runs into trouble as the time-varying aspects of the building block behavior become important. Especially for PLLs with a fast feedback loop and a sampling PFD, the impact of the time-varying behavior on the system performance can become quite dramatic.

A second problem involving the CT-LTI PLL model lies in a lack of clear theoretical foundations. Contrary to true LTI systems, like filters, the CT-LTI PLL model cannot be derived formally as a small-signal approximation in the neighborhood of a DC operating point. As noted above, the operating point of a PLL is a periodically time-varying one. This rises questions as to how to determine when we can safely assume the CT-LTI model to be correct.

4.2.2.2 Discrete-time, linear time-invariant model

It is already known for some time that PLLs implemented using sampling-PFDs, e.g. based on digital logic and a charge pump, exhibit discrete-time behavior [Gard80, Hein88]. The instantaneous phase error is sampled once (or twice) per period of the reference signal. Sampling is triggered by the zero-crossings of the input signal. Fig. 4.4 illustrates the functional model that serves as a starting point for analysis. Sampling is modeled as a multiplication of the phase error with a train of Dirac impulses.

A number of authors [Gard80, Hein88] have suggested to treat these PLL implementations as discrete-time systems. To do so, they restrict the points in time for which they computate the PLL signal values to the sampling instances. These DT-LTI methods essentially amount to a lifting technique [Bam92]. They reveal constraints on the design of a PLL's open-loop characteristic that cannot be derived using a CT-LTI approximation.

However, discrete-time (z-domain) models [Gard80, Hein88] do not fully recognize the mixed continuous-time/discrete-time nature of a PLL. Moreover, these methods are not very well suited as a framework for symbolic computations, at least not for arbitrary loop characteristics. This limits their help in gaining understanding and in making decisions on design parameters. Finally, this discrete-time approach does not allow a unified treatment of PLL behavior that, amongst others, easily encompasses all possible PFD implementations.

4.2.2.3 Numerical simulation of PLLs

When the previous (CT-LTI and DT-LTI) approximations fail, it is of course always possible to analyze a PLL's behavior by means of numerical simulation. This can either be done using a general-purpose simulation algorithm (SPICE or a VHDL-AMS simulator) or using dedicated methods [Perr02]. Note that use of dedicated algorithms can help us to reduce simulation times from hours to minutes.

The advantage of numerical simulations lies in the inherent accuracy of the results. By construction, simulations capture time-varying behavior and even nonlinear effects[3]. The disadvantage is that numerical simulations require (sensible) numerical values for all design parameters. Somehow, a designer must come up with these values in advance. Hence, numerical simulation algorithms are mainly useful in a verification stage or when finetuning a design. However, for the initial stages of a design process, one (also) desires methods that allow us to express input-output relations in a symbolic manner.

4.2.2.4 Contributions of this work

This work presents a method for constructing s-domain small-signal PLL models that capture the time-varying aspects of a PLL's behavior. The method is applicable to arbitrary types of PLL architectures, implemented using either a sampling- or mixing-type PFD. The models are stated in terms of the HTM formalism. Hence, it is a frequency-domain approach, extending ideas and intuitions on LTI systems to include a PLL's time-varying behavior. This method can be used to obtain both numerical results and symbolic expressions. Furthermore, it is shown that the traditional CT-LTI model is obtained from the more complete time-varying one through rank reduction of the open-loop gain HTM. This establishes solid mathematical underpinnings for the CT-LTI model with its shortcomings clearly identified. Finally, it is outlined how a HTM-based analysis unifies the derivation of both continuous-time and discrete-time PLL models. The theory is illustrated by means of several examples.

[3] This of course when the building block models are sufficiently accurate.

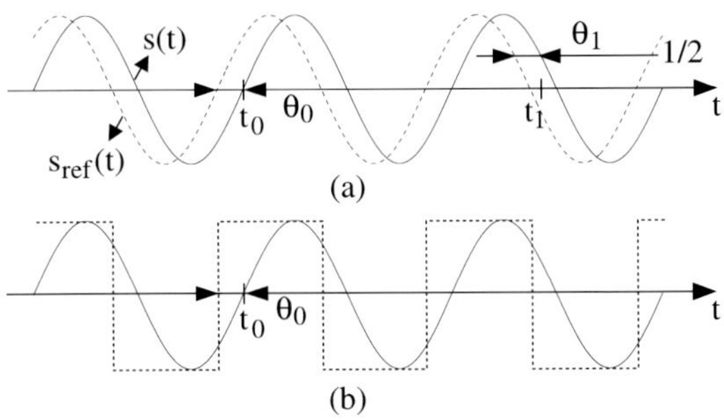

Figure 4.5: (a) The phase θ_0 associated with a zero-crossing of $s(t)$ is the time to some zero-crossing of some the reference signal $s_{ref}(t)$. Likewise, the phase θ_1 is defined via the events whereby $s(t)$ and $s_{ref}(t)$ cross the level $1/2$. (b) When $s(t)$ and $s_{ref}(t)$ are not entirely similar, it only makes sense to define phases that are, for example, associated with zero-crossings.

4.2.3 Signal phases and phase-modulated signal models

The fundamental task of a PLL is to align the phase of the (frequency-divided) VCO to that of the input signal. In order to elaborate mathematical models for these type of systems, we therefore first need a sound grasp on the concept of *the phase* of an oscillating signal. This then brings us to phase-modulated signal models and their use in describing PLL small-signal behavior.

4.2.3.1 Signal phase

By nature, a phase, or phase difference, is not associated with a continuous signal as a whole, but with discrete events generated by those signals. Such an event could, for example, be a zero-crossing. Defining the phase of an event then goes as follows:

Definition (phase): *The phase θ of an event generated by a signal $s(t)$ is the time to the occurrence of a reference event generated by a reference signal $s_{ref}(t)$. This time can be both positive and negative valued.*

Fig. 4.5(a) illustrates the definition above for $s(t)$ and $s_{ref}(t)$ both being sine waves. The events generated by $s(t)$ and $s_{ref}(t)$ correspond to the signals crossing a certain level. For example, θ_0 is the difference in time between a zero-crossing of $s(t)$ and a zero-crossing of $s_{ref}(t)$. Likewise, θ_1 is the phase associated with $s(t)$ and $s_{ref}(t)$ crossing the signal level $1/2$. Fig. 4.5(b) illustrates that, when $s_{ref}(t)$ and $s(t)$ are not entirely similar, e.g. when $s(t)$ is a sine wave and $s_{ref}(t)$ is a square wave, it only makes sense to define phases associated with, for example, zero-crossings. Observing the figure, it is seen to be senseless to define a phase associated with the crossing of the signal level $1/2$. As a final note, we mention it to be possible to associate a particular phase value θ with an absolute instance t of time. For example, in Fig. 4.5(a), the phase value θ_0 can be associated with t_0, the time at which $s_{ref}(t)$ crosses zero and generates the reference event. This unambiguously defines the concept of a phase $\theta(t)$ that varies with time.

4.2.3.2 Phase-modulated signal models

The type of signals $s(t)$ that appear in dealing with PLL small-signal behavior can best be described as oscillations whose phase varies with respect to some large-signal steady-state operating point. Hence, in modeling the small-signal behavior, we first need to specify the operating point (or reference) signals $s_{ref}(t)$ with respect to which small-signal phases are defined. Once $s_{ref}(t)$ is fixed, we assume that $s(t)$ can be written as

$$s(t) \approx s_{ref}\left(t+\theta(t)\right) \quad . \tag{4.13}$$

The right-hand side of (4.13) constitutes *a phase-modulated signal model.* Here, the process $\theta(t)$ is introduced to capture all kinds of small-signal phase variations. Observing (4.13), it is readily appreciated that, instead of directly computing $s(t)$, we can equally well compute $\theta(t)$ and use the signal model (4.13) to retrieve $s(t)$ afterwards. This is one of the key ideas underlying the construction of PLL small-signal models.

Phase-modulated signals occur at three positions in the PLL architecture depicted in Fig. 4.2. There is the input signal $X_{in}(t)$, the oscillation $X_{vco}(t)$ at the output of the VCO and the oscillation $X_{1/N}(t)$ at the output of the frequency divider. Typically, the input signal is deliberatly generated to fit the phase-modulated signal model (4.13). In order for the other two signals to fit (4.13), we need to assume that the PLL as a whole settles much, i.e. an order of magnitude, slower than the free-running (frequency-divided) VCO. This condition is mostly satisfied. If not, the poles induced by the settling behavior of the (frequency-divided) VCO tend to jeopardize the PLL's stability.

To describe the phase-modulated signals mentioned above, we first need to determine the reference signals with respect to which their phases will be defined. For the input signal, we set

$$X_{in}(t) = S_{in}\left(t+\theta_{in}(t)\right) \quad . \tag{4.14}$$

Here, $S_{in}(t)$ is the periodic reference with respect to which the input phase $\theta_{in}(t)$ is defined. Typically, it is a sinusoid or a square wave.

Selecting the references $S_{vco}(t)$ and $S_{1/N}(t)$ that are used to measure the phase changes of the VCO's output and of the output of the frequency divider is a little bit more complicated. One possible choice are the PLL's steady-state waveforms that occur when the (unmodulated) signal $S_{in}(t)$ is applied to the input. The advantage of such a choice lies in some nice structural properties characterizing the PLL's input-output HTM. The disadvantage lies in the fact that it complicates modeling the VCO behavior. A more common choice for $S_{vco}(t)$ ($S_{1/N}(t)$) involves the steady-state output of the VCO (frequency divider) with the VCO input being driven by a control voltage with constant value. This value is hereby chosen to equal the DC component of the PLL's effective steady-state control voltage. This ensures that the frequency-divided VCO reference signal $S_{1/N}(t)$ oscillates at the same frequency as $S_{in}(t)$. Fig. 4.6 shows a setup that can be used to determine the reference waveforms $S_{vco}(t)$ and $S_{1/N}(t)$.

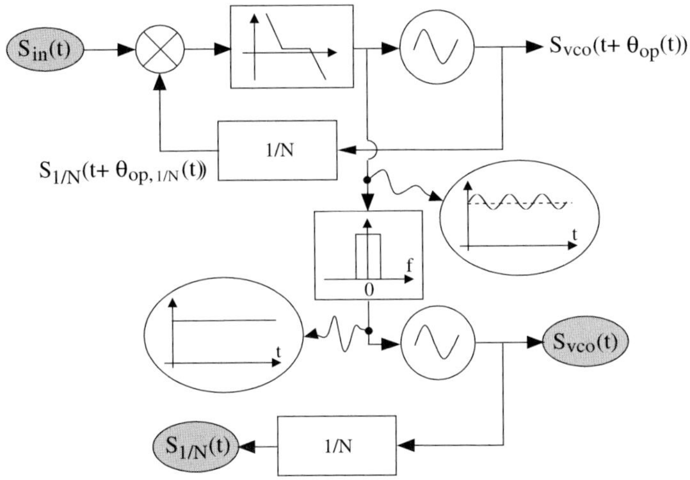

Figure 4.6: Setup for determining the reference waveforms $S_{vco}(t)$ and $S_{1/N}(t)$. These waveforms are chosen to equal the steady-state output of the VCO (frequency divider) with the VCO input being driven by a control voltage with constant value. This value is hereby chosen to equal the DC component of the PLL's effective periodic steady-state control voltage. This ensures that $S_{1/N}(t)$ oscillates at the same frequency as $S_{in}(t)$.

With respect to the above reference waveforms, the actual VCO and frequency-divided VCO signals are described as

$$X_{vco}(t) = S_{vco}\left(t + \theta_{op}(t) + \theta(t)\right) \tag{4.15}$$

$$X_{1/N}(t) = S_{1/N}\left(t + \theta_{op,1/N}(t) + \theta_{1/N}(t)\right) \quad . \tag{4.16}$$

Here, $\theta(t)$ and $\theta_{1/N}(t)$ are small-signal phase variations that arise in response to small-signal phase variations $\theta_{in}(t)$ at the input. Furthermore, $\theta_{op}(t)$ and $\theta_{op,1/N}(t)$ are periodic operating point phase variations due to periodic variations in the VCO's control voltage when the PLL is operating in steady state (i.e. $\theta_{in}(t) = 0$). Although these operating-point phase changes are typically small, they can be of importance in small-signal computations. This especially holds for PLLs based on mixing-type PFDs.

By now, the reader is probably wondering what's so important about these signal models to spend this much time on their construction. The reason for this lies in the fact that for some types of PLLs, especially those using a mixing PFD, correct signal models are of utmost importance in obtaining valid results. Being sloppy in constructing the signal models may give rise to non-physical behavior and results that disagree with numerical simulations and measurements.

Example (Determining signal models): Fig. 4.7 shows a PLL with a mixing PFD and sinusoidal input and output signals. For this setup, the open-loop gain of the associated

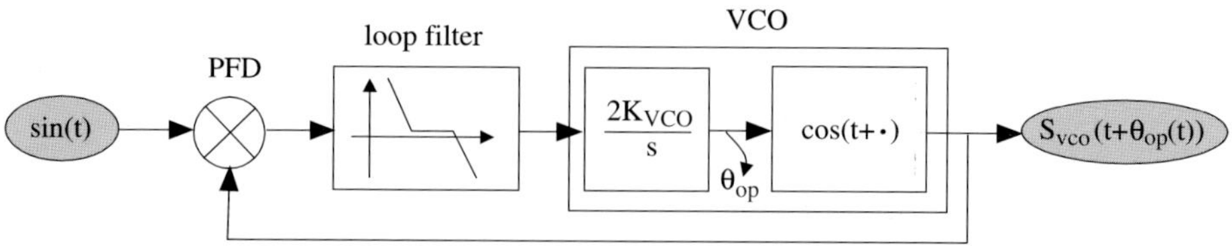

Figure 4.7: PLL architecture with mixing-type PFD and sinusoidal input and output oscillations. Note that the oscillation frequency is normalized to $2\pi f_0 = \omega_0 = 1$.

CT-LTI model equals

$$G(s) = \frac{K_{VCO}}{s} H_{LF}(s) \tag{4.17}$$

with $H_{LF}(s)$ the loop-filter characteristic. The closed-loop PLL transfer then equals

$$H(s) = \frac{G(s)}{1+G(s)} \quad . \tag{4.18}$$

As far as the reference waveforms are concerned, it is clear that for a constant (and zero) VCO control voltage

$$S_{vco}(t) = \cos(t) \quad . \tag{4.19}$$

The VCO's steady-state signal will, however, slightly deviate from this perfect cosine. This is because of small operating-point phase variations. These phase variations are due to the fact that the mixer's (large-signal) steady-state output approximately equals

$$Y_{PFD}(t) \approx \sin(t)\cos(t) = \frac{1}{2}\sin(2t) \quad . \tag{4.20}$$

Hence, the VCO's control voltage, and therefore the VCO's steady-state output phase, contains higher-order harmonics. It can be shown that, up to first order,

$$\theta_{op}(t) \approx \mathrm{Re}\left\{\frac{H(2j)}{j} e^{j2t}\right\} \quad . \tag{4.21}$$

According to (4.15), the modulated VCO signal can then be described as

$$X_{vco}(t) \approx \cos\left(t + \mathrm{Re}\left\{\frac{H(2j)}{j} e^{j2t}\right\} + \theta(t)\right) \quad . \tag{4.22}$$

Again, for certain types of PLL architectures, the operating-point phase variation $\theta_{op}(t)$ cannot be neglected during further computations. ▲

Having dicussed PLL signal modeling in depth, we are ready to construct HTM-based models that capture the time-varying behavior of each of the PLL building blocks.

4.2.4 HTM-based PLL building block models

In computing a PLL's overall input-output HTM, we first need to determine the HTMs corresponding to the PLL building blocks shown in Fig. 4.2. The next couple of sections respectively describe PFD, loop filter, VCO and frequency divider small-signal models together with the HTMs that accompany them. Note that in all the examples that follow, the PLL equations are assumed to be normalized such that the input oscillation frequency $2\pi f_0 = \omega_0 = 1^4$.

4.2.4.1 Phase-frequency detector (PFD)

In a first step, we discuss modeling the PFD small-signal behavior. As a reminder, the PFD measures the phase error between the input and the frequency-divided VCO signal. It turns out that, often, this PFD is the main responsible for a PLL's time-varying behavior. Two common PFD implementations are based on mixing and sampling [Lee98]. Mixing-PFDs operate by multiplying the input signal with the frequency-divided VCO signal. Sampling-PFDs produce a sequence of pulses the width of which equals the instantaneous phase error. In what follows, we derive HTMs for both types of PFDs.

Mixing PFDs: A mixing PFD, as illustrated in Fig 4.8, multiplies the output of the frequency divider $X_{1/N}(t) = S_{1/N}(t + \theta_{op,1/N}(t) + \theta_{1/N}(t))$ with the input signal $X_{in}(t) = S_{in}(+\theta_{in}(t))$. In what follows, we assume the mixing operation to be ideal. Using, for example, the algorithm later on outlined in section 4.3, it would also be possible to derive models for more complex structures. This, however, only complicates the analysis without contributing to the discussion below.

With both $\theta_{in}(t)$ and $\theta_{1/N}(t)$ being small-signal excursions, we find the mixer output to equal

$$
\begin{aligned}
Y_{PFD}(t) &= S_{in}(t+\theta_{in}(t)) \cdot S_{1/N}(t+\theta_{op}(t)+\theta_{1/N}(t)) && (4.23)\\
&\approx S_{in}(t) \cdot S_{1/N}(t+\theta_{op}(t)) + \\
&\quad \left(\frac{\mathrm{d}S_{in}}{\mathrm{d}\theta_{in}}(t) S_{1/N}(t+\theta_{op}(t))\right)\theta_{in}(t) + && (4.24)\\
&\quad \left(S_{in}(t)\frac{\mathrm{d}S_{1/N}}{\mathrm{d}\theta_{1/N}}(t+\theta_{op}(t))\right)\theta_{1/N}(t) \ .
\end{aligned}
$$

In the lower equality, the upper term represents the steady-state part of the mixer's output while the lower terms are due to the small-signal phase variations $\theta_{in}(t)$ and $\theta_{1/N}(t)$. Note that this latter part takes the form

$$y_{PFD}(t) = h^{in}_{PFD}(t)\theta_{in}(t) - h_{PFD}(t)\theta_{1/N}(t) \tag{4.25}$$

with $y_{PFD}(t)$ denoting the small-signal excursion of the mixer's output. Hence, the mixer small-signal behavior can be modeled as a linear combination of $\theta_{1/N}(t)$ and

[4]Such a normalization involves a change of the time variable $t_{norm} = \omega_0 t$. Note that proper normalization is important as it often helps us to simplify the resulting analysis while also improving the interpretability of the results.

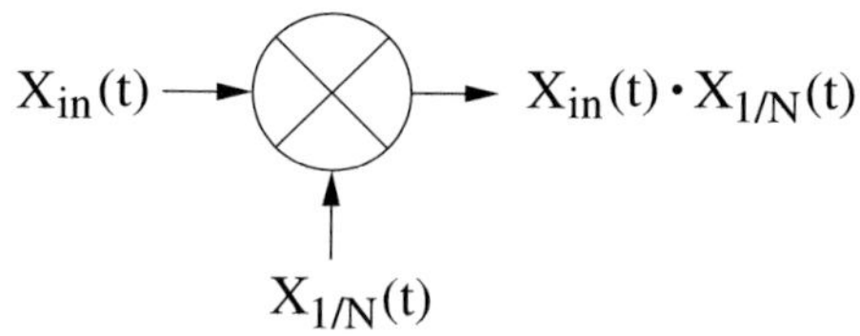

Figure 4.8: A multiplying mixer as a phase frequency detector.

$\theta_{in}(t)$ with coefficients that vary periodically over time. The minus sign in (4.25) is introduced to emphasize the role of the mixer as a phase comparator.

Using the HTM formalism, we can express (4.25) in the Laplace domain as

$$\mathbf{Y}_{PFD}(s) = \tilde{\mathbf{H}}_{PFD}^{in}(s)\Theta_{in}(s) - \tilde{\mathbf{H}}_{PFD}(s)\Theta_{1/N}(s) \quad . \tag{4.26}$$

Here, $\Theta_{in}(s)$ and $\Theta_{1/N}(s)$ are vectors of equivalent baseband components as defined by (4.1) and (4.2). The HTMs $\tilde{\mathbf{H}}_{PFD}^{in}(s)$ and $\tilde{\mathbf{H}}_{PFD}(s)$ are obtained by respectively inserting the Fourier coefficients of $h_{PFD}^{in}(t)$ and $h_{PFD}(t)$ in (4.9).

Example (Sinusoidal oscillations): Again, consider the system depicted in Fig. 4.7. Note that in this case, $X_{1/N}(t) = X_{vco}(t)$ as there is no frequency divider. After a little algebra, we then obtain

$$\begin{aligned} y_{PFD}(t) &= \frac{1}{2}\left[\cos(\theta_{op}(t)) + \cos(2t + \theta_{op}(t))\right]\theta_{in}(t) \qquad (4.27)\\ &\quad -\frac{1}{2}\left[\cos(\theta_{op}(t)) - \cos(2t + \theta_{op}(t))\right]\theta_{1/N}(t) \\ &\approx \frac{1}{2}\left[(1 + \cos(2t)) - \sin(2t)\,\theta_{op}(t)\right]\theta_{in}(t) \qquad (4.28)\\ &\quad -\frac{1}{2}\left[(1 - \cos(2t)) + \sin(2t)\,\theta_{op}(t)\right]\theta_{1/N}(t) \end{aligned}$$

with $\theta_{op}(t)$ given by (4.21). The lower (first-order) approximation holds for $|\theta_{op}(t)| \ll 1$. This typically holds true as long as the feedback loop does not become exceedingly fast.

Straightforward conversion of (4.28) into HTM language yields

$$\tilde{\mathbf{H}}_{PFD}^{in}(s) \approx \frac{1}{2}\begin{bmatrix} \ddots & \ddots & \ddots & \ddots & & \\ \ddots & 1 & 0 & 1/2 & 0 & \\ \ddots & 0 & 1 & 0 & 1/2 & \ddots \\ \ddots & 1/2 & 0 & 1 & 0 & \ddots \\ & 0 & 1/2 & 0 & 1 & \ddots \\ & & \ddots & \ddots & \ddots & \ddots \end{bmatrix} - \Delta\tilde{\mathbf{H}}_{op}(s) \tag{4.29}$$

and

$$\tilde{\mathbf{H}}_{PFD}(s) \approx \frac{1}{2}\begin{bmatrix} \ddots & \ddots & \ddots & \ddots & & \\ \ddots & 1 & 0 & -1/2 & 0 & \\ \ddots & 0 & 1 & 0 & -1/2 & \ddots \\ \ddots & -1/2 & 0 & 1 & 0 & \ddots \\ & 0 & -1/2 & 0 & 1 & \ddots \\ & & \ddots & \ddots & \ddots & \ddots \end{bmatrix} + \Delta\tilde{\mathbf{H}}_{op}(s) \qquad (4.30)$$

as the HTMs that characterize the behavior of the mixing PFD. Here,

$$\Delta\tilde{\mathbf{H}}_{op}(s) = \frac{1}{4}\begin{bmatrix} \ddots & \ddots & \ddots & \ddots & \\ & \mathrm{Re}\{H(2j)\} & 0 & 0 & \\ \ddots & 0 & \mathrm{Re}\{H(2j)\} & 0 & \ddots \\ & 0 & 0 & \mathrm{Re}\{H(2j)\} & \ddots \\ & \ddots & \ddots & \ddots & \ddots \end{bmatrix} \qquad (4.31)$$

with $H(s)$ specified by (4.18). Note that the name $\Delta\tilde{\mathbf{H}}_{op}(s)$ was chosen to emphasize the fact that it represents an increment that is typically smaller than the leading terms in (4.29) and (4.30). Note, however, that it still may be of importance in analyzing the details of the time-varying components of a PLL's behavior. ▲

A similar analysis as the one above is possible when the input signal and/or the VCO signal are square waves. This is equivalent with the mixer in Fig. 4.8 being implemented as an EXOR gate.

Sampling PFD: In sampling PFDs, the phase error $\theta_{in}(t) - \theta_{1/N}(t)$ between the input and the frequency-divided VCO signal is measured (sampled) once per period of the input signal. Typical sampling circuits, like the charge-pump-based topology in Fig. 4.9 [Lee98], code this error as the width of a sequence of digital pulses. These pulses are triggered by the zero-crossings of the input and the frequency-divided VCO signal. Hence, they occur at equidistant instances in time. If the width of such a pulse is small compared to the PLL's settling time constant[5], it will have the same effect as a Dirac impulse with a weight that equals the area underneath the original pulse. This equivalence is depicted in Fig. 4.10. Note that the pulses in the upper plot were normalized to have magnitude 1. The effective charge-pump current I_{cp} can, for example, be taken into account in the loop filter model.

[5] Note that this condition is always satisfied in the small-signal limit. PLL small-signal analysis, in principle, is concerned with phase deviations θ whose magnitude in the limit goes to zero. Hence, by choosing $|\theta|$ small enough, the pulses produced by the PFD will always be sufficiently narrow for the equivalence in Fig. 4.10 to hold.

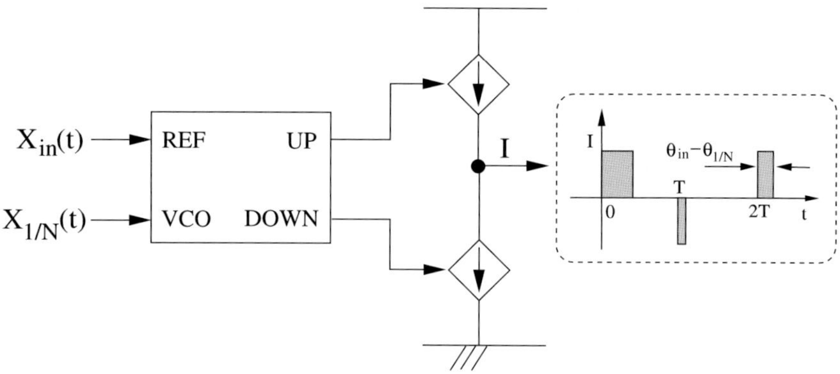

Figure 4.9: Sampling phase-frequency detector. The input and (frequency-divided) VCO phase are compared using sequential digital logic. This logic steers a charge pump which produces a sequence of current pulses whose width is proportional to the instantaneous phase error $\theta_{in}(t) - \theta_{1/N}(t)$.

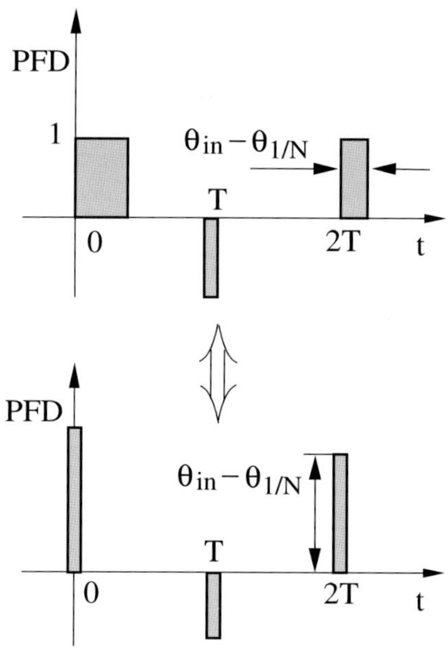

Figure 4.10: If the PFD produces pulses whose width is small compared to the time constant of the loop filter-VCO combination, they will have the same effect as Dirac impulses with a weight that equals the area underneath the original pulses.

Assuming the equivalence in Fig. 4.10 to hold, the input-output relation of a sampling PFD can be modeled as a multiplication of the phase error $\theta_{in} - \theta_{1/N}$ with a Dirac impulse train, or

$$y(t) = \left(\sum_{m=-\infty}^{+\infty} \delta(t - mT)\right) \left(\theta_{in}(t) - \theta_{1/N}(t)\right) \quad (4.32)$$

$$= \frac{\omega_0}{2\pi} \left(\sum_{m=-\infty}^{+\infty} e^{j\omega_0 mt}\right) \left(\theta_{in}(t) - \theta_{1/N}(t)\right) \quad (4.33)$$

with $T = 2\pi/\omega_0$ the sampling period. This multiplication can be stated in terms of HTMs as

$$\mathbf{Y}(s) = \tilde{\mathbf{H}}_{PFD}(s) \left(\Theta_{in}(s) - \Theta_{1/N}(s)\right) \quad (4.34)$$

with

$$\tilde{\mathbf{H}}_{PFD}(s) = \frac{\omega_0}{2\pi} \begin{bmatrix} & \vdots & \vdots & \vdots & \\ \cdots & 1 & 1 & 1 & \cdots \\ \cdots & 1 & 1 & 1 & \cdots \\ \cdots & 1 & 1 & 1 & \cdots \\ & \vdots & \vdots & \vdots & \end{bmatrix} \quad (4.35)$$

$$= \frac{\omega_0}{2\pi} \mathbf{1} \cdot \mathbf{1}^T . \quad (4.36)$$

Here, $\mathbf{1} = \begin{bmatrix} \cdots & 1 & 1 & 1 & \cdots \end{bmatrix}^T$. Note that (4.34) is a special case of (4.26) with $\tilde{\mathbf{H}}^{in}_{PFD}(s) = \tilde{\mathbf{H}}_{PFD}(s)$.

Equation (4.36) reveals an important property corresponding to the HTM $\tilde{\mathbf{H}}_{PFD}(s)$ of a sampling-PFD: it is a rank-one matrix. This should come as no suprise, since sampling maps all input signal content to the frequency band $[-\omega_0/2, \omega_0/2]$, a phenomenon known as aliasing. This baseband spectrum is then periodically repeated over the remainder of the frequency axis. So, knowledge of the output's signal content in one frequency band implies knowledge of its content in all other frequency bands. This explains the rank-one property of the HTM corresponding to the sampling operator.

4.2.4.2 Loop filter

The PFD output is fed to the loop filter. The transfer function of this loop filter is an important design variable in obtaining good PLL stability and settling behavior. Typically, the loop filter is implemented as a time-invariant analog filter with transfer function $H_{LF}(s)$.

Using (4.8), the HTM corrsponding to the time-invariant loop filter $H_{LF}(s)$ is found to

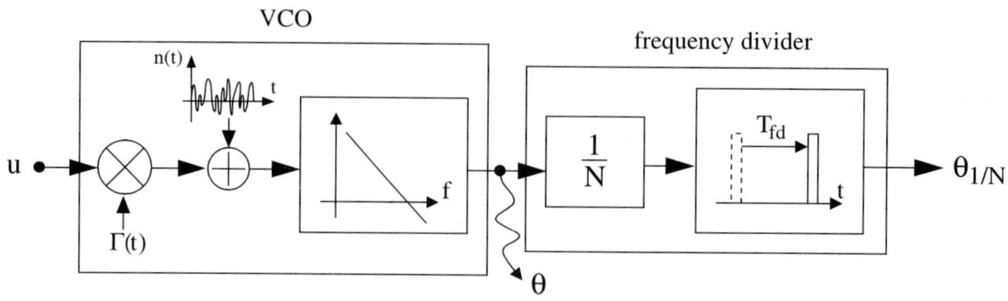

Figure 4.11: VCO-frequency divider model. The left-hand block represents a (noisy) VCO model. The right-hand block models the behavior of the frequency-divider.

equal

$$\tilde{\mathbf{H}}_{LF}(s) = \begin{bmatrix} \ddots & & & & \\ & H_{LF}(s-j\omega_0) & & 0 & \\ & & H_{LF}(s) & & \\ & 0 & & H_{LF}(s+j\omega_0) & \\ & & & & \ddots \end{bmatrix}. \tag{4.37}$$

Although filter nonlinearities can cause time-varying small-signal behavior, and therefore off-diagonal components in the loop filter HTM, we assume such effects to be negligible. This is justified by the fact that many loop filters are implemented as higly linear passive or opamp-RC based networks [Lee98]. Note, however, that this assumption of time invariance is not mandatory.

4.2.4.3 VCO–frequency divider

Finally, we discuss the VCO–frequency divider combination. The frequency of the VCO is steered by the control signal coming from the loop filter. This frequency is then divided by a factor N. Fig. 4.11 illustrates the functional model used in this text to capture the behavior of the VCO and frequency divider. In what follows, we discuss the underpinnings of this model in more detail.

VCO: By means of a procedure similar to the one for modeling the PLL behavior as a whole, it is possible to obtain a HTM that captures all aspects of the VCO's dynamic behavior. Each of the VCO building blocks is linearized in the neighborhood of the steady-state operating point[6]. Using the rules (4.6)-(4.7) for computing HTMs of composed systems, we then obtain a HTM for the VCO as a whole. However, when the settling behavior of the VCO is fast as compared to that of the PLL, we can make

[6]Note that this linearized model is valid since a PLL in lock prevents the VCO phase from drifting. As discussed in [Dem98, Dem00a], phase drifting renders a linearized VCO model invalid.

use of results from oscillator phase noise theory [Dem98, Dem00a, Haji98][7] to obtain a more compact VCO model. This model accounts for the long-term phase behavior while neglecting all VCO transients. Again, it is only valid when the VCO setlling behavior is sufficiently fast compared to that of the PLL as a whole. This condition, however, is satisfied for most practical PLL designs[8].

Let $U(t)$ be the input, produced by the loop filter, that steers the VCO. Furthermore, we assume that $U(t)$ can be decomposed in a large-signal DC component U_0, a component $u_{op}(t)$ that captures the small periodic operating-point variations around U_0 and a small-signal contribution $u(t)$ induced, for example, by the small-signal input phase variation $\theta_{in}(t)$. In summary, we assume

$$U(t) = U_0 + u_{op}(t) + u(t) \quad . \tag{4.38}$$

Since both $u_{op}(t)$ and $u(t)$ are small, we can use the results presented in [Dem98, Dem00a] to show that the VCO's small-signal phase variation $\theta(t)$ is governed by[9]

$$\begin{aligned}\frac{d\theta}{dt} &= \Gamma(t+\theta_{op}(t)+\theta(t))\,u(t) \\ &\quad + \left[\Gamma(t+\theta_{op}(t)+\theta(t)) - \Gamma(t+\theta_{op}(t))\right] u_{op}(t)\end{aligned} \tag{4.39}$$

Here, $\Gamma(t)$ is the VCO's impulse sensitivity function (ISF) associated with the input source $u(t)$. The ISF is periodic with a period equal to that of the VCO (before frequency division). Hence, if the frequency-divider divides by N, this means that the period of $\Gamma(t)$ equals T/N where T is the period of the PLL input signal $X_{in}(t)$. For most VCOs, the ISF associated with the control signal $u(t)$ is nearly constant, i.e. $\Gamma(t) \approx \Gamma_0$. In that case, (4.39) reduces to the traditional LTI model (4.12) with $\Gamma_0 = K_{VCO}$.

In a stable PLL, the control signal $u(t)$ is such as to keep the VCO phase deviations $\theta(t)$ small, i.e. $|N\theta/T| \ll 1$. It is therefore justified to linearize (4.39) near $\theta(t) = 0$. This results in the closed-form expression

$$\theta(t) \approx \int_{-\infty}^{t} \Gamma(\tau+\theta_{op}(\tau))\,u(\tau)d\tau \tag{4.40}$$

$$\approx \int_{-\infty}^{t} \Gamma(\tau)\,u(\tau)d\tau \quad . \tag{4.41}$$

[7] These results are also obtained as a special case of the results presented in section 5.7 of chapter 5.

[8] Having a VCO that settles as slow as the PLL as a whole bears stability problems with it. The VCO setlling behavior then introduces extra poles near the PLL's unity-gain frequency, degrading the phase margin.

[9] This relation follows from the fact that, for all small-signal excursions $u(t)$,

$$d(\theta_{op}+\theta)/dt = \Gamma(t+\theta_{op}(t)+\theta(t))\,(u_{op}(t)+u(t))$$

or

$$d\theta/dt = \Gamma(t+\theta_{op}(t)+\theta(t))\,(u_{op}(t)+u(t)) - d\theta_{op}/dt \quad .$$

By considering the PLL's steady-state behavior, i.e. $\theta(t) = u(t) = 0$, we obtain that

$$d\theta_{op}/dt = \Gamma(t+\theta_{op}(t))\,u_{op}(t) \quad .$$

Substituting this into the previous equation then yields the desired result.

Note that the latter approximation only holds valid when $\mathrm{d}\Gamma(t)/\mathrm{d}t \cdot \theta_{op}(t)$ is sufficiently small, something which is typically true. Hence, observing (4.41), we find that $\theta(t)$ is related to $u(t)$ via an LPTV operator consisting of a multiplication with $\Gamma(t)$ followed by an integration. This model is depicted by the left-hand block in Fig. 4.11. The HTM corresponding to this operator equals

$$\tilde{\mathbf{H}}_{VCO}(s) = \begin{bmatrix} & \ddots & \ddots & \ddots & & \ddots \\ \ddots & \ddots & \frac{\Gamma_0}{s-j\omega_0} & \underbrace{0}_{N-1\text{ bands}} & \frac{\Gamma_{-1}}{s-j\omega_0} & 0 \\ \ddots & \frac{\Gamma_1}{s} & 0 & \frac{\Gamma_0}{s} & 0 & \frac{\Gamma_1}{s} \\ & 0 & \frac{\Gamma_1}{s+j\omega_0} & \underbrace{0}_{N-1\text{ bands}} & \frac{\Gamma_0}{s+j\omega_0} & \ddots & \ddots \\ \ddots & & \ddots & \ddots & \ddots & \ddots \end{bmatrix} \tag{4.42}$$

and is readily computed using (4.7), (4.9) and (4.41). Here, the Γ_k are the Fourier coefficients of $\Gamma(t)$, or

$$\Gamma(t) = \sum_{k=-\infty}^{+\infty} \Gamma_k e^{jk\omega_{vco}t} = \sum_{k=-\infty}^{+\infty} \Gamma_k e^{j(kN)\omega_0 t} \ . \tag{4.43}$$

Since the VCO, and hence the ISF $\Gamma(t)$, oscillates at $\omega_{vco} = N\omega_0$ and since all HTMs are constructed with respect to the carrier waves $\psi_m(t) = e^{jm\omega_0 t}$, $\tilde{\mathbf{H}}_{VCO}(s)$ contains a great number of super- and sub-diagonal bands filled with zeros. This readily follows from the fact that, when expanding $\Gamma(t)$ with respect to $\psi_m(t) = e^{jm\omega_0 t}$, only the coefficients associated with the carriers $\psi_{kN}(t) = e^{j(kN)\omega_0 t}$ can assume non-zero values. As was already mentioned, often $\Gamma(t) \approx \Gamma_0$ in which case $\tilde{\mathbf{H}}_{VCO}(s)$ reduces to a diagonal matrix. This means that the high-frequency components of the small-signal control voltage $u(t)$ do (almost) not influence the oscillator's instantaneous frequency.

Note that, as illustrated in Fig. 4.11, the VCO's phase-noise behavior can be captured by introducing a noise source just before the integrator[10]. We can then use the cyclo-stationary noise theory (see section 3.5.3 in the previous chapter) to analyze the impact of the VCO's phase noise on the PLL's spectral purity.

Frequency divider: Frequency dividers are often implemented by means of digital logic, e.g. flipflops and NAND gates. If the VCO oscillation frequency hits the RF range, then we need, as part of the frequency divider, a prescaler that can handle these high frequencies [Demu02]. Due to their digital nature, frequency divider operations are typically triggered by the zero-crossings of the incoming VCO signal. A detailed time-varying model would therefore involve an interconnection of sampling and filtering operations. A similar argument can be made for prescaler architectures, e.g. the phase-switching architecture in [Cran98].

[10] Introducing the noise before the integrator is needed to account for the cumulative nature of oscillator phase noise.

However, as was the case for modeling the VCO behavior, such a detailed frequency divider model would be overly complex, introducing a multitude of irrelevant behavioral details[11]. In practice, the frequency-divider model depicted in the right-hand block of Fig. 4.11 captures most relevant behavior. Here, the frequency-divider divides the incoming VCO frequency by N. This result is delayed by a time T_{fd} before it is passed to the PFD. Typically, if present at all, the delay time T_{fd} is proportional to the period of the VCO signal. Hence, for a given input frequency, it can be considered constant. The resulting frequency divider model can be described as a time-invariant filter

$$H_{FD}(s) = \frac{\theta_{1/N}(s)}{\theta(s)} = \frac{e^{-sT_{fd}}}{N} \quad . \tag{4.44}$$

The corresponding HTM $\tilde{\mathbf{H}}_{FD}(s)$ is a diagonal matrix similar to the one in (4.37).

4.2.5 PLL closed-loop input-output HTM

Having constructed the building block models, we can now tie them together in order to obtain the HTM that describes the PLL's overall behavior. In what follows, we focus on the HTM that characterizes the tranfer from the input phase $\theta_{in}(t)$ to the output VCO phase $\theta(t)$. Computing HTMs associated with other input sources, e.g. with the VCO noise source in Fig. 4.11, is accomplished in a similar manner.

Using the relations (4.6)-(4.7), the input-output HTM for the PLL feedback topology in Fig. 4.2 is found to equal

$$\tilde{\mathbf{H}}(s) = \left[\left(\mathbf{I} + \tilde{\mathbf{G}}(s)\right)^{-1} \tilde{\mathbf{H}}_{VCO}(s) \cdot \tilde{\mathbf{H}}_{LF}(s) \cdot \tilde{\mathbf{H}}^{in}_{PFD}(s) \right] \quad . \tag{4.45}$$

Here, we used the expression (4.26) to describe the PFD behavior. Furthermore,

$$\tilde{\mathbf{G}}(s) = \tilde{\mathbf{H}}_{VCO}(s) \cdot \tilde{\mathbf{H}}_{LF}(s) \cdot \tilde{\mathbf{H}}_{PFD}(s) \cdot \tilde{\mathbf{H}}_{FD}(s) \tag{4.46}$$

is the PLL's open-loop gain HTM. The input-output relation (4.45) is valid for all types of PLLs, at least as far as their small-signal behavior is concerned. Note that this equation is the HTM generalization of the CT-LTI model (4.10).

Evaluating and/or approximating (4.45) is achieved by means of the techniques outlined in the previous chapter. For example, for PLLs with a mixing PFD, the power series method in section 3.4.3.2 tends to yield good results. The use of rank-reduced approximations (see section 3.4.3.3) for the open-loop gain HTM $\tilde{\mathbf{G}}(s)$, however, allows us to obtain results that are valid for any PFD implementation. Even more, for PLLs based on sampling PFDs, this method yields an exact closed-form expression for the PLL's closed-loop input-output HTM. In what follows, we first consider *approximate* expressions for (4.45) based on the rank reduction of $\tilde{\mathbf{G}}(s)$. These approximations hold for all types of PLL implementations. They establish the link with traditional CT-LTI models that are commonly used for PLL analysis and design. For PLLs with a sampling PFD, section 4.2.5.3 presents an *exact* input-output relation. It establishes the link with the discrete-time PLL models presented in [Gard80, Hein88].

[11] It is only interesting to construct a detailed model for circuit-level verification. In this case, one can use general-purpose numerical algorithms, build on the results in section 3.4.4 of the previous chapter, to extract the exact HTM corresponding with a particular frequency divider circuit-level implementation.

4.2.5.1 A general rank-reduced approximation:

By combining the different building block models, it is readily shown that the elements of $\tilde{\mathbf{G}}(s)$ are structured as

$$\tilde{G}_{n,m}(s) = \frac{\tilde{K}_{n,m}(s)}{s + jn\omega_0} \quad . \tag{4.47}$$

This shows that, at least for $|s| < \omega_0/2$, $\tilde{\mathbf{G}}(s)$ will be dominated by the rows closest to $n = 0$. This in turn suggests to try and approximate $\tilde{\mathbf{G}}(s)$ by its projection on the rows with indices $|n| \leq R$ with R the rank-reduction parameter. Mathematically, this is written as

$$\begin{aligned} \tilde{\mathbf{G}}(s) &\approx \mathbf{P}\left(\mathbf{P}^T \tilde{\mathbf{G}}(s)\right) & (4.48) \\ &\approx \mathbf{P} \cdot \mathbf{V}(s)^T \quad . & (4.49) \end{aligned}$$

Here, $\mathbf{V}(s).\mathbf{P} \in \mathbb{C}^{\infty \times (2R+1)}$ with

$$\mathbf{P} = \underbrace{\begin{bmatrix} \vdots & \vdots & \vdots \\ 0 & 0 & 0 \\ 1 & & 0 \\ \vdots & \ddots & \vdots \\ 0 & \cdots & 1 \\ 0 & 0 & 0 \\ \vdots & \vdots & \vdots \end{bmatrix}}_{2R+1 \text{ columns}} \quad . \tag{4.50}$$

In words, (4.48) and (4.50) express that, for an input signal with low-frequency equivalent baseband components, the output of the PLLs time-varying open-loop gain is dominated by the signal content contained in the lowest frequency bands. Note from the general theory outlined in section 3.4.3.3 that the approximation (4.48) only holds valid for those values of $s \in \mathbb{C}$ for which $\left\|\left(\mathbf{I} - \mathbf{P}\mathbf{P}^T\right)\tilde{\mathbf{G}}(s)\right\| \ll 1$[12]. Using a similar argument, we equally find

$$\begin{aligned} \tilde{\mathbf{H}}_{VCO}(s) \cdot \tilde{\mathbf{H}}_{LF}(s) \cdot \tilde{\mathbf{H}}_{PFD}^{in}(s) &\approx \mathbf{P}\left(\mathbf{P}^T \cdot \tilde{\mathbf{H}}_{VCO}(s) \cdot \tilde{\mathbf{H}}_{LF}(s) \cdot \tilde{\mathbf{H}}_{PFD}^{in}(s)\right) \\ &\approx \mathbf{P} \cdot \mathbf{V}_{in}(s)^T \quad . & (4.51) \end{aligned}$$

With (4.49), the factor $\left(\mathbf{I} + \tilde{\mathbf{G}}(s)\right)^{-1}$ in the input-output relation (4.45) can be elaborated using the Sherman-Morisson-Woodbury formula (see also section 3.4.3.3 of the previous chapter). This yields

$$\begin{aligned} \left(\mathbf{I} + \tilde{\mathbf{G}}(s)\right)^{-1} &\approx \left(\mathbf{I} + \mathbf{P} \cdot \mathbf{V}(s)^T\right)^{-1} \\ &\approx \mathbf{I} - \mathbf{P}\left(\mathbf{I} + \mathbf{G}_P(s)\right)^{-1} \mathbf{V}(s)^T \quad . & (4.52) \end{aligned}$$

[12] Due to the presence of poles at DC in a typical design of the loop filter $H_{LF}(s)$, this approximation may fail for values of s very close to 0. In such a case, rank-reduced approximations are still possible, but it is somewhat more involved to construct them. However, since this only makes a difference for signal transfers from or to frequency bands beyond $\omega_{vco} = N\omega_0$, it is unimportant for most practical purposes.

Here,

$$\mathbf{G}_P(s) = \mathbf{V}(s)^T\mathbf{P} = \mathbf{P}^T \cdot \tilde{\mathbf{G}}(s) \cdot \mathbf{P} \tag{4.53}$$

represents the $(2N+1)\times(2N+1)$ submatrix of $\tilde{\mathbf{G}}(s)$ centered around its origin $(n,m) = (0,0)$. Using (4.51) and (4.52), the input-output relation (4.45) turns into

$$\tilde{\mathbf{H}}(s) \approx \mathbf{P}\left(\mathbf{I}+\mathbf{G}_P(s)\right)^{-1}\mathbf{V}_{in}(s)^T \ . \tag{4.54}$$

Note from the structure of **P** in (4.50) that this rank-reduced approximation neglects all transfers of signal content to frequency bands beyond $R\omega_0$.

Example (Rank-one approximation): The most elementary rank-reduced approximation to $\tilde{\mathbf{G}}(s)$ is a matrix of rank one, i.e. we set $R=0$. In this case, the projector $\mathbf{PP}^T$ only takes the 0-th row of $\tilde{\mathbf{G}}(s)$ into account. We illustrate the results that are obtained by means of such an approximation for a PLL with a VCO that has a constant ISF, i.e. $\Gamma(t) = \Gamma_0 = K_{VCO}$. For the PFD, we select the mixing-PFD model specified by the HTMs (4.29)-(4.31). After a little algebra, we then find

$$\mathbf{G}_P(s) = \frac{K_{VO}}{s}H_{LF}(s)\frac{e^{-sT_{fd}}}{N} \in \mathbb{C}^{1\times 1} \tag{4.55}$$

and

$$\mathbf{V}_{in}(s)^T = \frac{K_{VO}}{s}H_{LF}(s)\left[\ \cdots\ 0\ \ \tfrac{1}{2}\ \ 0\ \ 1\ \ 0\ \ \tfrac{1}{2}\ \ 0\ \cdots\ \right] . \tag{4.56}$$

Substituting (4.55) and (4.56) into (4.54) finally yields

$$\tilde{\mathbf{H}}(s) \approx \begin{bmatrix} \vdots \\ 0 \\ 1 \\ 0 \\ \vdots \end{bmatrix} \frac{\frac{K_{VO}}{s}H_{LF}(s)}{1+\frac{K_{VO}}{s}H_{LF}(s)\frac{e^{-sT_{fd}}}{N}} \left[\ \cdots\ 0\ \ \tfrac{1}{2}\ \ 0\ \ 1\ \ 0\ \ \tfrac{1}{2}\ \ 0\ \cdots\ \right] . \tag{4.57}$$

Observing the input-output relation above, we note that the HTM element that models the phase transfers from baseband to baseband, i.e.

$$\tilde{H}_{0,0}(s) = \frac{\frac{K_{VO}}{s}H_{LF}(s)}{1+\frac{K_{VO}}{s}H_{LF}(s)\frac{e^{-sT_{fd}}}{N}} , \tag{4.58}$$

corresponds with the closed-loop transfer function as it is obtained by means of traditional CT-LTI analysis. However, (4.57) also predicts a transfer of high-frequency input phase variations (varying at about twice the carrier frequency) to the low-frequency phase content at the output. These transfers are due to the fact that input phase variations at twice the carrier frequency contribute at DC component at the output of the mixing PFD. They are captured by the HTM element

$$\begin{aligned} \tilde{H}_{0,2}(s) &= \frac{1}{2}\frac{\frac{K_{VO}}{s}H_{LF}(s)}{1+\frac{K_{VO}}{s}H_{LF}(s)\frac{e^{-sT_{fd}}}{N}} && (4.59) \\ &= \frac{1}{2}\tilde{H}_{0,0}(s) && (4.60) \end{aligned}$$

and are unaccounted for by traditional CT-LTI analysis. However, as (4.60) shows that $\tilde{H}_{0,2}(s)$ is of the same order of magnitude as $\tilde{H}_{0,0}(s)$, one must be careful in neglecting them. ▲

4.2.5.2 The continuous-time LTI PLL model and its limitations

The example above highlights the link between the traditional CT-LTI PLL model and the time-varying PLL models developed in this chapter:

CT-LTI PLL approximation: *The closed-loop PLL transfer function model obtained by means od traditional CT-LTI analysis corresponds to the HTM element $\tilde{H}_{0,0}(s)$ of* a rank-one approximation to (4.45).

The conditions for the contnuous-time LTI model to yield valid results are therefore twofold:

1. The approximations (4.49) and (4.51) must hold for $\mathbf{P}$ a (rank-one) vector, i.e. for $R = 0$. Since the accuracy of the approximations drop as $|s| \to \omega_0/2$, the LTI model is only valid at sufficiently low frequencies.
2. Since the LTI model only captures transfers from baseband to baseband, the inputs, e.g. $\theta_{in}(t)$, should not contain any high-frequency components. Hence, topics like noise folding, i.e. high-frequency noise components ending up at baseband, are left unaccounted by CT-LTI analysis.

As a consequence of the first condition, stability as predicted by the LTI model is only valid as long as the unity-gain frequency ω_{UG} of the open-loop gain, i.e. the relevant frequency for stability computations, is well below $\omega_0/2$. It will be demonstrated in section 4.2.6 that for fast feedback loops, i.e. $\omega_{UG}/\omega_0 > 0.05$, time-varying effects do need to be taken into account for stability predictions to be accurate.

4.2.5.3 PLL input-output HTM for sampling-PFDs

While (4.54) is an *approximation* that holds for all types of PFDs, *exact* expressions can be derived for PLLs built on sampling PFDs. This is done by means of similar techniques as in the previous section and by making use of the fact that, for sampling PFDs, $\tilde{\mathbf{H}}_{PFD}(s) = (\omega_0/2\pi)\mathbf{1}\cdot\mathbf{1}^T$ is by nature rank one. As a reminder, $\mathbf{1} = \begin{bmatrix} \cdots & 1 & 1 & 1 & \cdots \end{bmatrix}^T$. If we define

$$\mathbf{V}(s) = \frac{\omega_0}{2\pi}\tilde{\mathbf{H}}_{VCO}(s)\cdot\tilde{\mathbf{H}}_{LF}(s)\cdot\mathbf{1} \tag{4.61}$$

and

$$\mathbf{W}(s)^T = \mathbf{1}^T\tilde{\mathbf{H}}_{FD}(s) \ , \tag{4.62}$$

then we can write

$$\tilde{\mathbf{G}}(s) = \tilde{\mathbf{V}}(s)\cdot\mathbf{W}(s)^T \ . \tag{4.63}$$

Using the Sherman-Morisson-Woodbury formula, we then find

$$\left(\mathbf{I}+\tilde{\mathbf{G}}(s)\right)^{-1} = \left(\mathbf{I}+\mathbf{V}(s)\cdot\mathbf{W}^T\right)^{-1} \tag{4.64}$$

$$= \mathbf{I}-\frac{\mathbf{V}(s)\cdot\mathbf{W}(s)^T}{1+\lambda(s)} \tag{4.65}$$

with

$$\lambda(s)=\mathbf{W}(s)^T\mathbf{V}(s)=\mathbf{1}^T\left[\tilde{\mathbf{H}}_{FD}(s)\cdot\tilde{\mathbf{H}}_{VCO}(s)\cdot\tilde{\mathbf{H}}_{LF}(s)\right]\mathbf{1}\ . \tag{4.66}$$

In words, $\lambda(s)$ equals the sum of all elements of $\tilde{\mathbf{H}}_{FD}(s)\cdot\tilde{\mathbf{H}}_{VCO}(s)\cdot\tilde{\mathbf{H}}_{LF}(s)$. Substituting (4.65) in (4.45) and using the fact that $\tilde{\mathbf{H}}_{VCO}(s)\cdot\tilde{\mathbf{H}}_{LF}(s)\cdot\tilde{\mathbf{H}}^{in}_{PFD}(s)=\mathbf{V}(s)\cdot\mathbf{1}^T$, the input-output relation now becomes

$$\tilde{\mathbf{H}}(s)=\left(\frac{1}{1+\lambda(s)}\mathbf{V}(s)\cdot\mathbf{1}^T\right)\ . \tag{4.67}$$

This relation is exact as long as (4.36) accurately models the sampling-PFD's behavior.

4.2.5.4 The discrete-time LTI PLL model and its limitations

Observing (4.67), it is seen that the stability of a PLL with sampling PFD is governed by a single scalar quantity $\lambda(s)$. Using its definition (4.66) and the structural properties of HTMs as discussed in section 3.3.4, it is readily shown that $\lambda(s)$ is periodic in the direction of the imaginary axis, or $\lambda(s+j\omega_0)=\lambda(s)$. This implies that $\lambda(s)$ corresponds to a discrete-time LTI system with z-domain transfer function $\lambda_D(z)=\lambda\left(\omega_0\ln(z)/2\pi\right)$. Hence, the stability of a PLL with sampling PFD is determined by that of a discrete-time system. This is in correspondence with discrete-time (LTI) PLL models in [Gard80, Hein88].

However, the discrete-time models in [Gard80, Hein88] do not fully recognize the mixed-signal aspects of PLLs with sampling PFDs. Mapping the PLL on a discrete-time LTI system, disregards the continuous-time nature of the PLL's output signal. This is, for example, important for spectral computations and noise analysis. These issues are readily handled using the HTM framework where sampling is recognized as a continuous-time rank-one periodically time-varying operator. Moreover, HTM-based PLL analysis is easier and more straighforward than is the conversion to a discrete-time system. This is particularly the case for the derivation of symbolic expressions.

4.2.6 Example 1: PLL with sampling PFD

We illustrate and verify our analysis of time-varying PLL behavior for a system implementing a sampling PFD, lacking a frequency divider, i.e. $N=1$, and containing a VCO with constant ISF, i.e. $\Gamma(t)=\Gamma_0$. HTM-based computations are compared with results extracted from time-marching simulations in Matlab/Simulink^{TM}. The Matlab/Simulink simulation model is shown in Fig. 4.12. It implements the sampling PFD using flip-flops and therefore encodes the phase error through the *width* of the pulses it produces. This corresponds to the behavior of an actual PFD circuit realization. This allows us to test the accuracy of our approximations.

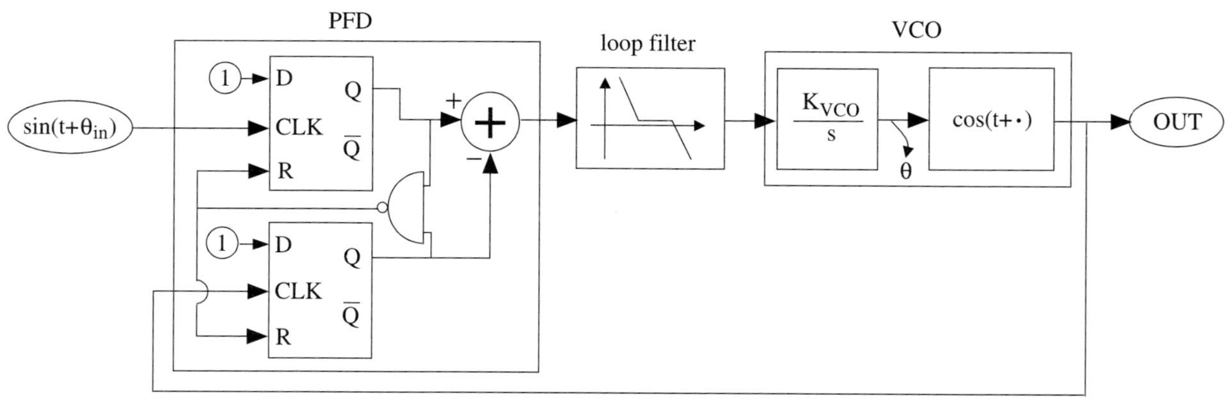

Figure 4.12: Matlab/Simuling model used for time-marching simulations. Note that the sampling PFD is implemented using flip-flops. It therefore encodes the phase error through the *width* of the pulses that it produces. This corresponds to the behavior of an actual circuit implementation.

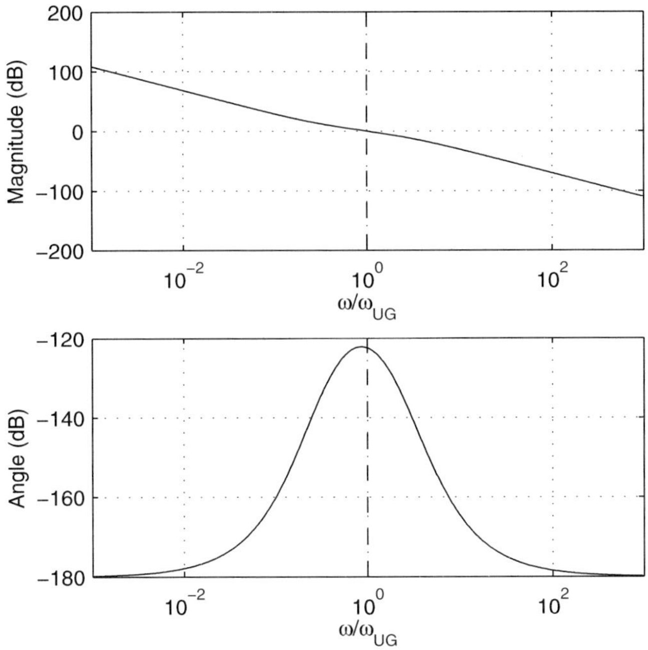

Figure 4.13: Typical characteristic for $A(j\omega)$ containing three poles (two at DC) and one zero. Note that the frequency-axis is normalized with respect to the unity-gain frequency ω_{UG} of $A(s)$.

4.2.6.1 Input-output HTM

With $\Gamma(t) = \Gamma_0$, the HTM (4.43) —capturing the VCO behavior— is diagonal and represents an LTI system with transfer characteristic $H_{VCO}(s) = \Gamma_0/s$. The open-loop gain of the CT-LTI approximation then becomes

$$A(s) = \frac{\omega_0}{2\pi}\frac{\Gamma_0}{s}H_{LF}(s) \ . \tag{4.68}$$

The factor $\omega_0/2\pi$ in front arises from the sampling-PFD model (4.35). Fig. 4.13 shows a typical gain characteristic. It contains three poles (the first two at DC) and one zero. This characteristic will be used for all further numerical computations.

Using (4.61) and (4.62), we find the PLL's open loop gain HTM to equal

$$\tilde{\mathbf{G}}(s) = \mathbf{V}(s) \cdot \mathbf{W}(s)^T = \begin{bmatrix} \vdots \\ A(s - j\omega_0) \\ A(s) \\ A(s + j\omega_0) \\ \vdots \end{bmatrix} \begin{bmatrix} \cdots & 1 & 1 & 1 & \cdots \end{bmatrix} \ . \tag{4.69}$$

Substituting this into (4.66) and (4.67), the input-output HTM $\tilde{\mathbf{H}}(s)$ is found to equal

$$\tilde{\mathbf{H}}(s) = \frac{1}{1+\lambda(s)} \begin{bmatrix} \vdots \\ A(s-j\omega_0) \\ A(s) \\ A(s+j\omega_0) \\ \vdots \end{bmatrix} \begin{bmatrix} \cdots & 1 & 1 & 1 & \cdots \end{bmatrix} . \tag{4.70}$$

Here,

$$\lambda(s) = \sum_{m=-\infty}^{+\infty} A(s+jm\omega_0) \tag{4.71}$$

is the PLL's effective open-loop gain. The HTM element $\tilde{H}_{0,0}(s)$, modeling the closed-loop transfers from baseband to baseband, is given by

$$\tilde{H}_{0,0}(s) = \frac{A(s)}{1+\lambda(s)} \approx \frac{A(s)}{1+A(s)} . \tag{4.72}$$

The latter approximation, corresponding to traditional LTI analysis, is valid as long as $|A(s)| \gg \left|\sum_{m\neq 0} A(s+jm\omega_0)\right|$. If this no longer holds, e.g. for feedback loops with a large unity-gain frequency ω_{UG}, time-varying effects become important and the other terms of $\lambda(s)$ must be taken into account. Note that (4.70) also allows us to study signal transfers to frequency bands other than baseband.

Fig. 4.14 illustrates the impact of an increasing ω_{UG}/ω_0 on $\tilde{H}_{0,0}(s)$. The solid lines are obtained by evaluating (4.72) while the marks follow from spectral analysis of time-marching simulation results in Matlab/Simulink. They are within 2%. This shows the accuracy ofthe methods here developed. Note, however, that evaluating (4.72) is only a matter of seconds while it takes several minutes for the time-marching simulations to complete.

4.2.6.2 Stability behavior

Observing $\tilde{H}_{0,0}(s)$ in Fig. 4.14, it is seen that the effective bandwidth shifts to the right as ω_{UG}/ω_0 increases. Also, peaking at the passband's edge becomes worse. Fig. 4.15explains this behavior in terms of the effective open-loop gain $\lambda(s)$. The increase of the closed-loop bandwidth corresponds to an increase of $\omega_{UG,eff}$, the (effective) unity-gain frequency of $\lambda(s)$. More important is the phase margin of $\lambda(s)$ which is rapidly degrading for increasing values of ω_{UG}/ω_0. For $\omega_{UG}/\omega_0 = 0.1$, this phase margin is already 9% worse than predicted by CT-LTI analysis. This makes clear that CT-LTI analysis does not give reliable results when ω_{UG}/ω_0 grows large.

Expressions for this degarding phase margin are obtained by assuming that, for ω near ω_{UG}, $\lambda(j\omega)$ is dominated by $A(j\omega)$, i.e. $\left|\sum_{m\neq 0} A(j\omega+jm\omega_0)\right| \ll |A(j\omega)|$. With $A(j\omega) = |A(j\omega)|\, e^{\phi_A(j\omega)}$, the phase of $\lambda(j\omega) = |\lambda(j\omega)|\, e^{\phi_\lambda(j\omega)}$ then approximately equals

$$\phi_\lambda(j\omega) \approx \phi_A(j\omega) + \mathrm{Im}\left\{ \frac{\sum_{m\neq 0} A(j\omega+jm\omega_0)}{A(j\omega)} \right\} . \tag{4.73}$$

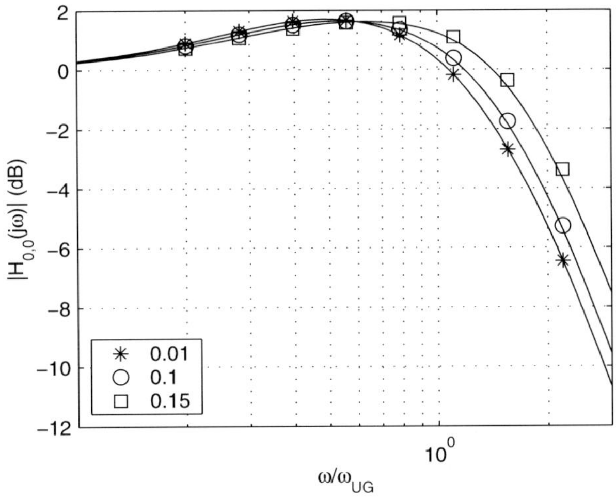

Figure 4.14: HTM element $\tilde{H}_{0,0}(s)$ capturing baseband-to-baseband signal transfers for a PLL with sampling-PFD. The element is plotted for $\omega_{UG}/\omega_0 = 0.01, 0.1$ and 0.15. The solid lines are obtained by evaluating (4.70). The marks are extracted from Matlab/Simulink simulations.

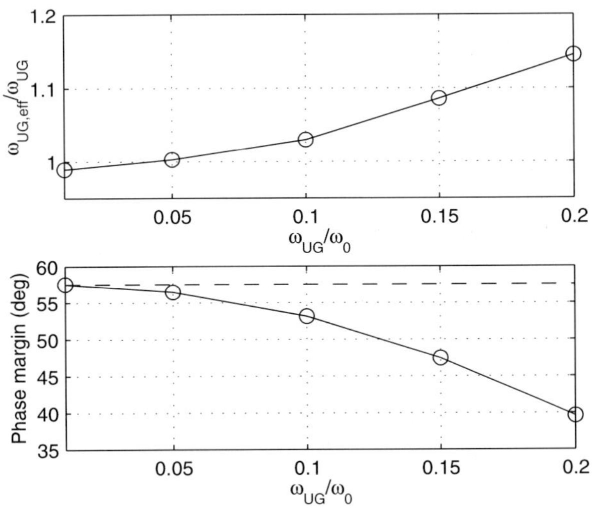

Figure 4.15: The normalized effective unity-gain frequency (upper plot) and the effective open-loop phase margin (lower plot) versus ω_{UG}/ω_0 as obtained by means of the methods developed in this work. In the lower plot, the dashed horizontal line indicates the phase margin as predicted by CT-LTI analysis.

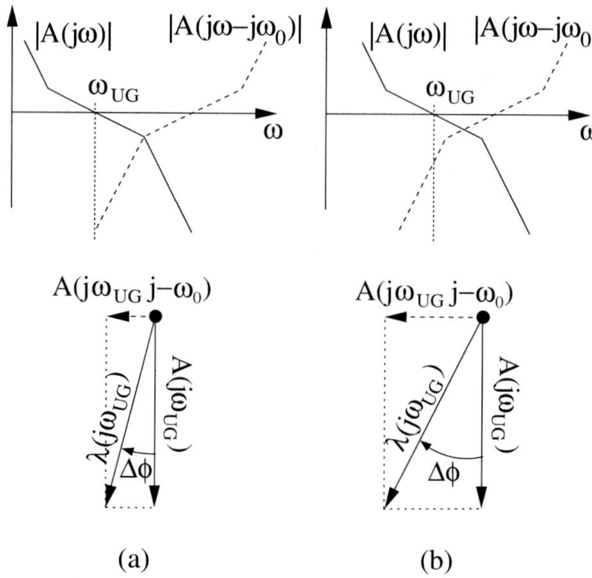

Figure 4.16: Impact of the position of ω_{UG} on the stability of PLLs with sampling-type PFDs. (a) For low ω_{UG} , aliasing terms have only a low impact on the phase margin which is then largely determined by the phase margin of $A(j\omega)$. (b) As ω_{UG} increases, the impact of the aliasing grows, degrading the phase margin. Note that in the figures above, for sake of simplicity, we approximated $\sum_{m\neq 0} A(j\omega + jm\omega_0) \approx A(j\omega - j\omega_0)$ for ω near ω_{UG}.

The second term In the expression above explains the problems that arise when the unity-gain frequency of $A(j\omega)$ approaches $\omega_0/2$. As illustrated in Fig. 4.16, for an increasing ω_{UG}, the second term in (4.73) starts to yield a significant contribution on top of the phase shift generated by $A(j\omega)$. This pushes the total phase shift of the open-loop gain $\lambda(j\omega)$ towards $-\pi$, resulting in a less stable feedback loop.

4.2.6.3 Noise folding in PLLs

Stability, however, is not the only thing that is going bad as ω_{UG} increases. For fast feedback loops, phase noise behavior is also worse than predicted by CT-LTI analysis. This is due to noise folding: high-frequency noise components are downconverted, contributing to the low-frequency phase noise content. Noise folding is illustrated in Fig. 4.17 and is caused by the time-varying behavior of the PLL building blocks. Noise folding is responsible for high-frequency spurs appearing in the PLL's close-in phase noise spectrum.

Noise folding is readily analyzed by means of the HTM-based noise analysis techniques outlined in section 3.5.3 of the previous chapter. Consider, for instance, the VCO noise process $n(t)$ (capturing its phase-noise behavior) injected at the position

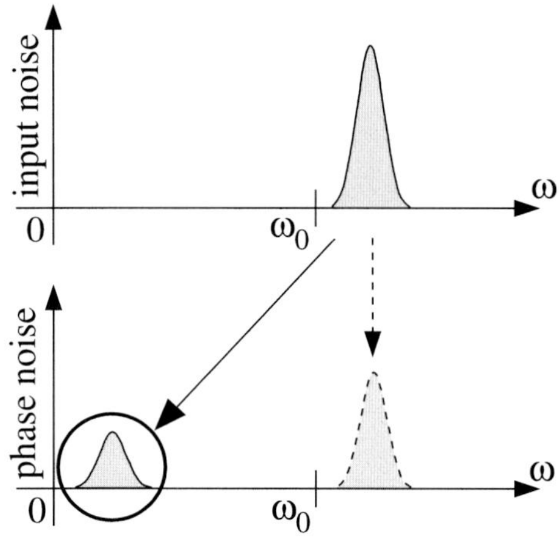

Figure 4.17: Noise folding: high-frequency noise components, e.g. spurs, are downconverted, contributing to the PLL's low-frequency phase noise content.

indicated in Fig. 4.11. The HTM $\tilde{\mathbf{H}}_{noise}(s)$ that characterizes the transfer of noise to the output phase $\theta(t)$ can be shown to equal

$$\tilde{\mathbf{H}}_{noise}(s) = \left(\mathbf{I} - \frac{1}{1+\lambda(s)} \begin{bmatrix} \vdots \\ A(s-j\omega_0) \\ A(s) \\ A(s+j\omega_0) \\ \vdots \end{bmatrix} \begin{bmatrix} \cdots & 1 & 1 & 1 & \cdots \end{bmatrix} \right) \times \begin{bmatrix} \ddots & & & & \\ & \frac{1}{s-j\omega_0} & & & \\ & & \frac{1}{s} & & \\ & & & \frac{1}{s+j\omega_0} & \\ & & & & \ddots \end{bmatrix} . \tag{4.74}$$

The PLL's output phase noise (at least the part due to the action of the VCO noise) is now characterized by

$$\mathbf{S}_{\Theta\Theta}(j\omega) = \tilde{\mathbf{H}}_{noise}(j\omega)\mathbf{S}_{\mathbf{nn}}(j\omega)\tilde{\mathbf{H}}_{noise}(j\omega)^* \ . \tag{4.75}$$

Here, $\mathbf{S}_{\mathbf{nn}}(j\omega)$ and $\mathbf{S}_{\Theta\Theta}(j\omega)$ represent harmonic power spectral density (HPSD) matrices respectively that correspond to the noise at the VCO's input and the (phase) noise at the VCO's output.

If $n(t)$ is stationary and characterized by the power spectral density (PSD) function $S_{nn}(j\omega)$, then, by evaluating (4.75), it can be shown that the PLL's baseband phase

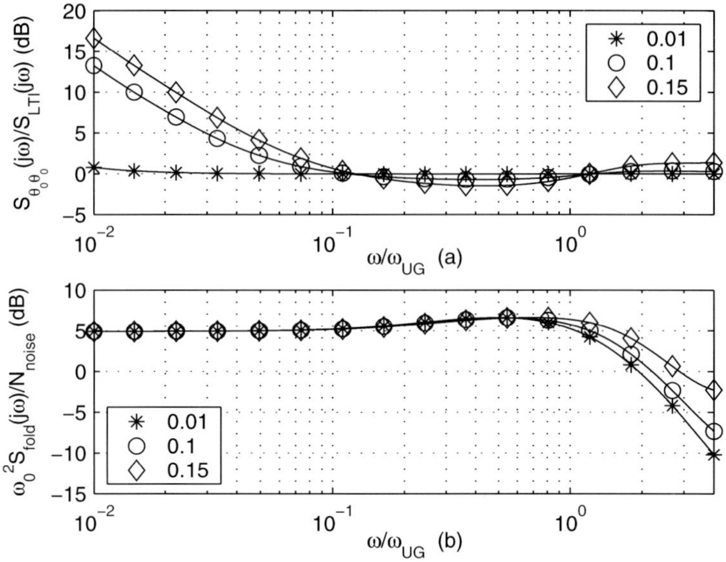

Figure 4.18: Baseband noise PSD for PLLs using sampling PFDs. (a) Ratio $S_{\theta_0\theta_0}(j\omega)/S_{LTI}(j\omega)$ between the noise PSD predicted by respectively time-varying and time-invariant analysis. (b) Normalized contributions due to high-frequency noise components folding into baseband.

noise content[13] is characterized by

$$\begin{aligned} S_{\theta_0\theta_0}(j\omega) &= \left|1-\frac{A(j\omega)}{1+\lambda(j\omega)}\right|^2 \frac{S_{nn}(j\omega)}{\omega^2} \\ &\quad + \sum_{m\neq 0} \left|\frac{A(j\omega)}{1+\lambda(j\omega)}\right|^2 \frac{S_{nn}(j\omega+jm\omega_0)}{|\omega+m\omega_0|^2} \ . \end{aligned} \tag{4.76}$$

Here, the first term captures that part of the phase noise as it is (approximately) obtained by means of traditional CT-LTI analysis. The lower terms, however, are unaccounted for by CT-LTI analysis. They model the noise folding caused by the sampling nature of the PFD. High-frequency input noise components, e.g. spurs, are downconverted to baseband. Note that all extra terms are positive. Hence, the resulting baseband noise is, therefore, always worse than the one predicted by LTI techniques.

Fig. 4.18 illustrates the impact of noise folding for a white input noise process $n(t)$, i.e. $S_{nn}(j\omega) = N_{noise}$. Fig. 4.18(a) plots $S_{\theta_0\theta_0}(j\omega)/S_{LTI}(j\omega)$, the ratio of the PLL's (baseband) phase noise PSD predicted by respectively time-varying and time-invariant

[13]The baseband phase noise component is characterized by the element $S_{\theta_0\theta_0}(j\omega)$ of the HPSD matrix $\mathbf{S}_{\Theta\Theta}(j\omega)$, i.e. the element with indices $(0,0)$. See section 3.5.3 in chapter 3 for notational conventions concerning the elements of HPSD matrices.

analysis. Here,

$$S_{LTI}(j\omega) = \left|\frac{1}{1+A(j\omega)}\right|^2 \frac{S_{nn}(j\omega)}{\omega^2} \ . \tag{4.77}$$

This ratio is drawn for several values of ω_{UG}/ω_0. Fig. 4.18(b) shows the output phase noise contributions coming from high-frequency noise components folding into baseband. This corresponds to the lower terms in (4.76). Observing Fig. 4.18(a), it is seen that, for ω_{UG}/ω_0 growing large, the PSD predicted by time-varying analysis starts to deviate from the one predicted by time-invariant analysis. This especially holds true for frequencies near zero. Fig. 4.18(b) clarifies the reason for this. While (4.77) predicts the noise PSD to be 0 near $\omega = 0$, time-varying analysis predicts a non-zero noise level due to high-frequency components folding into baseband. These components are (silently) neglected when using traditional time-invariant techniques.

4.2.7 Example 2: PLL with mixing PFD

The time-varying behavior of a mixing PFD turns out to have less of an impact on a PLL's small-signal performance than that of a sampling PFD. For the setup in Fig. 4.7 and with the mixing PFD HTMs specified by (4.29)-(4.31), the most important HTM elements characterizing the transfers from $\theta_{in}(t)$ to the baseband component of $\theta(t)$ are found to equal

$$\begin{aligned} \tilde{H}_{0,0}(s) \approx\ & H_{LTI}(s) + \frac{H_{LTI}(s)}{4}\left(H_{LTI}(s+2j\omega_0) - H_{LTI}(2j\omega_0)\right) \\ & + \frac{H_{LTI}(s)}{4}\left(H_{LTI}(s-2j\omega_0) - H_{LTI}(-2j\omega_0)\right) \\ & + \frac{H_{LTI}(s)^2}{4}\left(H_{LTI}(s+2j\omega_0) - H_{LTI}(2j\omega_0)\right) \\ & + \frac{H_{LTI}(s)^2}{4}\left(H_{LTI}(s-2j\omega_0) - H_{LTI}(-2j\omega_0)\right) \end{aligned} \tag{4.78}$$

$$\tilde{H}_{0,2}(s) \approx \frac{1}{2}H_{LTI}(s) + \frac{1}{2}H_{LTI}(s)H_{LTI}(s+2j\omega_0) \ . \tag{4.79}$$

These expressions were obtained using the power series method discussed in section 3.4.3.2 of the previous chapter.

In fig. 4.19, we plot HTM-element $\tilde{H}_{0,0}(s)$ that characterizes transfers from baseband to baseband. We do so for several values of the unity-gain frequency ω_{UG} of the LTI open-loop gain $A(s)$ (see Fig. 4.13). Here, the solid lines are obtained from (4.78) while the marks follow from the spectral analysis of time-marching simulation results in Matlab/Simulink. The figure illustrates that, even for high-speed loops, the HTM-element $\tilde{H}_{0,0}(s) \approx H_{LTI}(s)$, i.e. it deviates little from results predicted by time-invariant analysis. Note, however, that there is also a significant transfer of signal content from twice the carrier frequency to baseband. The latter transfer is modeled by $\tilde{H}_{0,2}(s)$ and is not predicted by LTI analysis.

Similarly as was accomplished for sampling-PFD-based PLLs, we could go on and analyze the stability and noise behavior for PLLs built using mixing PFDs. This, how-

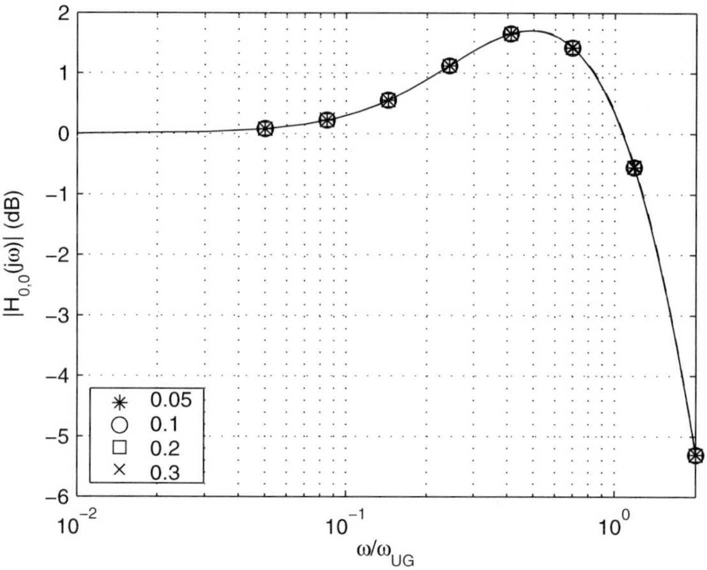

Figure 4.19: Baseband to baseband signal transfer for a PLL implemented with a mixing PFD. The HTM element $\tilde{H}_{0,0}(s)$ is plotted for $\omega_{UG}/\omega_0 = 0.1, 0.2, 0.3, 0.4$ and 0.5. The solid lines are obtained by evaluating (4.78). The marks are extracted from Matlab/Simulink simulations.

ever, does not contribute new elements to our HTM-based analysis of a PLL's behavior. We therefore do not discuss this any further.

4.2.8 Conclusions

The small-signal behavior of all current PLL implementations is inherently time-varying. As a PLL's steady-state operating point varies periodically with time, the linearization of its behavior in the neighborhood of this periodic operating point will result in an LPTV system. HTMs provide a natural framework for a frequency-domain description of this time-varying PLL small-signal behavior.

It turns out that time-varying effects, especially for PLLs with a fast feedback loop and a sampling PFD, can have a significant impact on PLL system performance. Examples discussed in this section concern stability degradation and noise folding. Traditional techniques for PLL analysis, mostly grounded on continuous-time linear time-invariant feedback theory, fail to predict these phenomena. In most cases, they are over-optimistic with regard to the predicted PLL behavior. This may result in erroneous design decisions.

HTM-based analysis of small-signal PLL behavior does not suffer from this drawback. It is exact: there are no hidden approximations or assumptions. The only approximations are the ones explicitly introduced during the analysis. This, however, does not mean that approximations aren't useful. They often help simplifying results, focusing attention to what is really important. For example, traditional (and often useful) PLL models are found to be particular (rank-one) approximations to the exact time-varying models. This provides these traditional models with solid mathematical underpinnings while their shortcomings are clearly identified.

In summary, small-signal PLL analysis is an excellent example to illustrate the use of HTMs in analyzing real-life system behavior. It emphasizes the practical relevance of the HTM framework in supporting more accurate and reliable system analysis (and hence design).

4.3 Automated symbolic LPTV system analysis

As holds for all pencil and paper analysis, HTM-based computations, especially symbolic ones, can be tedious and error prone. This section therefore introduces an algorithm, called *SymbolicHTM*, for semi-automated HTM-based symbolic analysis of LPTV system behavior. In general, symbolic analysis involves the generation of expressions that describe the behavior of a system (circuit) starting from a topological description stated in terms of the system's (circuit's) design variables [Gie91]. Examples of topological descriptions are circuit netlists and block diagrams. Of course, generating expressions is not a goal by itself. As symbolic analysis often aims to increase our understanding on the system's or circuit's behavior, expressions should be kept as simple as possible.

The SymbolicHTM algorithm presented in this section takes a block diagram as input and produces symbolic expressions for the elements of the system's overall input-output HTM. It demonstrates that traditional techniques for automated symbolic analysis of LTI systems [Fern98, Gie91, Lin91] can be extended to handle LPTV systems as well. Moreover, for symbolic LPTV system analysis, the SymbolicHTM algorithm is more efficient than techniques based on weakly nonlinear time-invariant symbolic circuit analysis [Wamb98a]. This is because of the fact that SymbolicHTM exploits the (time-varying) linear nature of a multiplication with a periodic carrier wave. We demonstrate the algorithm by generating expressions for the input-output HTM elements of two different downconversion stages.

In what follows, we first review prior art on symbolic circuit and system analysis. Next, we discuss the construction of the block diagrams that are needed as input (section 4.3.3). This is followed by a description of the data structures (section 4.3.4) and the computational flow (section 4.3.5) of the Symbolic HTM algorithm. We conclude this section with experimental results obtained from the application of SymbolicHTM to two different downconversion stages (section 4.3.7 and 4.3.8).

4.3.1 Prior art

Symbolic analysis has a long standing history [Fern98, Gie91, Lin91]. Traditionally, it focuses on the behavior of LTI networks with extensions towards symbolic weakly-nonlinear circuit analysis [Wamb98a]. Expressions are derived using analytical techniques (either matrix-based or graph-based). More recently, a fitting-based approach for constructing symbolic expressions was suggested as an alternative to the analytical methods [Dae03].

4.3.1.1 LTI network analysis

The first research boom in the field of symbolic circuit analysis focused its efforts on LTI networks [Fern98, Gie91, Lin91]. These LTI networks were obtained by replacing nonlinear network elements, e.g. transistors, by a small-signal model. The network behavior is then characterized by symbolic expressions for input-output transfer functions, input/output impedances, etc.

Early on, it became clear that the number of terms in these symbolic expressions grows exponentially with the network's complexity. In order to deal with this "explosion of terms", people introduced techniques for simplifying the expressions, both after [Gie91], during [Wamb98b] and before [Dae02] generation. A second track of research involved hierarchical circuit analysis, resulting in sequences of expressions [Star80].

Although techniques for LTI symbolic network analysis work fine on relatively small circuits, they lack the ability to capture intermodulations between different signal components. As such, their range of application excludes mixing behavior, harmonic distortion, etc.

4.3.1.2 Weakly nonlinear circuit analysis

In an effort to overcome the limitations of LTI network analysis, attention was focused on the behavior of weakly nonlinear circuits [Verha01, Wamb98a]. This has resulted in methods based on Volterra series and perturbation techniques, methods that operate by an iterative application of linear techniques.

However, weakly nonlinear network analysis is not able to distinguish between linear time-varying and truely nonlinear behavior. For example, although one would be tempted to consider a multiplication $y(t) = e^{jk\omega_0 t} \times u(t)$ a nonlinear operation, it was argued in section 3.3.1 of the previous chapter that the principle of superposition still holds, making it a linear operation. Hence, the symbolic computation of the HTM-elements using techniques for nonlinear symbolic LTI network analysis hence does not fully exploit the linear nature of LPTV systems. This makes them suboptimal for this kind of applications.

4.3.1.3 Fitting-based approaches

All of the methods discussed so far are analytical. This implies that they try to solve the system or circuit equations through determinant expansion techniques, series expansions, spanning tree enumerations, etc. There is no predefined template for the expressions that result, i.e. the form of the expressions is developed at run time.

An alternative technique that was recently proposed builds on the existence of a predefined and parametrized template expression [Dae03]. Using numerical simulation and data regression, one tries to determine suitable values for the template parameters. This method suits very well for performance and strongly nonlinear system modeling.

Basically, they can model anything that can be simulated. The disadvantage, of course, lies in the need to find a suitable expression template a priori.

4.3.1.4 Contributions of this work

This section presents an algorithm to construct symbolic expressions for the HTM elements of LPTV systems. The method is called *SymbolicHTM*. The input is a system model in the form of a block diagram. The output consists of symbolic expressions for the system's HTM-elements in terms of the parameters characterizing the behavior of its building blocks. The algorithm is based on analytical methods. There is no need for a predefined expression template.

The input block diagram can, for example, be extracted from a circuit netlist using techniques similar to the ones for behavioral modeling of LTI networks [Leyn01]. For circuits consisting of a large number of nodes, more compact expressions are obtained by modeling the often large LTI subcircuits using multiport representations. With SymbolicHTM expressing the results in terms of symbols representing the multiport parameters, lengthy and uninterpretable expressions are avoided. The symbolic computation of these multiport parameter characteristics is then accomplished afterwards, using well-established techniques for (LTI) symbolic circuit analysis [Fern98, Gie91, Lin91]. Hence, In the hierarchy of symbolic analysis, SymbolicHTM makes up a layer on top of the existing tools for symbolic circuit analysis.

4.3.2 Symbolic LPTV system analysis: outlining the flow

Fig. 4.20 outlines the top-level flow used for the symbolic characterization of LPTV system behavior. At the heart lies the symbolic analysis engine that computes symbolic expressions for the input-output HTM elements given a description of the system's topology and building blocks. In this work, the input is required to be in the form of a block diagram (or, equivalently, a signal flow graph). As shown in Fig. 4.20, this may require an extra preprocessing step to convert, for example, a circuit netlist into a block diagram description. Finally, when numerical values for the building block parameters are available, they can be used to try and simplify the resulting expressions by discarding terms that have negligible contribution.

In what follows, we discuss each of the steps of the SymbolicHTM algorithm in greater detail. Section 4.3.3 briefly discusses constructing the input block diagram. Next, sections 4.3.4 and 4.3.5 deal with the operations of the engine for generating the expressions. Finally, use of the algorithm is illustrated in sections 4.3.7 and 4.3.8 for two types of downconversion stages.

4.3.3 Input model construction

SymbolicHTM takes a block diagram of the system to be analyzed as input. The composing blocks can be LTI systems, represented by a symbol $H(s)$ for their transfer characteristic, multiplications with a periodic signal $p(t)$, represented by the symbols P_k for its Fourier coefficients, or even other LPTV systems, represented by the symbols for the elements of the 0-th column of the corresponding HTM (see section 4.3.4).

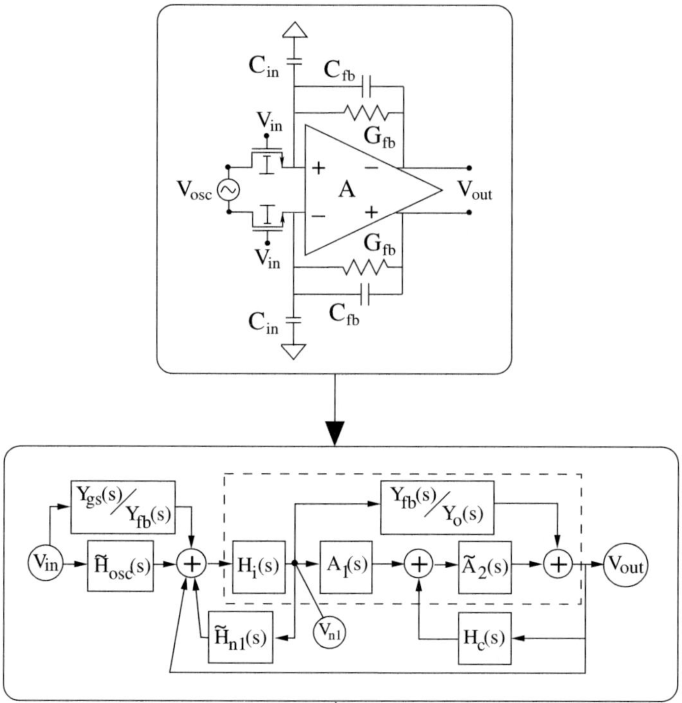

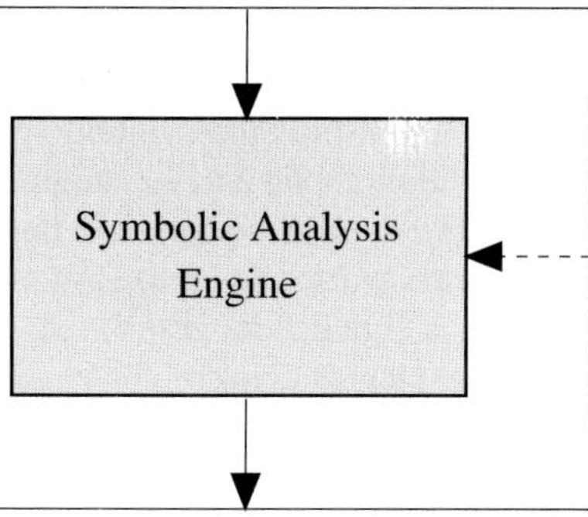

Symbol	Value
G_{fb}	0.1 mS
C_{fb}	1.6 pF
C_{in}	16 pF
C_c	2 pF
C_{out}	10 pF
A_0	55 dB
V_{osc}	0.15 V

$$\tilde{H}_{0,1}(s) = \frac{1}{2}\frac{\beta V_{osc} G_0(s)}{\left(1 - G_0(s) - \frac{H_c(s)(g_{m2,0}+sC_c)}{Y_o(s)}\right) Y_{fb}(s)}$$

$$\begin{aligned}\tilde{H}_{0,2}(s) = {} & \frac{1}{2}\frac{\beta V_{osc} G_{-1}(s+j\omega_0)}{L(s) Y_{fb}(s+j\omega_0)} \\ & + \frac{\beta V_{osc} G_0(s+j\omega_0) G_{-1}(s+j\omega_0)}{L(s) L(s+j\omega_0) Y_{fb}(s+j\omega_0)} \\ & - \frac{1}{2}\frac{\gamma \beta^2 V_{osc} V_{n1,-1} G_0(s) H_i(s+j\omega_0)}{L(s) Y_{fb}(s) Y_{fb}(s+j\omega_0)} \\ & + \dots\end{aligned}$$

Figure 4.20: Top-level flow for generating symbolic expressions for the HTM elements characterizing LPTV system behavior. This work focuses on the symbolic analysis engine that generates the expressions starting from a block-diagram system description.

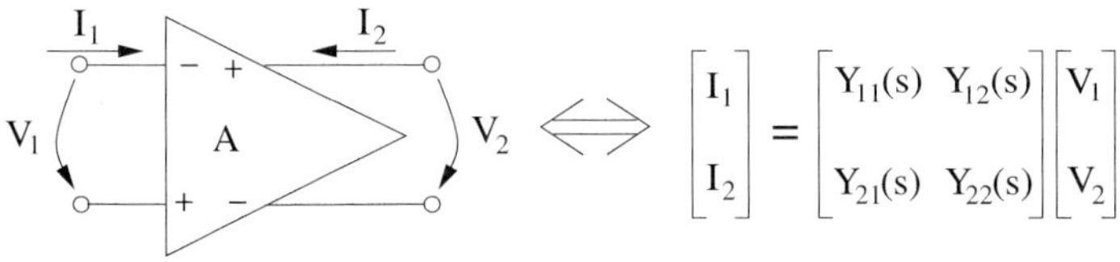

Figure 4.21: Subcircuits consisting of a large number of nodes, like opamps, are replaced by symbolic stamps, representing the subcircuit behavior using (time-varying) multiport parameters

SymbolicHTM then expresses the HTM elements of the overall input/output characteristics in terms of these symbols.

The extraction of the block diagram from a circuit netlist can be automated using existing techniques for behavioral modeling [Leyn01]. However, in order to keep the resulting symbolic expressions interpretable, for complex circuits it is necessary to work in an hierarchical way. As illustrated in Fig. 4.21, this means that subcircuits consisting of a large number of nodes, like opamps, are replaced by symbolic stamps, representing the subcircuit, for example, using (time-varying) multiport parameters. Generation of this symbolic stamp can be done separately and is a one-time effort. This procedure is similar to the way how algorithms for traditional symbolic LTI network analysis replace transistors by a stamp containing small-signal parameters like transconductances, resistors and capacitors. With the SymbolicHTM algorithm expressing the system's input/output HTM-elements in terms of the stamp parameters, lengthy and uninterpretable expressions are avoided. This supports hierarchical analysis, where, for example, the detailed dependence of the subcircuit's multiport parameters on component parameters is extracted afterwards using techniques for classical symbolic LTI network analysis [Fern98, Gie91, Lin91].

Furthermore, before starting to generate symbolic expressions, it is possible to simplify the input block diagram by applying similar techniques as used in LTI network analysis [Dae02, Leyn01]. For example, a branch $\tilde{\mathbf{H}}(s) = \tilde{\mathbf{H}}_1(s) + \tilde{\mathbf{H}}_2(s) + \tilde{\mathbf{H}}_3(s) \ldots$ containing a number of parallel paths can be simplified by eliminating those paths $\tilde{\mathbf{H}}_m(s)$ having negligible contribution. Technically, this implies that we remove all paths for which $\left\|\tilde{\mathbf{H}}_m(s)\right\| \ll \left\|\tilde{\mathbf{H}}(s)\right\|$. We refer to appendix A for more details on HTM norms. This simplification requires of course numerical data, or other prior knowledge on the system, to be available.

4.3.4 Data structures

In storing the HTM, it is not needed to store every matrix element separately. From the structural relation (3.53) derived in section 3.3.4, it is seen that it is sufficient to only store the nonzero elements $\tilde{H}_{\cdot,0}(s)$ of the HTM's 0-th column. When these elements are known over the entire frequency range $\omega \in [-\infty, \infty]$, one is able to compute all other elements of the HTM. SymbolicHTM, therefore, stores HTMs by means of expressions for the elements of their 0-th column.

As a note aside, we mention that this particular choice in storing HTMs best suits the power series expansion method that SymbolicHTM is using to perform HTM inversions. Algorithms performing the inversions, for example using rank-reduced approximations, may require alternate ways for storing HTMs and their elements.

4.3.5 Computational flow of the SymbolicHTM algorithm

Fig. 4.22 shows the global outline of the computational flow of the Symbolic HTM algorithm. It starts from a system block diagram model and consists of three steps. In a first (preprocessing) step, the input model's feedback loops are cut by introducing a set of loop variables. During the second step, which comprises the actual symbolic computations, these loop variables are eliminated using symbolic Gaussian elimination in order to obtain expressions for the input-output HTM elements. In a final (postprocessing) step, these expressions can be simplified to obtain more compact and better readable expressions. The latter is only possible if numerical data are available to control the truncation error [Gie91, Gie94]. In what follows, we explain each of these steps in more detail.

4.3.5.1 Breaking the loops

In a first step, all loops in the block diagram are cut. This results in the introduction of a set of loop variables. Given a graph-like representation, loops can be detected using graph traversal algorithms [Thul92]. They are cut at the point where they first leave the feedforward path. This process results in a set of equations

$$\begin{cases} \mathbf{X}_1(s) &= \sum \tilde{\mathbf{G}}_{1,k}(s)\mathbf{X}_k(s) + \tilde{\mathbf{G}}_{1,in}\mathbf{U}(s) \\ \mathbf{X}_2(s) &= \sum \tilde{\mathbf{G}}_{2,k}(s)\mathbf{X}_k(s) + \tilde{\mathbf{G}}_{2,in}\mathbf{U}(s) \\ &\vdots \\ \mathbf{Y}(s) &= \sum \tilde{\mathbf{G}}_{y,k}(s)\mathbf{X}_k(s) + \tilde{\mathbf{G}}_{y,in}\mathbf{U}(s) \end{cases} \tag{4.80}$$

with the $\mathbf{X}_k(s)$ representing the loop variables and the $\tilde{\mathbf{G}}_{l,\cdot}(s)$, $l = 1, 2, \ldots$, being the HTMs characterizing the open-loop behavior. These open-loop HTMs are computed as series and/or parallel connections of the building-block HTMs, using the principles outlined in section 4.1. As a heuristic for obtaining better readable expressions, for each loop equation l, a symbolic name can be assigned to the HTM corresponding to the intersection of the graphs of the open-loop HTMs $\tilde{\mathbf{G}}_{l,\cdot}(s)$. This means that one introduces auxiliary symbols to represent the HTM corresponding to the part in common to the HTMs $\tilde{\mathbf{G}}_{l,\cdot}(s)$. Expressing the open-loop HTMs $\tilde{\mathbf{G}}_{l,\cdot}(s)$ in terms of these auxiliary symbols yields hierarchically structured and better readable expressions.

All of this is illustrated in Fig. 4.23 for the case of a simple feedback system. The top graph represents the graph of the overall system. Each branch represents a (possibly time-varying) building block. There is one feedback loop, hence leading to the introduction of a loop variable $\mathbf{X}_1$ and its corresponding loop equation. The bottom part of Fig. 4.23 shows the graphs corresponding to the open-loop HTMs $\tilde{\mathbf{G}}_{1,\cdot}(s)$. Here, the

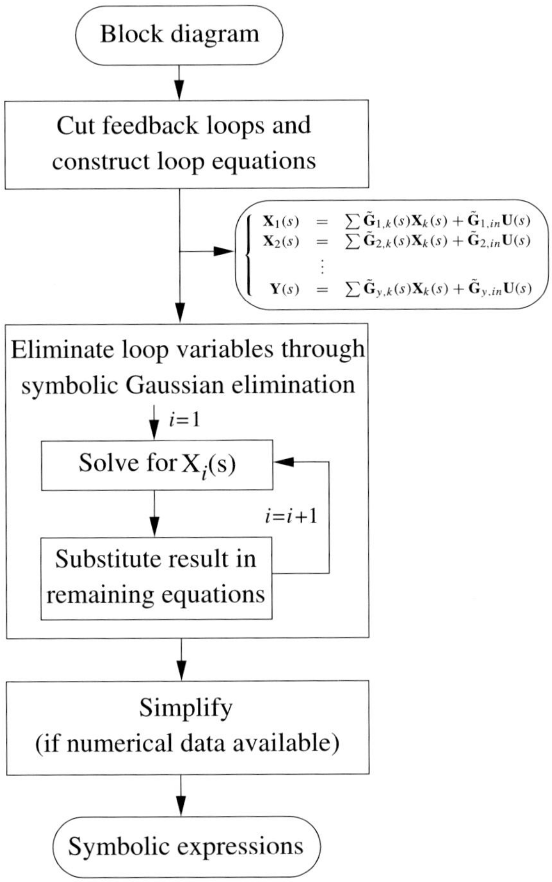

Figure 4.22: Outline of the computational flow of the Symbolic HTM algorithm. First, all loops are cut and the loop equations constructed. Next, the loop variables are eliminated using symbolic Gaussian elimination. When numerical data are available, it can be used to simplify the resulting expressions by eliminating irrelevant terms.

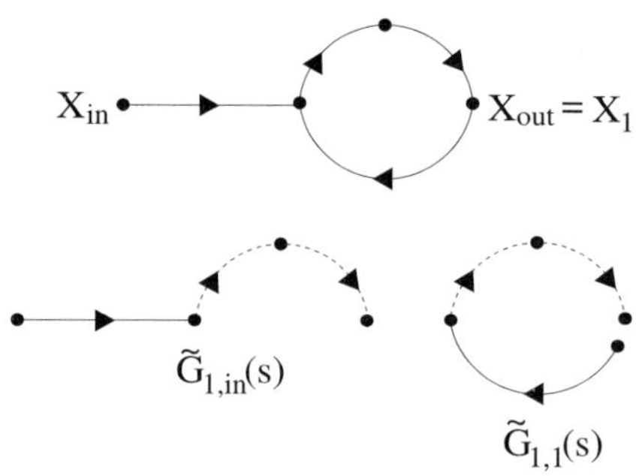

Figure 4.23: Graphs representing a simple feedback system (top) and its open-loop HTMs (bottom). Each branch represents a (possibly time-varying) building block. The dashed lines mark the parts in common to both open-loop HTMs.

dashed part is in common to both graphs and its corresponding HTM can be represented using auxiliary symbols in order to make further expressions better readable.

4.3.5.2 Eliminating the loop variables

Next, the loop variables $\mathbf{X}_k(s)$ from (4.80) are eliminated using symbolic Gaussian elimination. For $\mathbf{X}_1(s)$, this implies that we solve

$$\left(\mathbf{I}-\tilde{\mathbf{G}}_{11}(s)\right)\mathbf{X}_1(s) = \sum_{k>1} \tilde{\mathbf{G}}_{1,k}(s)\mathbf{X}_k(s) + \tilde{\mathbf{G}}_{1,in}\mathbf{U}(s) \tag{4.81}$$

and substitute the result back into the equations for $\mathbf{X}_2(s), \mathbf{X}_3(s), \ldots$ in (4.80). The result is a reduced system with one equation and one variable removed. We proceed in a likewise manner to eliminate the other loop variables until we end up with the desired input/output relation.

The inversion of $\left(\mathbf{I}-\tilde{\mathbf{G}}_{11}(s)\right)$, implied in equation (4.81), is performed using the power series expansion (3.80) discussed in section 3.4.3.2. This expansion is truncated after the first R terms. Here, R is provided by the user, unless numerical data are available. In the latter case, its value can be determined based upon error control, as is done in symbolic approximation techniques used for analyzing LTI networks [Fern98, Gie91, Gie94, Wamb98b].

During computations, a number of simple measures can be taken to avoid wasting time on unimportant contributions. Firstly, we can limit the number of tones, i.e. the number of HTM elements, taken into account during computations. For each LPTV subsystem and interconnection of subsystems, we assume that the HTM elements $\tilde{G}_{0,k}(s) \approx 0$ for $|k| > K$ with K a parameter that controls the SymbolicHTM algorithm. This assumption allows us to skip a detailed computation of these HTM elements. Secondly, in performing the necessary HTM inversions by means of (3.80), we introduce a dummy variable μ in the expansion, writing

$$\tilde{\mathbf{H}}(s)^{-1} \approx \sum_{r=0}^{R-1} \mu^r \left(\tilde{\mathbf{H}}_{LTI}(s)^{-1}\Delta\tilde{\mathbf{H}}(s)\right)^r \tilde{\mathbf{H}}_{LTI}(s)^{-1} \ . \tag{4.82}$$

Here, for example, $\tilde{\mathbf{H}}(s) = \mathbf{I} + \tilde{\mathbf{G}}_{11}(s) = \tilde{\mathbf{H}}_{LTI}(s) + \Delta\tilde{\mathbf{H}}(s) = \left(\mathbf{I} + \tilde{\mathbf{G}}_{11,LTI}(s)\right) + \Delta\tilde{\mathbf{G}}_{11}(s)$. The dummy variable μ is carried along during all subsequent inversion and substitution processes. As a heuristic for obtaining only the relevant terms, it is assumed that all contributions which are $O\left(\mu^R\right)$ are unimportant and can hence be neglected. Choosing suitable values for both K and R can be done automatically, based on error control, when numerical data are available. If not so, they need to be chosen manually. This may require some educated guessing. The validity of the results that are obtained for the chosen values of K and R then, of course, needs to be verified further on in the design process, when numerical data becomes available.

4.3.5.3 Symbolic simplification

In a final step, the expressions for the input/output HTM elements can be simplified by eliminating all terms with negligible energy, as is done in traditional linear symbolic analysis [Fern98, Gie91, Gie94, Wamb98b]. Given the HTM element

$$\tilde{H}_{m,n}(s) = \tilde{H}_{m,n;1}(s) + \tilde{H}_{m,n;2}(s) + \tilde{H}_{m,n;3}(s) + \ldots \tag{4.83}$$

the most straightforward approach is to eliminate all terms $\tilde{H}_{m,n,k}(s)$ for which

$$\left\|\tilde{H}_{m,n;k}(s)\right\| \ll \left\|\tilde{H}_{m,n}(s)\right\| \tag{4.84}$$

Practically, this elimination is done by ranking the terms in equation (4.83) in order of decreasing magnitude $\left\|\tilde{H}_{m,n;k}(s)\right\|$. Terms are then eliminated, starting with the smallest ones, until the error being introduced exceeds some specified threshold. For the norm $\left\|\tilde{H}_{m,n;k}(s)\right\|$, one can for example use

$$\left\|\tilde{H}_{m,n;k}(s)\right\| = \sqrt{\sum_l w_l \left\|\tilde{H}_{m,n;k}(j2\pi f_l)\right\|^2} \tag{4.85}$$

Here, the set of frequencies $\{f_l\}$ is chosen to cover the frequency range of interest. The weight coefficients w_l allow us to (de-)emphasize frequency bands with respect to each other. Note that this simplification procedure is only possible if numerical data are available to estimate and control the simplification error.

4.3.5.4 Computational complexity

The complexity of the Symbolic HTM algorithm is roughly proportional to the total number of all terms generated during the second (elimination) step. This number is (in worst case) of the order $O\left(N(2K+1)^{R+N-1}T^{R+N-1}\right)$, where N is the number of time-varying feedback loops, K is a measure for the number of tones taken into account (as defined above), T is the average number of terms necessary for modeling the open-loop

HTM elements and $R-1$ is the order in μ up to which the contributions are computed. Note that N and T are mainly functions of the size of the block diagram being analyzed (as determined by the number of nodes of the corresponding signal flow graph). It is through these parameters that the block diagram size (or equivalently, the circuit size) influences the computational complexity.

4.3.6 SymbolicHTM: advantages and limitations

The great advantage of the Symbolic HTM algorithm lies in the automated symbolic characterization of LPTV system behavior. HTM-based analysis is made accessible by partial elimination of tedious and error-prone pencil and paper computations.

While being a strength, automation also implies a disadvantage. Automated analysis often relies on heuristics and is not necessarily as smart as a human may be. Results can, therefore, become more complicated than needed. SymbolicHTM, for example, relies on the assumption that all terms that are $O\left(\mu^R\right)$ are not important while all other terms are. If this assumption fails, it may either result in low-accuracy expressions or in expressions that are too complicated to handle. A further condition for SymbolicHTM to be applicable lies in the accuracy of (4.82) in computing the inverse of a HTM. Although it often works fine, this power-series approximation does not apply to all types of systems. For example, SymbolicHTM does not yield good results when applied to PLLs implemented using sampling-type PFDs[14].

Finally, care must be taken in order to avoid an explosion of the number of terms in the expressions that are generated. Computational complexity considerations in the previous section show that this number of terms grows exponentially with both the circuit size N and the order of the power-series expansion R. Large values of N and R hence lead to complex expressions that are hard to interpret. Term explosion is a problem inherent to all algorithms for symbolic circuit analysis that are based on analytical methods. Avoiding term explosion is accomplished by keeping the input model complexity as low as possible, for example, by the introduction of extra hierarchy. Furthermore, it is required that (4.82) converges sufficiently fast, i.e. R can be chosen sufficiently small in order to prevent term explosion.

4.3.7 Application 1: linear downconversion mixer

By nature, mixers are intended to operate as LPTV systems: they are designed to transfer information from one frequency band (carrier) to another. However, in almost any mixer, both wanted and unwanted signal transfers occur. Proper characterization of the mixer's behavior requires quantifying all of these transfers.

In a first application, we use the SymbolicHTM algorithm to characterize the behavior of the linear downconversion mixer shown in Fig. 4.24. This is accomplished by generating symbolic expressions for the HTM elements $\tilde{H}_{0,1}(s)$, $\tilde{H}_{0,0}(s)$ and $\tilde{H}_{0,2}(s)$. These expressions are verified by comparison with numerical simulations for typical

[14]This, however, could be overcome by allowing SymbolicHTM to also use rank-reduced techniques to perform HTM inversions. In that case, however, we need a heuristic that decides on which method to use to invert a particular HTM. This again requires expertise of the field.

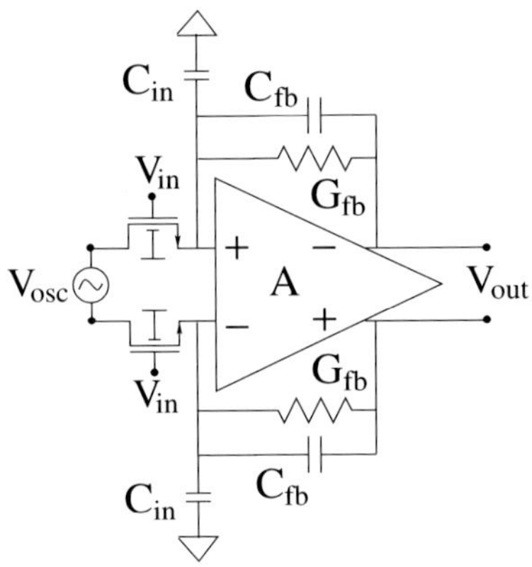

Figure 4.24: Linear downconversion mixer topology. The amplifier is realized as a two-stage Miller-OTA.

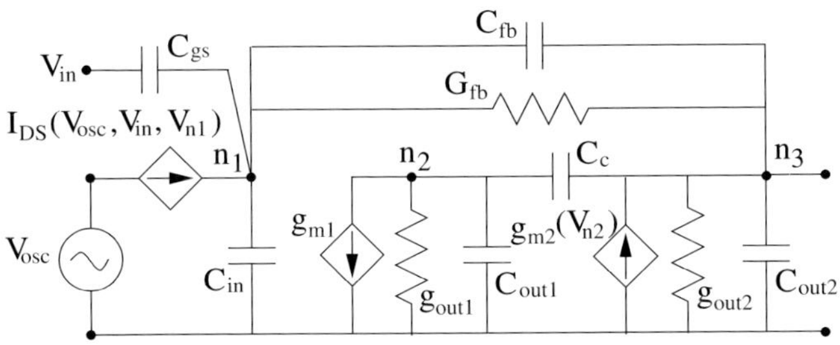

Figure 4.25: Single-sided equivalent circuit of the linear downconversion mixer. The transistors have been replaced by (nonlinear) voltage-controlled current sources.

values of the design parameters. The mixer topology, introduced in [Crol95], was modeled using a two-stage Miller-OTA. In what follows, the mixer building block models are discussed, some remarks concerning the construction of the loop equations (4.80) are made and the resulting symbolic expressions are verified through numerical simulations. As a notational convention, we use uppercase for the large-signal currents/voltages and lowercase for the (time-varying) small-signal components. The Laplace transforms of the latter are also denoted in uppercase but are recognized by their argument s or $j\omega$.

4.3.7.1 Modeling the mixer components

In order to construct a block diagram representation for the mixer topology shown in Fig. 4.24, we transform it to the single-sided equivalent circuit shown in Fig. 4.25. Table 4.1

Symbol	Value	Symbol	Value
C_c	2 pF	G_{fb}	0.1 mS
C_{fb}	1.6 pF	g_{m1}	2.8 mS
C_{gs}	70 fF	$g_{m2,0}$	3.8 mS
C_{in}	16 pF	$\frac{g_{m1}}{g_{out1}}$	30
C_{out1}	60 fF	$\frac{g_{m2,0}}{g_{out2}}$	20
C_{out2}	10 pF	β	0.5 mS/V
V_{osc}	0.15 V	$V_{in,DC} - V_T$	0.2 V

Table 4.1: Numerical values for some important mixer design parameters.

lists the values for some of the most important parameters as they were set for numerical evaluation of the mixer's behavior. In the equivalent model, transistors are replaced by (nonlinear) voltage-controlled current sources. The single-sided input transistor current model is based upon the level-1 equations for a MOST in triode region, or

$$I_{DS} = \beta\left(V_{osc} - V_{n1}\right)\left(V_{in} - V_T\right) - \frac{\gamma\beta}{2}\left(V_{n1}^2 - V_{osc}^2\right) \quad . \tag{4.86}$$

Here, V_T is the threshold voltage and γ a parameter introduced to capture the mismatch between the two input transistors in the original balanced configuration in Fig. 4.24. $V_{osc}(t)$ is the oscillator signal, $V_{in}(t)$ is the input signal as applied to the gates of the input transistors and $V_{n1}(t)$ is the voltage at the OTA input terminal n_1. Furthermore, due to nonlinear behavior, the effective small-signal transconductance g_{m2} of the second OTA stage depends on the voltage $V_{n2}(t)$ on the node n_2, as is indicated in Fig. 4.25. As a consequence, the periodic large-signal swing of this voltage causes a periodic modulation of the small-signal transconductance g_{m2}. The voltage swing V_{n1} at the input of the first OTA stage is small enough to allow us to neglect this effect for the transconductance of this first stage.

The mixer block diagram was then extracted from the nodal equations of the topology in Fig. 4.25 and is shown in Fig. 4.26. The time-varying subblocks are indicated with a tilde on top of the block name. The two main sources of time-varying behavior are due to the modulation of the small-signal parameters of the input transistors and of the transconductance of the second OTA stage. As the input transistors are concerned, use of equation (4.86) yields the small-signal current model

$$i_{DS} = \beta\left(V_{osc}(t) - V_{n1}(t)\right) v_{in} - \beta\left(V_{in}(t) - V_T\right) v_{n1} - \gamma\beta V_{n1}(t) v_{n1} \quad . \tag{4.87}$$

Here, the capital letters represent the (time-varying) operating points while the small letters represent the small-signal components. The first term in (4.87) captures the desired mixing behavior, corresponding to the block $\tilde{\mathbf{H}}_{osc}(s)$ in Fig. 4.26. The last term in (4.87) adds a time-varying component to the OTA input admittance. This corresponds to the feedback path $\tilde{\mathbf{H}}_{n1}(s)$. More precisely, $\tilde{\mathbf{H}}_{osc}(s)$ corresponds to the series connection of a multiplication with $\beta\left(V_{osc}(t) - V_{n1}(t)\right)$ and an LTI filtering operation with transfer function $\frac{1}{Y_{fb}(s)}$. Furthermore, $\tilde{\mathbf{H}}_{n1}(s)$ is the series connection of $-\gamma\beta V_{n1}(t)$ and

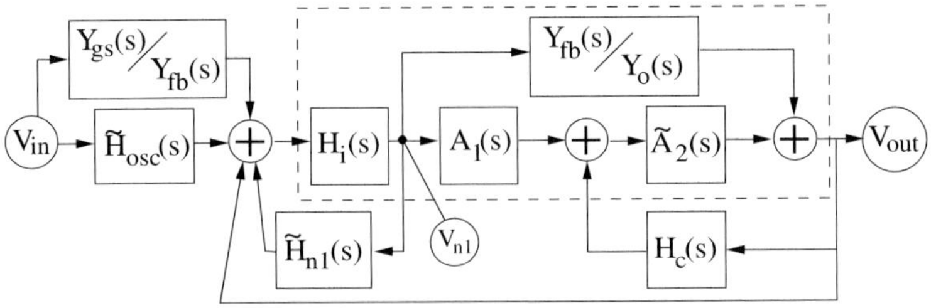

Figure 4.26: Block diagram of the linear downconversion mixer. Time-varying blocks are marked with a tilde on top of their name. Blocks without a tilde are time-invariant.

$\frac{1}{Y_{fb}(s)}$. In the above, $V_{osc}(t) = V_{osc}\cos(\omega_0 t)$ is modeled as a pure cosine. Note that, in the expressions that follow, the symbol V_{osc} is always used to refer to the oscillation amplitude. When referring to the complete oscillation waveform, we'll explicitly denote the time dependency using $V_{osc}(t)$.

The second source of time-varying behavior is, as was already explained before, due to the modulation of the small-signal transconductance of the second OTA stage, or

$$g_{m2}(t) = \sum_k g_{m2,k} e^{jk\omega_0 t} \quad . \tag{4.88}$$

This modulated small-signal transconductance is symbolically represented by the Fourier coefficients $g_{m2,k}$. In the block diagram of Fig. 4.26, it is contained within the time-varying gain block $\tilde{\mathbf{A}}_2(s)$, which represents the series connection of a multiplication with $g_{m2}(t)$ and filtering by $\frac{1}{Y_o(s)}$, the output impedance.

The remaining blocks in Fig. 4.26 are time-invariant and are determined by

$$Y_{fb}(s) = G_{fb} + sC_{fb} \tag{4.89}$$
$$Y_{gs}(s) = sC_{gs} \tag{4.90}$$
$$Y_o(s) = g_{out2} + s(C_{out2} + C_c) + Y_{fb}(s) \tag{4.91}$$
$$Y_{n1}(s) = g_{in} + s(C_{in} + C_{gs}) + Y_{fb}(s) \tag{4.92}$$
$$g_{in} = \beta(V_{in,DC} - V_T) \tag{4.93}$$
$$Y_{n2}(s) = g_{out1} + s(C_{out1} + C_c) \tag{4.94}$$
$$H_i(s) = \frac{Y_{fb}(s)}{Y_{n1}(s)} \tag{4.95}$$
$$H_c(s) = \frac{sC_c}{Y_{n2}(s)} \tag{4.96}$$
$$A_1(s) = \frac{g_{m1}}{Y_{n2}(s)} \tag{4.97}$$

Numerical values for the most important parameters occurring in these formulas were already listed in Table 4.1.

4.3.7.2 Computing the HTM elements of the overall mixer

In a first step of the symbolic HTM algorithm, the loop equations (4.80) are constructed. During this process, two loop variables and one auxiliary HTM $\tilde{\mathbf{G}}(s)$ are introduced. This auxiliary HTM corresponds to the part encircled by the dashed line in Fig. 4.26, and is generated as the intersection of the graphs corresponding to the open-loop HTMs of the second loop equation. It represents a system with

$$\begin{cases} \tilde{G}_{0,0}(s) = \frac{H_i(s)\left(A_1(s)\left(g_{m2,0}+sC_c\right)+Y_{fb}(s)\right)}{Y_o(s)} \\ \tilde{G}_{k,0}(s) = \frac{H_i(s)A_1(s)g_{m2,k}}{Y_o(s+jk\omega_0)} \quad \text{for } k \neq 0 \end{cases} \tag{4.98}$$

Next, symbolic expressions are generated for the mixer's input-output HTM elements $\tilde{H}_{0,0}(s)$, $\tilde{H}_{0,1}(s)$ and $\tilde{H}_{0,2}(s)$. These elements, respectively, capture the downconversion behavior of the signal contents around 0, f_0 and $2f_0$ to baseband. Computations were done up to first order in the dummy variable μ (see equation (4.82)), while taking 7 tones into account. They took about 3 seconds of CPU time to complete on a Sun Ultra30. In the absence of prior numerical data, these values were selected manually. In what follows, the resulting symbolic expressions are verified numerically for a mixer design with a bandwidth of 1 MHz and a local oscillator frequency $f_0 = 20\,\text{MHz}$. Numerical extraction of the HTM elements was done in Matlab™ using an algorithm similar to the one in [Royc99]. Comparison of the symbolic expressions generated by SymbolicHTM with numerically extracted HTM elements is done by integrating the difference between both over the interval $[0, f_0]$, with a weight function that emphasizes the characteristic in the 1 MHz passband by a factor of 10 dB.

The desired downconversion behavior from f_0 to baseband, determined by the HTM element $\tilde{H}_{0,1}(s)$, is dominated by the contribution of 0-th order in μ. It is given by

$$\tilde{H}_{0,1}(s) = \frac{1}{2}\frac{\beta V_{osc}\tilde{G}_{0,0}(s)}{\left(1-\tilde{G}_{0,0}(s)-\frac{H_c(s)\left(g_{m2,0}+sC_c\right)}{Y_o(s)}\right)Y_{fb}(s)} \tag{4.99}$$

with V_{osc} being the amplitude of the local oscillator signal and $\tilde{G}_{0,0}(s)$ given by (4.98). At low frequencies, this expression can be approximated as

$$\tilde{H}_{0,1}(s) \approx \frac{1}{2}\frac{\beta V_{osc}}{Y_{fb}(s)} = \frac{\beta V_{osc}}{2G_{fb}}\frac{1}{1+\frac{s}{G_{fb}/C_{fb}}} \tag{4.100}$$

It exhibits a lowpass behavior, as was to be expected. Comparison with numerical computations shows the full expression (4.99) to be accurate up to 0.01%.

For the HTM elements $\tilde{H}_{0,0}(s)$ and $\tilde{H}_{0,2}(s)$, things are more complicated. The expression for $\tilde{H}_{0,0}(s)$ contains 11 terms while the one for $\tilde{H}_{0,2}(s)$ contains 8 terms. In both cases, no term is really dominating. As an example, we list the first couple of terms of the HTM element $\tilde{H}_{0,2}(s)$:

$$\begin{aligned}\tilde{H}_{0,2}(s) &= \frac{1}{2}\frac{\beta V_{osc}\tilde{G}_{-1,0}(s+j\omega_0)}{L(s)Y_{fb}(s+j\omega_0)} \\ &+\frac{\beta V_{osc}\tilde{G}_{0,0}(s+j\omega_0)\tilde{G}_{-1,0}(s+j\omega_0)}{L(s)L(s+j\omega_0)Y_{fb}(s+j\omega_0)} \\ &-\frac{1}{2}\frac{\gamma\beta^2 V_{osc}V_{n1,-1}\tilde{G}_{0,0}(s)H_i(s+j\omega_0)}{L(s)Y_{fb}(s)Y_{fb}(s+j\omega_0)} \\ &+\ldots\ .\end{aligned} \tag{4.101}$$

Here, $L(s) = 1 - \tilde{G}_{0,0}(s) - \frac{H_c(s)(g_{m2,0}+sC_c)}{Y_o(s)}$, the $\tilde{G}_{k,0}(s)$ are as defined in (4.98) and the $V_{n1,k}$ are the Fourier coefficients corresponding to the time-varying operating point $V_{n1}(t)$ in equation (4.87). The full expressions for both $\tilde{H}_{0,0}(s)$ and $\tilde{H}_{0,2}(s)$ can be found in appendix C.

For purpose of analysis, the terms in the expressions for $\tilde{H}_{0,0}(s)$ and $\tilde{H}_{0,2}(s)$ can be grouped by combining those having a similar low-frequency behavior. In most cases, this concerns the frequency range of interest. Grouping is accomplished by using the fact that, typically, for $|s| \ll \omega_0$ and $k \neq 0$, the factors of the form $H(s+jk\omega_0)$ occurring in the terms above can be approximated as $H(s+jk\omega_0) \approx H(jk\omega_0)$, i.e. a constant. This leads us, for example, to group the first two terms in expression (4.101), since both are proportional to $\frac{1}{L(s)Y_o(s)}$ as their low-frequency behavior is concerned. Combining terms in this way, $\tilde{H}_{0,0}(s)$ contains 3 components and $\tilde{H}_{0,2}(s)$ 2. This provides a much better starting point for analyzing and interpreting the expressions. For example, for $|s| \ll \omega_0$, $\tilde{H}_{0,0}(s)$ looks like

$$\begin{aligned}\tilde{H}_{0,0}(s) &\approx C_1\frac{G_0(s)Y_{gs}(s)}{L(s)Y_{fb}(s)} + C_2\frac{\beta V_{osc}}{L(s)Y_o(s)} \\ &+C_3\frac{\beta V_{osc}G_0(s)}{L(s)Y_{fb}(s)}\end{aligned} \tag{4.102}$$

$$\begin{aligned}&\approx C_1\frac{Y_{gs}(s)}{Y_{fb}(s)} + C_2\frac{\beta V_{osc}}{g_{m2,0}}\frac{1}{A_1(s)H_i(s)} \\ &+C_3\frac{\beta V_{osc}}{Y_{fb}(s)}\end{aligned} \tag{4.103}$$

with the coefficients $C_1(V_{osc},\omega_0,\ldots)$, $C_2(V_{osc},\omega_0,\ldots)$ and $C_3(V_{osc},\omega_0,\ldots)$ frequency-independent functions of the building block parameters. Likewise, it holds for $\tilde{H}_{0,2}(s)$ that

$$\tilde{H}_{0.2}(s) \approx D_1\frac{\beta V_{osc}}{L(s)Y_o(s)} + D_2\frac{\beta V_{osc}G_0(s)}{L(s)Y_{fb}(s)} \tag{4.104}$$

$$\approx D_1\frac{\beta V_{osc}}{g_{m2,0}}\frac{1}{A_1(s)H_i(s)} + D_2\frac{\beta V_{osc}}{Y_{fb}(s)} \tag{4.105}$$

with, again, $D_1(V_{osc},\omega_0,\ldots)$ and $D_2(V_{osc},\omega_0,\ldots)$ frequency-independent functions of the building block parameters.

HTM element	Relative error
$\tilde{H}_{0,1}(s)$	-81.0 dB
$\tilde{H}_{0,0}(s)$	-53.8 dB
$\tilde{H}_{0,2}(s)$	-33.5 dB

Table 4.2: Relative accuracy of the symbolically computed downconversion mixer characteristics as compared to direct numerical computation.

The accuracy of these symbolic models —as compared to HTM elements numerically extracted by means of an algorithm similar to the one in [Royc99]— is listed in table 4.2. The results show that the expressions, generated up to first order in the dummy variable μ, allow us to model $\tilde{H}_{0,0}(s)$ and $\tilde{H}_{0,2}(s)$ up to accuracies of respectively 0.2% and 2%, with the expression for $\tilde{H}_{0,1}(s)$ being even more accurate. This is quite sufficient for most purposes. Fig. 4.27 plots $\tilde{H}_{0,0}(s)$ and $\tilde{H}_{0,2}(s)$ and their components for the design parameters chosen as in table 4.1. The solid bold line represents the exact overall transfer characteristic, while the marked lines correspond to the individual, symbolically generated components (after grouping).

Computing contributions up to higher order in μ rapidly leads to an explosion in the number of symbolic terms. Simplification techniques [Fern98, Gie91, Gie94, Wamb98b] become unavoidable to control this explosion. This, however, requires numerical data concerning the building block parameters. Often, such data is not available at the stage in the design process where symbolic system characterization is desirable. The results in table 4.2, however, indicate that the contribution of these higher-order terms to the overall characteristic is negligible, making it in most cases unnecessary to compute them. Up to first order, the number of terms is small enough for the symbolic models to be useful and interpretable.

4.3.8 Application 2: Receiver stage with feedback across the mixing element

Fig. 4.28 shows part of a receiver system consisting of a (bandpass) filter $H_1(s)$, a mixer and a (lowpass) filter $H_2(s)$. The output of the mixer stage is fed back to its input via an often unwanted (i.e. parasitic) feedback filter $H_{fb}(s)$. In the example of the linear mixer in Fig. 4.24, this parasitic feedback path runs through the gate-source capacitances of the input transistors. It causes downconversion to baseband of the signal content in the frequency bands centered around $f = 0$, $f = 2f_0$, etc. As a result, 2nd-order distortion components generated in previous stages of the receiver end up in the desired output frequency band.

For the system shown in Fig. 4.28, we used SymbolicHTM to compute the expressions for the input-output HTM elements for several levels of accuracy. Here, the level of accuracy is determined by the order of the dummy variable μ up to which results are computed, and the number of tones taken into account. The HTM elements were evaluated for $f_0 = 10\,\text{MHz}$, $H_1(s)$ a 4th-order maximally flat bandpass filter centered around f_0 with a 3 dB bandwidth of 1.5 MHz, $H_2(s)$ a 6th-order Chebyshev lowpass filter with

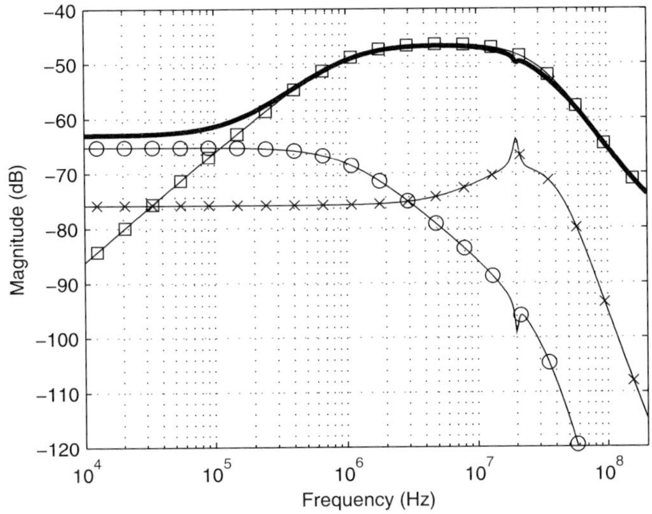

(a) HTM element $\tilde{H}_{0,0}(s)$

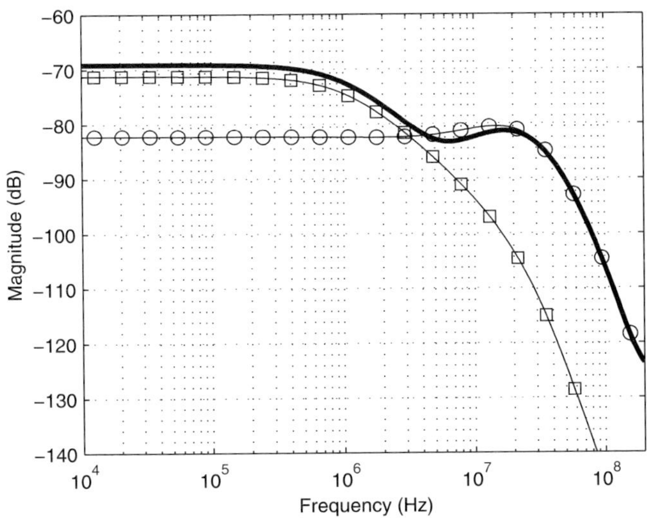

(b) HTM element $\tilde{H}_{0,2}(s)$

Figure 4.27: Exact transfer characteristic of $\tilde{H}_{0,0}(s)$ and $\tilde{H}_{0,2}(s)$ (solid bold lines) and their symbolically generated components (marked lines). (a) The squared, crossed and circled line respectively correspond to the first, second and third term in equation (4.102). (b) The circled and squared line respectively correspond to the first and second term in equation (4.104).

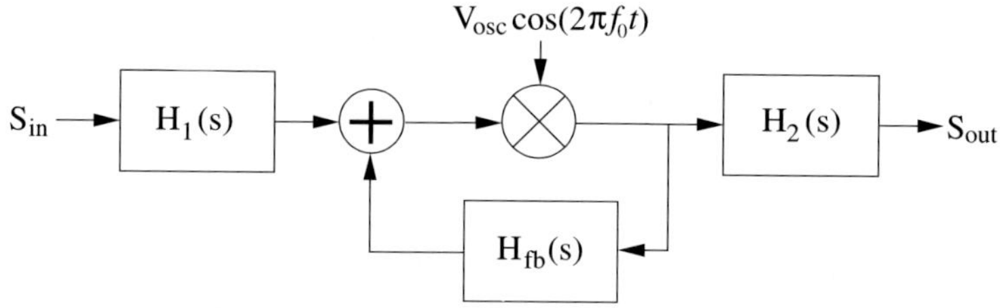

Figure 4.28: Receiver system consisting of a bandpass filter $H_1(s)$, a lowpass filter $H_2(s)$ and a downconversion mixer in between. The mixer output is fed back to its input via the (parasitic) feedback path $H_{fb}(s)$.

a passband ripple of 0.5 dB and its corner frequency at 10 kHz, and $H_{fb}(s)$ a highpass filter with its corner frequency at 100 kHz and a high-frequency attenuation of -15dB. This highpass feedback path corresponds with the fact that it typically arises due to capacitive current injections.

The desired downconversion behavior from $f = f_0$ to baseband is described by the HTM element $\tilde{H}_{0,1}(s)$ and is computed to be

$$\begin{aligned} \tilde{H}_{0,1}(s) &= \tfrac{V_{osc}}{2} H_2(s)H_1(s+j\omega_0) \\ &+ \tfrac{V_{osc}^3}{8} H_2(s)H_1(s+j\omega_0)H_{fb}(s)H_{fb}(s-j\omega_0) \\ &+ \tfrac{V_{osc}^3}{8} H_2(s)H_1(s+j\omega_0)H_{fb}(s)H_{fb}(s+j\omega_0) \end{aligned} \tag{4.106}$$

This result is accurate up to 2nd order in μ while taking 7 tones into account. In expression (4.106), the first term describes the intended signal transfer, while the other terms can be considered as linear distortion components due to the parasitic feedback filter $H_{fb}(s)$. The downconversion behavior from $f = 0$ and $f = 2f_0$ to baseband is respectively modeled by the HTM elements

$$\begin{aligned} \tilde{H}_{0,0}(s) &= \tfrac{V_{osc}^2}{4} H_2(s)H_1(s)H_{fb}(s-j\omega_0) \\ &+ \tfrac{V_{osc}^2}{4} H_2(s)H_1(s)H_{fb}(s+j\omega_0) \end{aligned} \tag{4.107}$$

and

$$\tilde{H}_{0,2}(s) = \frac{V_{osc}^2}{4} H_2(s)H_{fb}(s+j\omega_0)H_1(s+2j\omega_0) \tag{4.108}$$

From (4.107), it is observed that, with $H_1(s)$ being a bandpass filter centered around ω_0, $\tilde{H}_{0,0}(s) \sim H_1(s)$ vanishes almost completely at low frequencies. On the other hand, equation (4.108) shows that, for a given feedback path $H_{fb}(s)$, the order of magnitude of $\tilde{H}_{0,2}(s)$ will be largely determined by $H_1(2j\omega_0)$. In order to get $\tilde{H}_{0,2}(s)$ as low as possible, it is advantageous to design the upper (high-frequency) stopband characteristic of $H_1(s)$ as steep as possible. For a given filter order, this suggests an asymmetric filter design with the majority of the filter zeros located in the upper stopband. This will result in an increase of $\tilde{H}_{0,0}(s)$ as compared to a symmetric design, but as long as

HTM element	Relative error
$\tilde{H}_{0,1}(s)$	-78.0 dB
$\tilde{H}_{0,0}(s)$	-41.1 dB
$\tilde{H}_{0,2}(s)$	-35.9 dB

Table 4.3: Relative accuracy of the symbolically computed receiver HTM elements (4.106)-(4.108) with respect to results obtained by direct numerical extraction.

the bandwidth of the lowpass filter $H_2(s)$ is much smaller than ω_0, this won't matter much.

Fig. 4.29 plots the HTM elements with $H_1(s)$ respectively being a symmetric and an asymmetric bandpass filter. $H_1(s)$ being symmetric means that it has two zeros at $\omega = 0$ and two zeros at $\omega = \infty$. $H_1(s)$ being asymmetric means that it has one zero at $\omega = 0$ and three zeros at $\omega = \infty$. The solid lines plot the values as obtained by evaluating the expressions (4.106)-(4.108), while the marks indicate values as extracted out of Matlab/Simulink simulations. The crosses, circles and squares respectively represent the transfer functions $\tilde{H}_{0,1}(j\omega)$, $\tilde{H}_{0,0}(j\omega)$ and $\tilde{H}_{0,2}(j\omega)$. Comparing Fig. 4.29(a) and (b) shows that use of an asymmetric filter design indeed improves rejection of the signal content near $2\omega_0$, lowering $\tilde{H}_{0,2}(s)$ with 6 dB. As predicted, this comes at the expense of an increase of $\tilde{H}_{0,0}(s)$. Note that this improved $\tilde{H}_{0,2}(s)$ requires no extra hardware (the filter order remains unchanged) but only proper placement of the filter zeros. It illustrates the design insights gained by having the expressions (4.106)-(4.108) available.

The accuracy of the symbolic models (4.106)-(4.108) was compared to values extracted from numerical simulations. These simulations involved applying a single input tone at respectively $f_0 + f_s$, f_s and $2f_0 + f_s$ Hz, for f_s ranging from 200Hz to 1MHz. Analysis of the steady-state output signal then yields the output signal component at baseband (at frequency f_s). The result is listed in table 4.3. As can be seen, all symbolic expressions are at least within 2% accurate, with $\tilde{H}_{0,1}(s)$ even being 0.01% accurate.

Fig. 4.30 illustrates how the model accuracy varies with the order of μ and the number of tones taken into account in computing the input-output HTM. It was plotted for the HTM element $\tilde{H}_{0,1}(s)$. For other HTM elements, similar results were obtained. Fig. 4.30 shows that, for constant μ, there is a maximum number of tones beyond which no improvement in accuracy can be obtained by increasing the number of tones taken into account. This is due to the fact that, when limiting the number of terms in the approximation (4.82), effects due to the presence of higher-order tones do not enter the expressions for the HTM elements of interest. A likewise reasoning holds when keeping the number of tones constant and varying μ. As a rule of thumb, it is of no use to select the order of μ larger than half the number of tones taken into account during computations.

Fig. 4.31 shows the total number of terms generated by SymbolicHTM, summed over all computed HTM elements, versus the order of μ and the number of tones taken

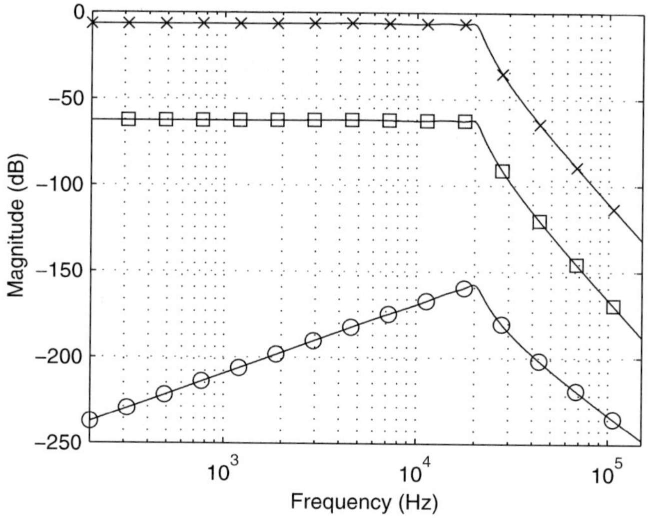

(a) Symmetric bandpass filter $H_1(s)$

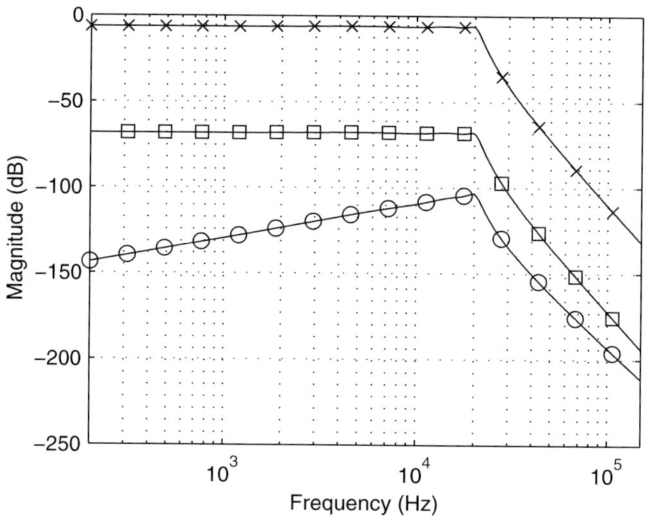

(b) Asymmetric bandpass filter $H_1(s)$

Figure 4.29: Input-output HTM elements of the overall receiver for (a) a symmetric and (b) an asymmetric bandpass filter $H_1(s)$. The crosses, circles and squares respectively represent the HTM elements $\tilde{H}_{0,1}(j\omega)$, $\tilde{H}_{0,0}(j\omega)$ and $\tilde{H}_{0,2}(j\omega)$.

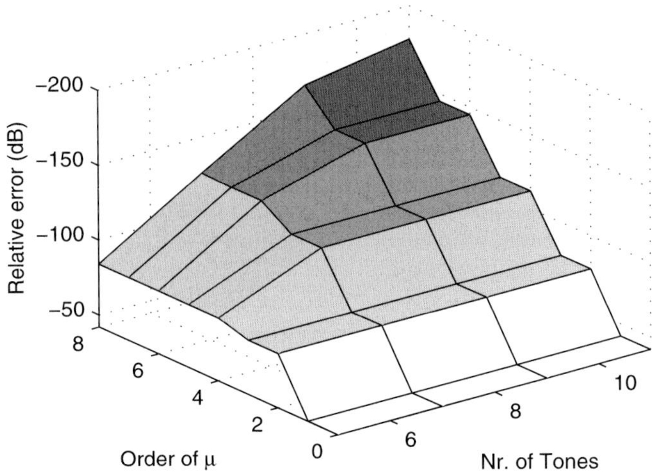

Figure 4.30: Accuracy of the symbolic expression for $\tilde{H}_{0,1}(s)$ versus the order of μ and the number of tones $(2K+1)$ taken into account in generating the expression. Note that the direction of the Z-axis has been reversed.

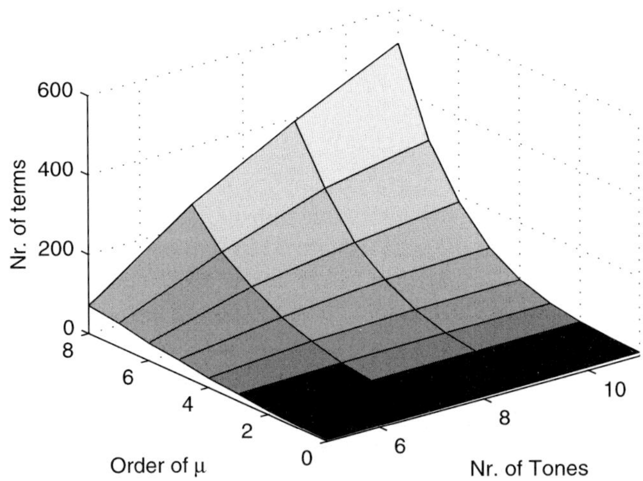

Figure 4.31: The total number of terms generated versus the order of μ and the number of tones $(2K+1)$ taken into account in generating the HTM expressions. The CPU time spent per term equals on average 4e-3 seconds.

into account. On average, the CPU time spent to generate a single term equals 4e-3 seconds. It is the number of terms which determines the computational cost of the SymbolicHTM algorithm. As can be seen, the number of generated terms quickly rises with the order of μ, especially when a large number of tones is taken into account. In this second example, it took 0.15 sec to compute results up to 2nd order in μ while taking 7 tones into account. This time increases up to a couple of seconds for both μ and K becoming large. Fig. 4.30, however, shows that these extra terms often do not contribute to the accuracy of the HTM elements of interest. This means that much time is spent on computing terms that are insignificant and hence of no interest. Also, a larger number of terms makes the symbolic results harder to interpret. However, combining results in Fig. 4.30 and 4.31, it is observed that a satisfactory degree of accuracy can be obtained for relatively simple expressions.

4.4 Conclusions and directions for further research

Both PLL analysis and the SymbolicHTM algorithm demonstrate that the use of the HTM formalism goes beyond shear theory. They show that it is possible to obtain detailed results —both numerical and symbolic— that characterize the behavior of real-life systems. Hereby, it is possible to automate part of the analysis. This contributes to the power and flexibility of HTMs as a frequency-domain framework for capturing LPTV system behavior. As such, it is made a valuable extension to the well-known techniques for LTI system analysis.

HTMs were proven to be an elegant framework for an exact mathematical treatment of (linearized) PLL behavior. HTMs support a detailed description of the time-varying characteristics inherent to all practical PLL implementations. Existing (LTI) techniques for PLL modeling were shown to be specific approximations of the exact time-varying description. For the first time, these traditional models were given solid mathematical groundings with their shortcomings clearly identified.

The systems to which HTM-based analysis applies, are by no means exhausted. For example, an oscillator's impulse sensitivity function (see section 4.2.4.3) can be obtained using HTM-based techniques. The HTMs involved turn out to have a rank-degenerate structure at DC. This can be exploited to streamline computations. A first analysis of oscillator behavior using HTM-based techniques shows promising results for both the efficient computation of steady-state and phase noise behavior.

Furthermore, having established methods that capture time-varying linearized PLL behavior, we can proceed by taking time-varying weak nonlinearities into account. In a PLL, for instance, the PFD and the oscillator make up sources of time-varying weakly nonlinear behavior. Analyzing time-varying weakly nonlinear behavior builds on perturbation techniques and linear time-varying system analysis. It thereby proceeds along the same lines of reasoning as used for time-invariant weakly nonlinear system analysis [Wamb98a]. Developing (symbolic) methods for time-varying weakly nonlinear system analysis would contribute yet another step towards a true top-down analysis (and hence design) of electronic systems and circuits.

Chapter 5

Modeling oscillator dynamic behavior

If man were restricted to collecting facts the sciences would be only a sterile nomenclature and he would never have known the great laws of nature. It is in comparing phenomena with each other, in seeking to grasp their relationships, that he is led to discover these laws...

—*Pierre-Simon Laplace*

Oscillators are key building blocks in almost all of today's communication systems. Because of their importance, we need a clear understanding of the fundamentals of their behavior. This understanding must be accompanied by methods for accurate analysis and for the generation of compact behavioral models. The dynamic behavior of many oscillators is characterized by a mixture of fast- and slow-varying time constants. On the one hand, we observe the system's fast-varying oscillations. On the other hand, the properties of these oscillations, e.g. amplitude and phase, vary slowly over time. This kind of behavior results in long simulation times when handled by means of traditional algorithms (e.g. SPICE). In this chapter, we focus on methods that allow us to separate an oscillator's fast- and slow-varying dynamics. These methods are used for the improvement of simulation efficiency, for behavioral model extraction and for the symbolic characterization of an oscillator's dynamic behavior. As such, they are a step towards improved oscillator design methods.

The methods presented in this chapter are based on perturbation analysis and averaging techniques. Oscillator perturbation analysis relies on partitioning the overall oscillator circuit into a core system and a set of perturbation components. The core system is solved by a multitude of different steady-state solutions. Each solution is called a *state* of the system. The perturbation components disturb the core, driving it from one state to another. This state-changing behavior captures the essence of the oscillator's dynamic behavior.

Detailed observation shows that the changes in the state of an oscillator are dominated by a slow-varying component. Typically, this component is the one of main interest. Using averaging techniques, it is possible to identify the part of the state-changing equations that generates this slow-varying behavior. These *averaged equations* can be solved very efficiently. They can be used as a starting point for oscillator phase-noise analysis and for the generation of compact behavioral models.

In what follows, section 5.1 introduces the basic concepts underlying the oscillator analysis elaborated in this chapter. Hereby, we use the earth's motion around the sun as an example. The mathematics underlying these concepts is addressed in sections 5.3

to 5.6. The theory is applied to oscillator phase noise analysis in section 5.7 and to the behavioral modeling of harmonic oscillators in section 5.8.

5.1 The story behind the math

Constructing of compact oscillator models requires a clear understanding of the mechanisms that govern oscillator dynamics. Oscillators exhibit a mixture of fast- and slow-varying behavior. Rapid variations are caused by, for instance, energy that oscillates between a set of energy storing elements, e.g capacitors and inductors. Energy exchanges with the surroundings cause slow-varying modulations. Understanding the interplay between fast- and slow-varying dynamics is key to understanding an oscillator's behavior.

In order to clarify the fundamental mechanisms that govern an oscillator's behavior, we consider one of the first oscillations ever studied: the motion of the earth around the sun. As is the case for an electronic oscillator, the earth's motion contains a periodic fast-oscillating component: its elliptic motion around the sun. Here, "fast" implies a time scale of the order of a year. Due to small perturbations in the solar system's gravitational field, the shape and period of the earth's elliptic orbit slowly change over time. Here, slow indicates a time span of ten-thousands of years. These slow-varying changes are governed by mechanisms similar to the ones that govern the transient dynamics of an electronic oscillator's amplitude and phase. As such, the same mechanisms that govern an oscillator's phase noise behavior caused the ice ages.

5.1.1 Earth: a big oscillator

One of the first oscillations ever to be studied in depth is the earth's motion around the sun. The modern point of view on this subject started with the work of Copernicus (1543), Kepler (1609), Galilei (1616) and Newton (1687). Lagrange (1784), Laplace (1799) and Poincaré (1892) made important contributions to the topic, for example, on the use of perturbation series techniques.

Fig. 5.1 shows the problem setup for computing the earth's trajectory around the sun. Gravity attracts the earth towards the sun with a force that is inversely proportional to the square of the distance of separation r. However, the presence of other planets, the non-spherical shape of both the sun and the earth and some other effects introduce an additional but small force component $\varepsilon\mathbf{f}(\mathbf{r},t)$. The total force exerted on the earth therefore equals

$$\mathbf{F}(\mathbf{r},t,\varepsilon) = \frac{K}{r^2}\hat{\mathbf{r}} + \varepsilon\mathbf{f}(\mathbf{r},t) \quad . \tag{5.1}$$

Here, $\hat{\mathbf{r}}$ is the unit vector pointing from the earth to the sun while $\mathbf{r} = r\hat{\mathbf{r}}$. Furthermore, ε denotes a small number indicating that the second term in (5.1) is, typically, much smaller than the first component. The term $\varepsilon\mathbf{f}(\mathbf{r},t)$ is, therefore, called a *perturbation term*. Given the driving force (5.1), the earth's motion can now be solved from

$$m\frac{\mathrm{d}^2\mathbf{r}}{\mathrm{d}t^2} = \mathbf{F}(\mathbf{r},t,\varepsilon) \tag{5.2}$$

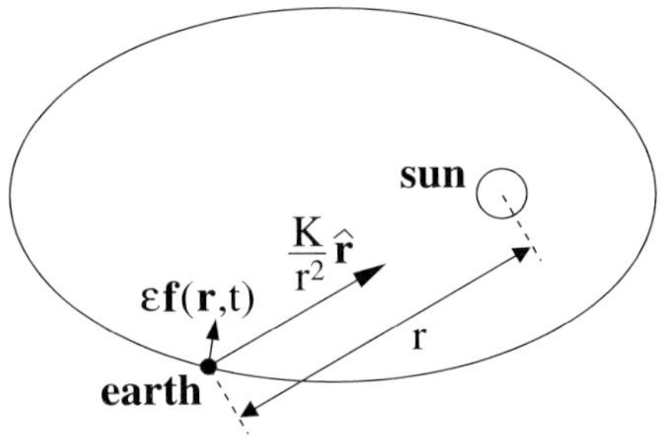

Figure 5.1: Gravity attracts the earth towards the sun with a force that is inversely proportional to the square of the distance of separation r. This $1/r^2$ term causes the earth to describe an ellipse around the sun. The presence of other planets and some other effects introduce an additional but small force component $\varepsilon\mathbf{f}$. On the long haul, this tiny force causes the shape of the earth's orbit to change.

where m is the earth's mass[1].

5.1.2 Unperturbed system behavior: neglecting small forces

The smallness of the term $\varepsilon\mathbf{f}$ in (5.1) suggests that it only has a small impact on the earth's behavior. In studying the earth's motion, we are therefore tempted, at least for the time being, to forget about this perturbation term. Hence, we use

$$\mathbf{F}_{newton} = \mathbf{F}(\mathbf{r},t,0) = \frac{K}{r^2}\hat{\mathbf{r}} \ . \tag{5.3}$$

as the force in the equation of motion (5.2). As shown in Fig. 5.1, this force causes the earth to describe an elliptic orbit around the sun. This behavior can be interpreted as a periodic oscillation with a period that equals a single year.

The orbit drawn in Fig. 5.1 is by no means unique. This non-uniqueness should be understood as the existence of a multitude of steady-state solutions, each satisfying (5.2) for $\varepsilon = 0$. If the earth is set off with different initial conditions, different orbits result, either located in a different plane or with a different shape. Fig 5.2 shows a number of different orbits, each solving (5.2) for $\varepsilon = 0$. Characterizing the earth's (possible) behavior therefore requires us to find *all possible* solutions to (5.2) for $\varepsilon = 0$.

Mathematically, for $\varepsilon = 0$, the equation (5.2) is called *unperturbed equation of motion*. This equation models the behavior of the system's *core*. In the current example, the core is associated with the earth, the sun and the $1/r^2$ attracting force between them. The steady-state solutions that solve the unperturbed equations of motion are called the *states* of the core system. Characterizing the core system behavior involves *finding*

[1] This is a somewhat simplified representation of the problem. A more accurate problem statement can be found in [Hest87].

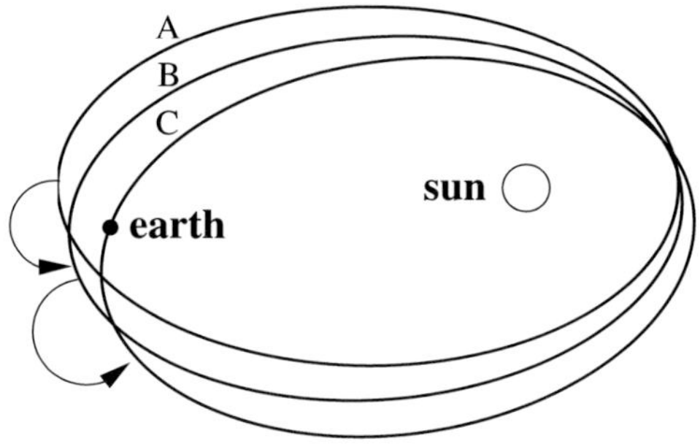

Figure 5.2: The unperturbed equation of motion is solved by a multitude of steady-state solutions. Depending upon the initial conditions, the earth could either settle on, for instance, orbit A, B or C. The perturbing force $\varepsilon\mathbf{f}$ drives the earth from one possible orbit to another.

all possible states. In other words, we are looking for all orbits that the earth could possibly settle on. In asymptotically stable systems, like electronic filters, there is only one state. The system's output quickly settles on a unique steady-state solution. The earth, however, could settle on a multitude of different orbits. Every ellipse with the sun at one of its focal points represents a valid steady-state solution[2] and, therefore, a state of the core system. In this example, such a state is characterized by the properties of the corresponding orbit, e.g. the plane of revolution and the eccentricity. Note that none of these orbits is absolutely stable, i.e. if earth is driven away from an orbit, e.g. because of a meteorite impact, there is no force restoring earth to this original orbit.

5.1.3 Perturbed system behavior: changes in the earth's orbit

What is so important about finding *all possible* orbits (states)? Isn't it sufficient to just identify the one that the earth is currently tracking? No, it isn't! The fact that the earth is, for example, currently tracking orbit A (see Fig. 5.2) does not imply that it will do so forever. Impact of a meteorite or any other external influence may drive the earth from orbit A to orbit B. Since orbit B is a valid solution to the unperturbed equation of motion, the earth would be perfectly happy on this second orbit and would make no attempt to return to the original orbit A. External influences can hence *drive the core system from one state to another.* When the external influences persist, this process continues, driving the earth, for instance to a third orbit (state) C.

The state-changing behavior discussed above represents the basic idea underlying oscillator perturbation analysis. Due to a small but non-zero ε in (5.2), the equation of motion, and therefore the earth's behavior, slightly deviates from that of the unperturbed core system. This can be dealt with by considering the perturbing force term $\varepsilon\mathbf{f}$

[2]The same holds for all parabolas and hyperbolas.

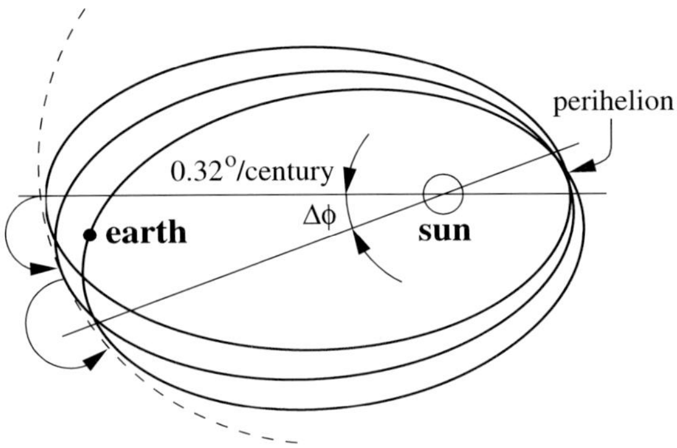

Figure 5.3: The joint forces exerted by the other planets in the solar system cause, amongst others, a precession of the earth's orbit around the sun. The position of the perihelion advances about 0.32° per century.

as an external influence that drives the core system from one state (orbit) to another. In Fig. 5.2, $\varepsilon\mathbf{f}$ causes the earth, for instance, to change from state (orbit) A to state (orbit) B. Since $\varepsilon\mathbf{f}$ is small, the equations that govern this state-changing behavior can be extracted using *perturbation series expansion techniques* [Kev81]. These equations describe how the characteristics of the earth's elliptic orbit, like plane of revolution and eccentricity, change over time.

The smallness of the perturbing force $\varepsilon\mathbf{f}$ has another important consequence: the rate at which the earth is changing states (orbits) is much slower than the speed with which the earth is traversing the orbits associated with these states. It takes centuries before a significant change in state (orbit) is observed. Hence, changes in the system's state occur on a time scale that is much slower than that of the earth's periodic revolution around the sun. It is therefore said that the revolution around the sun occurs on a *fast-varying time scale* while the earth's dynamic (state-changing) behavior occurs on a *slow-varying time scale*. Of course, from a human point of view, an oscillation with a period of a year isn't particularly fast. However, the words fast and slow must be interpreted in the context of the system that is investigated. In the context of the evolution of our solar system, changes taking place in a year's time are to be considered very fast.

In order to give a feeling as to the consequences of the perturbing force $\varepsilon\mathbf{f}$, we consider the impact of the joint forces exerted by the other planets in the solar system. As shown in Fig. 5.3, these forces induce a precession of the earth's orbit around the sun [Hest87]. The position of the earth's perihelion[3] advances about 0.32° per century. One also observes periodic changes in the eccentricity of the earth's orbit, making it first more and then again less elliptical. These changes in the shape and position of the earth's orbit are called the Milankovitch cycles [Hays76]. These cycles have a period of about 100,000 years. They are of prime importance for glaciation and the occurrence

[3]The perihelion is that point of the earth's orbit that is closest to the sun.

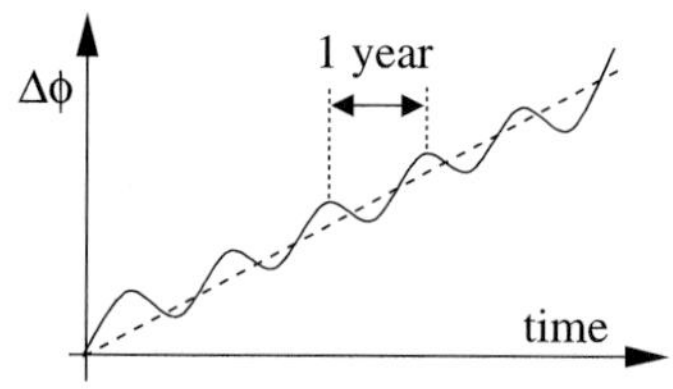

Figure 5.4: The position of the perihelion of the earth's orbit as a function of time (solid line). This position is expressed as an angle $\Delta\phi$ with respect to some reference position. On average, we observe a steady increase (dotted line) with time. In addition, there is a small fast-varying fluctuation.

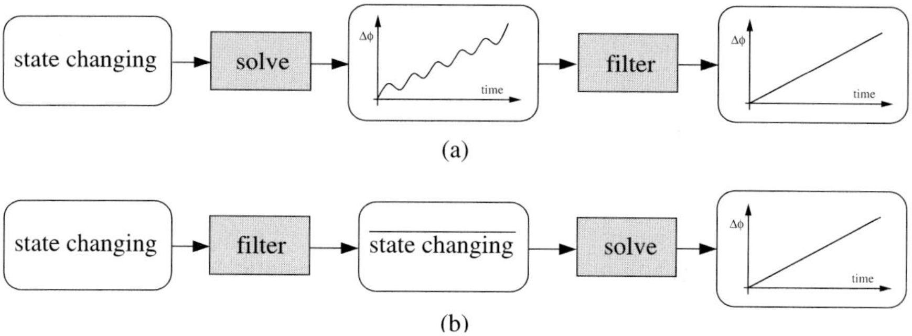

Figure 5.5: In obtaining the long-term (dashed) characteristic in Fig. 5.4, there are two ways to proceed: (a) first solve the original set of state-changing equations and remove what's unimportant afterwards by filtering; (b) first filter (average) the state-changing equations and then solve these averaged equations for the desired characteristic.

of the ice ages. Hence, small forces can have significant long-term consequences.

5.1.4 Averaging: focusing on what's important

Before linking the material above to electronic oscillators, there is one aspect about the earth's state-changing behavior that still needs to be addressed. Zooming in on the (extrapolated) position of the earth's perihelion with respect to time (expressed as an angle $\Delta\phi$ with respect to some reference position), we typically observe the behavior in Fig. 5.4. Here, the solid line represents the exact position, showing a slow but steady increase with small fast-varying fluctuation on top. In many cases, the latter fluctuation is of no interest and we would like to focus on the long-term progress. It is, therefore, the dashed line in Fig. 5.4 that is of interest. One way to obtain this dashed curve is illustrated in Fig. 5.5(a). Here, we first compute the perihelion's exact motion after which a lowpass filtering operation removes all fast-varying components. This,

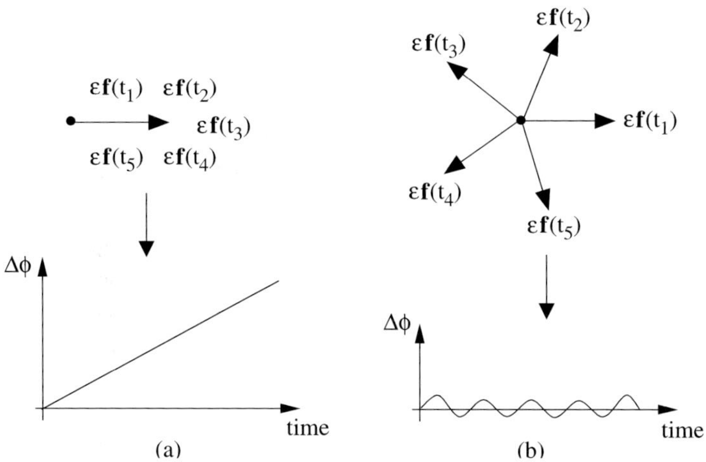

Figure 5.6: (a) A small force acting during a very long time in the same direction may, over time, cause significant changes to occur. (b) A small force that continuously changes direction counteracts itself and can never cause more than small fluctuations.

however, is very time-consuming as the computation of the perihelion's exact motion requires us to use a small simulation time step: we need a couple of samples per year in order to capture the perihelion's fast-varying fluctuations. Using this time step, we need to cover a simulation time interval over a million years, i.e. the duration of a couple of Milankovitch cycles. Running such a simulation is like having to bridge a thousand kilometers by foot (with steps of about one meter).

Fortunately, there is a more efficient way to solve this problem. As illustrated in Fig. 5.6(a), small forces can only cause substantial changes when they act in the same direction for a long time. Fig. 5.6(b) shows that force components that quickly vary in direction over time can never generate large changes as they counteract themselves. This suggests to first filter the equations governing the system's state-changing behavior. Hereby, we only retain the force components that act in the same direction over a sufficiently long time. Next the filtered equations can be solved for the long-term behavior. This procedure is illustrated in Fig. 5.5(b). The filtering process is also called *averaging* [Bogo61][4]. The resulting equations, indicated with a bar on top of them, are called the *averaged equations*. Since solutions of the averaged equations are, by construction, void of fast-varying components, they can be solved using a large integration time step and, therefore, in a much shorted CPU time. Hence, averaging allows us to focus attention to the important long-term system behavior while neglecting small fast-varying fluctuations.

[4]Averaging, in turn, is related to the theory of continuous transformation (or Lie) groups [Cohn65, Firk00].

5.1.5 How does electronic oscillator dynamics fit in?

Fig. 5.7(a) summarizes the procedure outlined above. Firstly, we partition the system into a core and a set of perturbation terms. Next, the core is characterized by identifying all its states, i.e. all possible steady-state solutions of the corresponding unperturbed system equations. The perturbation terms slowly drive the core system from one state to another. Finally, averaging helps us to identify the relevant part of the equations that govern the system's state-changing behavior.

Fig. 5.7(b) redraws Fig. 5.7(a) and shows how electronic oscillator analysis can be made to fit within the same picture. The steps are illustrated for a harmonic oscillator. In a harmonic oscillator, the principle of oscillation relies on the exchange of energy between an inductive and a capacitive element. Here, the core system corresponds to a lossless resonant tank. Feedback transistors and tank losses are considered perturbation terms. The unperturbed (lossless) resonant tank is solved by a double-infinite set of steady-state oscillations. Each oscillation corresponds to a state of the core system and is identified by a value for its amplitude and starting phase. The perturbation terms, i.e. the currents coming from the feedback transistors and the tank losses, slowly drive the resonant tank from one state to another. Hence, they cause the tank's amplitude and phase to change with time. Averaging helps us to extract the equations that govern the slow-varying components of these changes. These components are the only ones that can grow large.

5.1.6 Modeling oscillator behavior

The discussion above may have left the reader wondering what all of this has to do with the (behavioral) modeling of an oscillator's dynamic behavior. As stated before, system modeling starts with a thorough understanding of the system's behavior. Next, this understanding must be formalized by means of a mathematical model. As far as an oscillator's dynamic behavior is concerned, the equations that model the averaged state-changing behavior of the oscillator's core system turn out to provide an excellent starting point for the creation of compact behavioral models. These averaged equations capture the most essential part of an oscillator's dynamics. For the behavior of the earth, this involves the long-term changes in the position and shape of the earth's orbit around the sun. For a harmonic oscillator, this involves changes in amplitude and phase. Efficient behavioral modeling of oscillator dynamics can, therefore, be accomplished by means of suitable approximation of the averaged state-changing dynamics. This allows us to analyze, for instance, an oscillator's settling behavior and its phase noise behavior.

This concludes our overview of the basic concepts underlying the oscillator analysis presented in this chapter. In what follows, we discuss each of the steps in greater depth. Firstly, however, we review previous art on the matter.

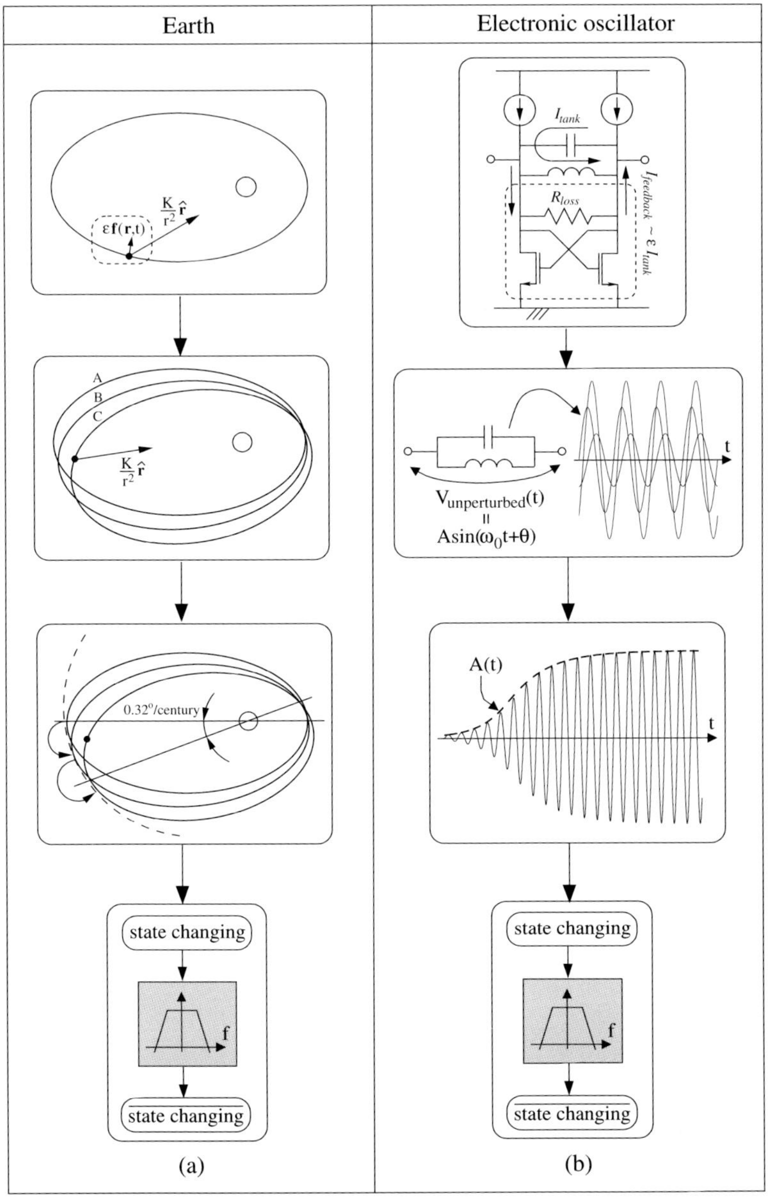

Figure 5.7: The behavior of both the earth and an electronic oscillator is governed by the same mechanisms. Both can be partitioned in a core system and a set of perturbation terms (encircled by the dashed line). Next, the core system's behavior is characterized by identifying all of its states (steady-state solutions). The perturbation terms slowly drive the system from one state to another. This causes slow-varying transient dynamics. Finally, averaging helps us to identify the relevant part of the equations that govern the oscillator's state-changing behavior.

5.2 Prior art

In what follows, we briefly review general autonomous system theory, oscillator phase noise theory and methods for numerical simulation.

5.2.1 General theory

Over the centuries, the analysis of nonlinear oscillators and their behavior has been subject to many publications. Historically among the first are the treatments on celestial mechanics [Hest87, Suz96]. With the advent of the industrial revolution and its machines, the theory of mechanical vibrations [Kryl47, Rand03, Sto50] became a popular area of research. Also, the abstract study of autonomous differential equations resulted in many useful techniques. Here, the work of Krylov, Bogoliubov and Mitropolsky [Bogo61] offers one of the most concise discussions on the matter. They introduce both singular perturbation theory and averaging. Other treatments in this area [Kev81, Murd88, Verhu96] build on similar ideas.

With regard to averaging, the methods developed by Krylov, Bogoliuv and Mitropolsky can be shown a special case of the transformation groups introduced by Sophus Lie [Cohn65, Firk00]. His theory of near-identity transformations provides a solid theoretical basis for higher-order averaging methods. Here, we also mention the theory of manifolds, tangent spaces and integration on manifolds [Abra88]. These concepts are important in the theory of differential algebraic equations and model-order reduction.

The analysis of the nonlinear van der Pol oscillator by Balthazar van der Pol [Vand20] is probably one of the first studies on the behavior of nonlinear electronic oscillators. Many authors have followed his lead [Bogo61, Buo02, Chua92, Guck80, Murd88, Rob83, Stav00, Sto50][5]. Hereby, describing function analysis [Gelb68, Kha96, Lee98] and singular perturbation theory [Kha96] are among the most frequently used techniques for analyzing an oscillator's behavior. However, these methods are often tailored to the needs of a particular application, i.e a particular set of differential equations. Hence, they lack generality. Handling a set of equations extracted from an oscillator circuit netlist remains an open problem.

5.2.2 Phase noise analysis

Among the work on nonlinear oscillator behavior, we highlight the oscillator phase noise theories developed in recent years [Cran98, Dem00a, Dem00b, Dem02, Demu02, Haji98, Kae90, Lees66]. These theories study the impact of noisy disturbances on an oscillator's phase behavior. Fig. 5.8(a) illustrates the well-known impact noise has on the oscillator's power spectral density (PSD). Phase noise causes the energy contained in the oscillating signal to spread over the frequency band. This spectral spreading is, typically, characterized by a flat close-in phase-noise PSD followed by an $1/f^3$ region (due to upconverted $1/f$ input noise) and an $1/f^2$ region (due to white input noise).

The most rigorous treatments on phase noise analysis are given in [Dem00a, Dem00b, Dem02, Kae90]. Here, the authors develop the viewpoint of phase noise as being

[5]Note that this list just intends to give some leads. It is by no means exhaustive.

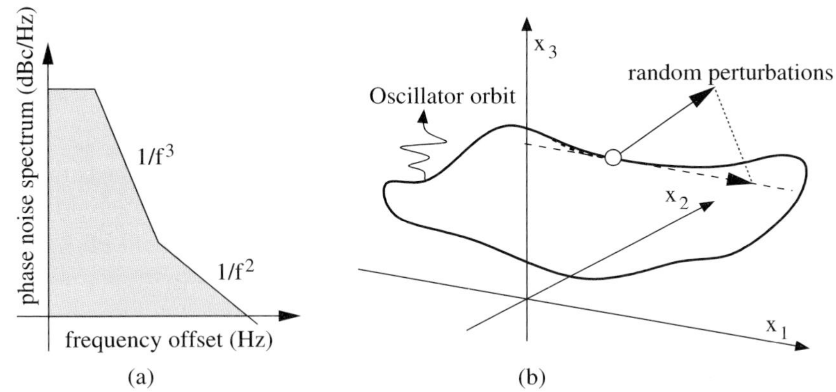

Figure 5.8: Appearance and operation of oscillator phase noise. (a) Phase noise causes the energy contained in an oscillating signal to spread all over the frequency band. (b) Phase noise is caused by random perturbations (noise) that make the oscillator (phase-)shift along its own orbit.

caused by random perturbations (noise) that make the oscillator (phase-)shift along its own orbit in the phase space. To compute the "force" that drives these phase shifts, they project the perturbations onto the tangent line to the orbit. This results in a phase noise equation that is structured as

$$\frac{d\theta}{dt} = \mathbf{v}^T(t+\theta)\mathbf{n}(t) \ . \tag{5.4}$$

Here, θ is the oscillator's phase, $\mathbf{n}(t)$ is the vector containing the perturbing noise sources while $\mathbf{v}^T(t)$ is the oscillator's *perturbation projection vector* (PPV) [Dem00a]. At this point, we do not go into further detail on (5.4). A more rigorous explanation is given in section 5.7 where (5.4) is shown a special case of the theory presented in this chapter. We only mention that when there is only a single noise source involved, i.e. $\mathbf{n}(t) = \mathbf{b} \cdot n(t)$, (5.4) reduces to

$$\frac{d\theta}{dt} = \Gamma(t+\theta)n(t) \ . \tag{5.5}$$

Here, $\Gamma(t) = \mathbf{v}^T(t) \cdot \mathbf{b}$ is called the *impulse sensitivity function* (ISF, a name introduced in [Haji98][6]) corresponding to $n(t)$.

The phase noise theories and methods in [Dem00a, Dem00b, Kae90] are very generic in a sense that they make little assumptions about the structure of the equations that

[6] In [Haji98], the author simply conjectures the approximate phase noise equation

$$\frac{d\theta}{dt} = \Gamma(t)n(t) \ .$$

This approximate phase noise equation neglects the right-hand-side's dependence on the phase θ that is present in the exact equation (5.5). The validity of this simplification is discussed in [Vana02d].

model the oscillator. As such, they can be applied directly to circuit-level netlists. However, by nature, these phase noise theories are only concerned with an oscillator's phase behavior, discarding amplitude variations and common-mode behavior. Furthermore, in [Dem00a, Dem02], the equation (5.4) is solved by constructing a (complicated) Fokker-Planck equation [Risk89]. In this chapter, we develop a more general mathematical framework for the analysis of an oscillator's behavior. It embeds the phase noise analysis in [Dem00a, Dem00b, Dem02] as a particular application. Moreover, the proof of these existing results is significantly simplified by means of the averaging methods here presented.

5.2.3 Numerical simulation

Finally, there is the possibility to study oscillator behavior by means of numerical simulation. Hereby, one can either use general-purpose algorithms, like SPICE [Nag75], or dedicated methods [Bra96, Nar99, Ngo96, Riz96]. SPICE, however, is troubled by the presence of the widely spaced time constants that characterize the behavior of many oscillators. Capturing the slow-varying transients requires a long overall simulation time interval. In contrast, the presence of fast oscillations forces the simulator to take a small simulation time step. Hence, covering the long simulation time interval comes at the expense of a huge number of time steps. It often takes minutes or even hours for a single simulation run to complete.

Recent developments brought about simulation methods that try to separate the computation of fast- and slow-varying behavior [Bra96, Nar99, Ngo96, Riz96]. These methods build on the multi-variate formulation of the circuit simulation problem as introduced in [Royc01]. However, although very efficient for the verification of transistor-level implementations, these algorithms do not provide a lead to construct compact oscillator models that can be used for system-level exploration and verification.

5.2.4 Contributions of this work

This chapter presents methods that allow us to efficiently handle the slow-varying behavior of oscillation amplitudes, phases, phase differences and common-mode levels. It presents a significant generalization of the work (on phase noise analysis) in [Dem00a, Dem00b, Dem02, Kae90]. To the author's knowledge, the theory presented in this chapter is, to a large extent, original. It is grounded onto two major components:

- Perturbation theory [Kev81, Kha96] provides an elegant framework to model the interactions between an oscillator's core system and a set of perturbation terms. The perturbation analysis here presented extends the projection technique in [Dem00a, Dem00b, Dem02, Kae90] to capture, besides the phase variations, the dynamics of the oscillator's amplitude and common-mode level. It is also able to handle the behavior of systems of coupled oscillators. The analysis results in a set of equations that govern the oscillator's state-changing behavior (see section 5.1.3). It requires proper partitioning of the oscillator circuit equations into a core system and perturbation terms. This work does not address general methods to do the partitioning. Instead, it operates on a case by case basis. Note, however, that such a "case" often spans a wide variety of circuits.

- Averaging allows us to extract the equations that govern the slow-varying components of the oscillator's state-changes. This work is restricted to the use of first-order averaging transformations. In most cases, this yields results that are sufficiently accurate. The averaging techniques presented in this text adapt the results in [Bogo61, Verhu96] to handle the presence of noise. This allows them to deal with matters like oscillator phase noise.

The advantages of this method over other techniques to analyze oscillator dynamic behavior, e.g. plain transient SPICE simulations, are:

- We are able to exploit the presence of the widely separated time constants (steady-state versus settling behavior) that typically characterize an oscillator's behavior. While this often poses a bottleneck to traditional (SPICE-like) circuit simulation algorithms, it enables the methods presented in this work to explicitly separate the oscillator's fast- and slow-varying signal components. This allows us to use a much larger simulation time step than possible when solving the original set of circuit equations. This in turn results in shorter simulation times.

- On a related issue, the averaged equations can be used to construct compact behavioral models. These models can be solved efficiently using a large simulation time step. They are readily implemented in almost any environment for system-level simulation (Matlab/Simulink, VHDL-AMS, etc.) and are especially useful for running lengthy or repetitive system-level simulations.

- The method provides a powerful framework to handle often complex nonlinear behavior. Besides for numerical analysis, it can also be used to obtain symbolic expressions that describe the oscillator's dynamics. These expressions are of great help during those stages of a design when not all parameters have been given a suitable value yet. This chapter presents examples on how to derive symbolic expressions that characterize an oscillator's injection locking behavior and the settling behavior of an harmonic (LC-based) oscillator.

- From a theoretical point of view, the averaging methods here presented yield a simplification of the proof of existing results on oscillator phase noise analysis. This part of the work culminated in a semi-analytic method for the computation of an oscillator's entire phase noise spectrum [Vana03c].

In short, this method shows quite some good and novel results with regard to the (symbolic and numerical) analysis of an oscillator's dynamic behavior.

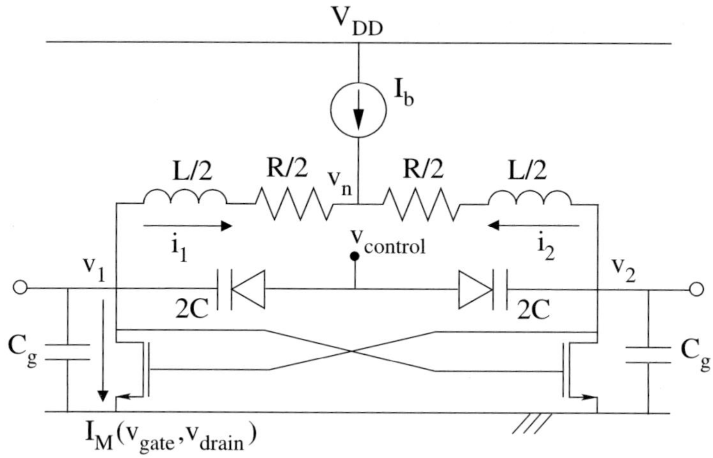

Figure 5.9: Harmonic oscillator circuit topology. Here, the resistors with values $R/2$ model the inductor losses, $I_M(v_{gate}, v_{drain})$ denotes the drain-source current of a MOS transistor and the capacitors C_g capture some small (transistor) capacitances.

5.3 Oscillator circuit equations

The behavior of an oscillator circuit is described by the system of differential-algebraic equations (DAEs) corresponding to the circuit's modified nodal equations:

$$\frac{\mathrm{d}}{\mathrm{d}t}\mathbf{q}(\mathbf{v})+\mathbf{j}(\mathbf{v})+\mathbf{b}(t)=0 \ . \tag{5.6}$$

Here, $\mathbf{v} \in \mathbb{R}^N$ is a vector storing the circuit's state variables, for example, the circuit's nodal voltages. Furthermore, $\mathbf{q}(\mathbf{v}) \in \mathbb{R}^N$ is a vector of capacitor charges and/or inductor fluxes while $\mathbf{j}(\mathbf{v}) \in \mathbb{R}^N$ is a vector of branch currents. The vector $\mathbf{b}(t) \in \mathbb{R}^N$ models the action of some weak (small) input sources, e.g. noise.

Example (Voltage-controlled harmonic oscillator): The voltage-controlled harmonic oscillator in Fig. 5.9 is described by the system of DAEs

$$\frac{L}{2}\frac{\mathrm{d}i_1}{\mathrm{d}t}-v_1+v_n+\frac{R}{2}i_1 = 0 \tag{5.7}$$

$$\frac{L}{2}\frac{\mathrm{d}i_2}{\mathrm{d}t}-v_2+v_n+\frac{R}{2}i_2 = 0 \tag{5.8}$$

$$2C\frac{\mathrm{d}v_1}{\mathrm{d}t}+i_1+I_M(v_2,v_1)+C_g\frac{\mathrm{d}v_1}{\mathrm{d}t} = 0 \tag{5.9}$$

$$2C\frac{\mathrm{d}v_2}{\mathrm{d}t}+i_2+I_M(v_1,v_2)+C_g\frac{\mathrm{d}v_2}{\mathrm{d}t} = 0 \tag{5.10}$$

$$i_1+i_2+I_b = 0 \ . \tag{5.11}$$

Here, the resistors with values $R/2$ model the inductor losses, $I_M(v_{gate}, v_{drain})$ denotes the drain-source current of the MOS transistors and the capacitors C_g capture some small (transistor) capacitances. By introducing the differential voltages and currents

$$i_d = (i_1 - i_2)/2 \quad (5.12)$$
$$i_{cm} = (i_1 + i_2)/2 \quad (5.13)$$
$$v_d = (v_1 - v_2)/2 \quad (5.14)$$
$$v_{cm} = (v_1 + v_2)/2 \ , \quad (5.15)$$

the equations (5.7)-(5.11) can be written as

$$2C\frac{\mathrm{d}v_d}{\mathrm{d}t} + i_d + I_{M,d}(v_{cm}+v_d, v_{cm}-v_d) + C_g\frac{\mathrm{d}v_d}{\mathrm{d}t} = 0 \quad (5.16)$$
$$\frac{L}{2}\frac{\mathrm{d}i_d}{\mathrm{d}t} - v_d + \frac{R}{2}i_d = 0 \quad (5.17)$$
$$2C\frac{\mathrm{d}v_{cm}}{\mathrm{d}t} + i_{cm} + I_{M,cm}(v_{cm}+v_d, v_{cm}-v_d) + C_g\frac{\mathrm{d}v_{cm}}{\mathrm{d}t} = 0 \quad (5.18)$$
$$\frac{L}{2}\frac{\mathrm{d}i_{cm}}{\mathrm{d}t} - v_{cm} + v_n + \frac{R}{2}i_{cm} = 0 \quad (5.19)$$
$$2i_{cm} + I_b = 0 \ . \quad (5.20)$$

In the equations above,

$$I_{M,d}(v_1, v_2) = (I_M(v_2, v_1) - I_M(v_1, v_2))/2 \quad (5.21)$$

and

$$I_{M,cm}(v_1, v_2) = (I_M(v_2, v_1) + I_M(v_1, v_2))/2 \quad (5.22)$$

respectively denote the differential and common-mode transistor currents. Furthermore, in order not to overload the example, we neglected the nonlinear characteristic of the resonant tank's capacitors (varactors). Also, in this example, we assumed the components of the vector $\mathbf{b}(t)$ to be zero. ▲

5.3.1 Normalizing the oscillator circuit equations

Although not strictly necessary, it is good practice to normalize the equations (5.6) such that all quantities involved are, approximately, one in order of magnitude. To do so, we normalize all voltages and currents with respect to some nominal value. These normalization constants can be user-provided. They can also be obtained by means of short SPICE-like trial simulations. Moreover, we also normalize the time variable

$$\tau = \omega_0 t \ . \quad (5.23)$$

Here, ω_0 is chosen to make derivatives with respect to τ one in order of magnitude. Normalizing the oscillator circuit equations later on facilitates partitioning them into a core system and a set of perturbation terms.

Example (Voltage-controlled harmonic oscillator): For the system of DAEs (5.16)-(5.20) that model the oscillator circuit in Fig. 5.9, all voltages are normalized with respect to V_{osc}, an estimate for the amplitude of the resulting differential-mode oscillation. The differential current $i_d(t)$ is then normalized with respect to

$$I_{osc} = (2\omega_0 C)\, V_{osc} = V_{osc}/\,(\omega_0 L/2) \quad . \tag{5.24}$$

The (small) common-mode current $i_{cm}(t)$ is best normalized with respect to I_{osc}/Q. Hereby,

$$\omega_0 = \frac{1}{\sqrt{LC}} \tag{5.25}$$

is the natural frequency of the oscillator's resonant tank, while

$$Q = \frac{1/R}{C\omega_0} = \frac{L\omega_0}{R} \tag{5.26}$$

equals its quality factor. With

$$x_d = \frac{v_d}{V_{osc}},\; x_{cm} = \frac{v_{cm}}{V_{osc}},\; x_n = \frac{v_n}{V_{osc}},\; y_d = \frac{i_d}{I_{osc}},\; y_{cm} = \frac{i_{cm}}{I_{osc}/Q} \quad , \tag{5.27}$$

being the normalized state variables, we find

$$\frac{\mathrm{d}x_d}{\mathrm{d}\tau} + y_d + \left[\frac{I_{M,d}}{I_{osc}} + \frac{C_g}{2C}\frac{\mathrm{d}x_d}{\mathrm{d}\tau}\right] = 0 \tag{5.28}$$

$$\frac{\mathrm{d}y_d}{\mathrm{d}\tau} - x_d + \left[\frac{y_d}{Q}\right] = 0 \tag{5.29}$$

$$\frac{\mathrm{d}x_{cm}}{\mathrm{d}\tau} + \left[\frac{y_{cm}}{Q} + \frac{I_{M,cm}}{I_{osc}} + \frac{C_g}{2C}\frac{\mathrm{d}x_{cm}}{\mathrm{d}\tau}\right] = 0 \tag{5.30}$$

$$-x_{cm} + x_n + \left[\frac{1}{Q}\frac{\mathrm{d}y_{cm}}{\mathrm{d}\tau} + \frac{y_{cm}}{Q^2}\right] = 0 \tag{5.31}$$

$$y_{cm} + \frac{Q}{2}\frac{I_b}{I_{osc}} = 0 \quad . \tag{5.32}$$

Here, the normalized time variable τ is as defined in (5.23). The reason for some terms to be enclosed in square brackets is explained in the next section. ▲

5.3.2 Partitioning the normalized circuit equations

Having normalized the oscillator's circuit equations, we partition them into a core system and a set of perturbations. The core system corresponds to the terms of the (normalized) circuit equations that are large in magnitude while the perturbation terms correspond to the small ones. As such, the core system captures that part of the oscillator that is the main responsible for the oscillations to occur. The perturbation terms cause the core system to change behavior over time over time.

Formally, the normalized and partitioned set of oscillator circuit equations can be written as

$$\frac{\mathrm{d}}{\mathrm{d}\tau}\mathbf{c}(\mathbf{x}) + \mathbf{g}(\mathbf{x}) + \varepsilon\mathbf{f}(\mathbf{x},\tau) = 0 \quad . \tag{5.33}$$

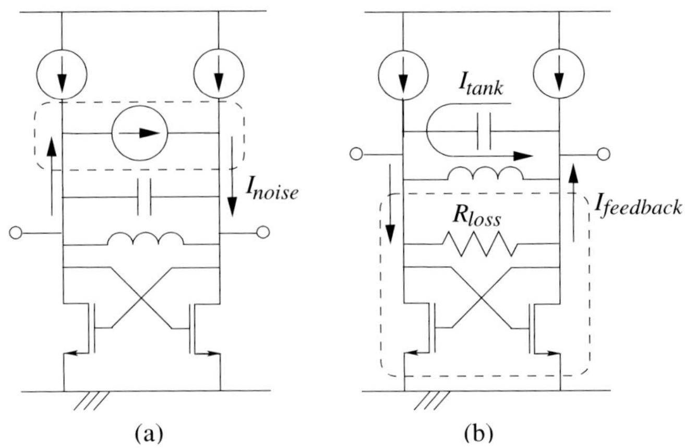

Figure 5.10: Two possible ways to partition an harmonic oscillator topology with the parts encircled by the dashed lines corresponding to the perturbation terms. (a) If one is mainly interested in the oscillator's phase noise behavior, only the noise source is treated as a perturbation term. (b) If one targets the oscillator's settling behavior, all currents due to the feedback transistors and tank losses are treated as perturbation terms.

Here, $\mathbf{x} \in \mathbb{R}^N$ represents the circuit's normalized state variables. The first and second term in (5.33) represent a system of autonomous DAEs that capture the oscillator's unperturbed core behavior. The term $\varepsilon\mathbf{f}(\mathbf{x}, \tau)$ represents the perturbation components. Here, ε is called a *perturbation variable*. It labels the corresponding term as small with respect to the other ones. Note that the perturbation term may depend explicitly on the (normalized) time variable τ. This is, for example, caused by the presence of a vector of non-zero input sources $\mathbf{b}(t) = \mathbf{b}(\tau/\omega_0)$ in (5.6). Since, in this text, these inputs are assumed to be weak , they can be treated as part of the perturbation term $\varepsilon\mathbf{f}(\mathbf{x}, \tau)$. As such, they introduce an explicit dependence on time.

Example (Harmonic oscillator): Fig. 5.10 shows two ways to partition an harmonic oscillator with the parts encircled by the dashed lines corresponding to the perturbation terms. The choice between both depends on the kind of behavior in which one is interested. In Fig. 5.10(a), one is interested in the oscillator's phase noise behavior, i.e. the oscillator phase shifts induced by noise sources. In that case, the core system corresponds to the noise-free oscillator. Only the noise sources are treated as perturbations. In Fig. 5.10(b), we target the harmonic oscillator's settling behavior. Here, the resonant tank —excluding tank losses— corresponds to the oscillator's core system. The feedback transistor currents and the tank losses are treated as perturbation terms. In order for them to be small compared to the current circulating in the resonant tank, the tank's Q-factor must be sufficiently large (about 10 in order of magnitude). ▲

Example (Voltage-controlled harmonic oscillator): For the voltage-controlled harmonic oscillator topology in Fig. 5.9, it is indicated in (5.28)-(5.32) how to perform the partitioning operation. Here, the terms in the square brackets mark the perturbation terms. In order for them to be small, it must hold that both $1/Q \ll 1$ and $C_g/2C \ll 1$. All others terms are assigned to the oscillator's core system. Note that this partitioning operation is *functional* and *not physical*. For example, in (5.28)-(5.32), the inductive part of the coil is classified as being part of the core system while the inductor losses are labeled as perturbation terms. Physically, however, it is impossible to separate a coil's inductive and resistive behavior. ▲

The partitioning operation discussed above is by no means trivial. It is hard to construct a general-purpose algorithm that accomplishes an appropriate partitioning for arbitrary oscillator topologies. In this work, no attempts are made to do so and all partitionings in the examples are done manually. However, in sections 5.7 and 5.8, we outline partitioning strategies that allow us to handle oscillator phase noise analysis and harmonic oscillator behavior. These partitioning strategies basically correspond with the ones in Fig. 5.10.

5.4 Characterizing the oscillator's unperturbed core

Next, we characterize the oscillator's unperturbed core system. This core system is described by the equations

$$\frac{\mathrm{d}}{\mathrm{d}\tau}\mathbf{c}(\mathbf{x})+\mathbf{g}(\mathbf{x})=0 \ . \tag{5.34}$$

These equations result when the perturbation term $\varepsilon\mathbf{f}(\mathbf{x},\tau)$ in (5.33) is neglected. Characterizing the core system requires us to find all states, i.e. all steady-state solutions $\mathbf{x}_s(\tau,\mathbf{p})$ that solve (5.34). In this work, a steady-state solution is defined as being periodic. Furthermore, we assume all periodic steady-state solutions to have the same period T[7]. Mathematically, this is expressed as

$$\frac{\mathrm{d}}{\mathrm{d}\tau}\mathbf{c}\left(\mathbf{x}_s\left(\tau,\mathbf{p}\right)\right)+\mathbf{g}\left(\mathbf{x}_s\left(\tau,\mathbf{p}\right)\right) \quad = \quad 0 \tag{5.35}$$

$$\mathbf{x}_s\left(\tau+T,\mathbf{p}\right) \quad = \quad \mathbf{x}_s\left(\tau,\mathbf{p}\right) \ . \tag{5.36}$$

Here, $\mathbf{p}\in\mathbb{R}^P$ is a constant parameter vector that identifies a particular steady-state solution, i.e. it identifies a particular state of the unperturbed core system. For this reason, $\mathbf{p}$ is called the system's *state vector*.

The set of all points in $\mathbb{R}^N$ that lie on one of the steady-state solutions $\mathbf{x}_s(\tau,\mathbf{p})$, i.e.

$$M=\left\{\mathbf{x}\,|\,\exists\mathbf{p}\in\mathbb{R}^P\wedge\tau\in\mathbb{R}:\mathbf{x}=\mathbf{x}_s(\tau,\mathbf{p})\right\} \ , \tag{5.37}$$

is assumed a P-dimensional stable manifold that quickly attracts all other solutions of

[7] In a more general sense, the term *steady-state solution* can be defined as a solution to (5.34) that is void of fast-varying transient behavior. For typical oscillators, these solutions tend to be periodic or quasi-periodic. Furthermore, the methods here presented can be extended to handle steady-state solutions with varying periods of oscillation (see for example [Vana02e]). However, the assumption of a constant period is often satisfied and helps to simplify analysis.

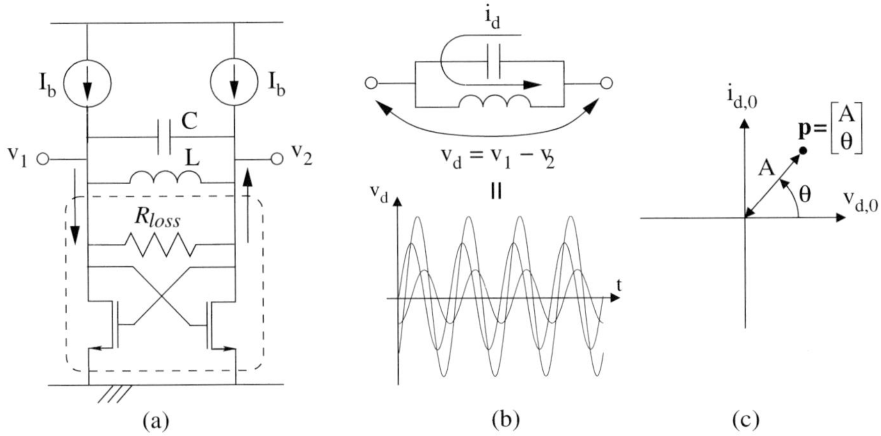

Figure 5.11: Characterizing a harmonic oscillator's core system: (a) The harmonic oscillator is partitioned into a core system and a set of perturbation components (enclosed by the dashed line). (b) The unperturbed core system is solved for all possible steady-state solutions (states). (c) Each solution (state) is identified by a particular value of the state vector $\mathbf{p} = \begin{bmatrix} A & \theta \end{bmatrix}^T$. Hence, it can be represented as a single point in a 2-dimensional space.

(5.34)[8]. Roughly, a manifold is a smooth surface [Abra88]. Solutions $\mathbf{x}(\tau)$ of (5.34) are attracted to the manifold if, for $\tau \to \infty$, $\mathbf{x}(\tau)$ approaches M infinitesimally closely.

Example (Harmonic oscillator): Fig. 5.11(b) and (c) illustrate the characterization of the core system associated with the harmonic oscillator in Fig. 5.11(a). The behavior of this oscillator is captured by the system of (unnormalized) DAEs

$$C\frac{\mathrm{d}v_d}{\mathrm{d}t} + i_d + \left[\frac{v_d}{R_{loss}} + \frac{I_M\left(\frac{v_{cm}-v_d}{2}, \frac{v_{cm}+v_d}{2}\right) - I_M\left(\frac{v_{cm}+v_d}{2}, \frac{v_{cm}-v_d}{2}\right)}{2}\right] = 0 \quad (5.38)$$

$$L\frac{\mathrm{d}i_d}{\mathrm{d}t} - v_d = 0 \quad (5.39)$$

$$I_M\left(\frac{v_{cm}-v_d}{2}, \frac{v_{cm}+v_d}{2}\right) + I_M\left(\frac{v_{cm}+v_d}{2}, \frac{v_{cm}-v_d}{2}\right) - 2I_b = 0 \quad (5.40)$$

where $v_d = v_1 - v_2$ and $v_{cm} = v_1 + v_2$. In (5.38)-(5.40), the terms in the square brackets denote the perturbation terms. This corresponds with the partitioning indicated in Fig. 5.11(a). As shown in Fig. 5.11(b), the oscillator's core system corresponds to a lossless resonant tank. This core system is described by the equations obtained by neglecting the perturbation terms in (5.38)-(5.40). These (unperturbed) equations are solved for $v_d(t) = A\cos(\omega_0 t + \theta)$ and $i_d(t) = A\sin(\omega_0 t + \theta)$, i.e. sine waves with an

[8] See appendix D on how to quantify the notion of "a manifold that quickly attracts all other solutions". Note that if M does not attract the other solutions in its neighborhood, this typically indicates that a bad job has been done in partitioning the original set of circuit equations.

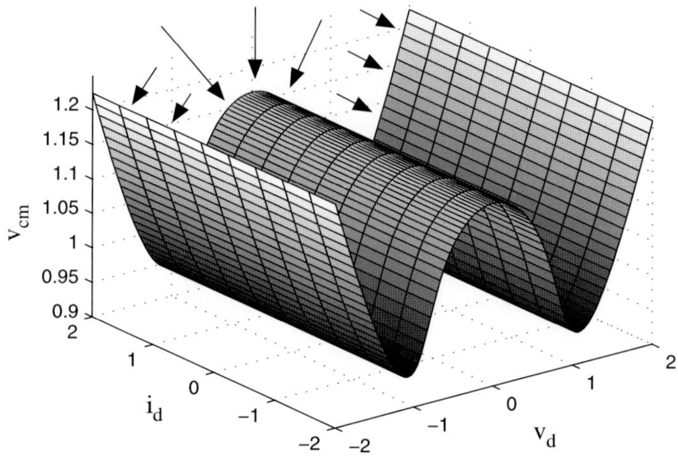

Figure 5.12: The set of all points that lie on one of the steady-state solutions of the harmonic oscillator's core system make up a 2-dimensional manifold (surface). All other solutions that solve the unperturbed core system's equations are quickly attracted towards this manifold.

arbitrary amplitude and phase. The algebraic equation (5.40) makes the third state variable, the common-mode voltage $v_{cm}(t)$, "track" the differential voltage $v_d(t)$. In in Fig. 5.11(c), these solutions are pictured as points in a 2-dimensional state-vector space. Fig. 5.12 plots the 2-dimensional manifold (surface) M —defined by (5.37)— consisting of all points that lie on a steady-state solution. This manifold is stable as it quickly attracts all other solutions. ▲

Example (Voltage-controlled harmonic oscillator): The (normalized) unperturbed core system associated with the voltage-controlled oscillator topology in Fig. 5.9 is described by the system of equations

$$\frac{dx_d}{d\tau} + y_d = 0 \tag{5.41}$$

$$\frac{dy_d}{d\tau} - x_d = 0 \tag{5.42}$$

$$\frac{dx_{cm}}{d\tau} = 0 \tag{5.43}$$

$$-x_{cm} + x_n = 0 \tag{5.44}$$

$$y_{cm} + \frac{Q}{2}\frac{I_b}{I_{osc}} = 0\ . \tag{5.45}$$

These equations are obtained by neglecting the perturbation terms in (5.28)-(5.32), i.e. the terms in the square brackets. The periodic steady-state solutions (states) of the core system are given by

$$\begin{aligned} x_d(\tau) &= A\cos(\tau+\theta) && (5.46) \\ y_d(\tau) &= A\sin(\tau+\theta) && (5.47) \\ x_{cm}(\tau) &= z && (5.48) \\ y_{cm}(\tau) &= -\frac{Q}{2}\frac{I_b}{I_{osc}} && (5.49) \\ x_n(\tau) &= z\ . && (5.50) \end{aligned}$$

All steady-state solutions taken together yield a 3-dimensional manifold. Hereby

$$\mathbf{p} = \begin{bmatrix} A & \theta & z \end{bmatrix}^T \in \mathbb{R}^3 \tag{5.51}$$

is the corresponding state vector. Note that, compared to the previous example, the dimension of the state vector $\mathbf{p}$ has increased by one by means of the addition of the common-mode variable z. This increase is due to the presence of the control voltage $v_{control}$ in Fig. 5.9. This control voltage induces the common-mode charging and discharging of the resonant tank capacitors. This, in turn, introduces an extra slow-varying (common-mode) process in the oscillator's transient behavior. Typical analyses, e.g. [Ham01], do not account for this slow-varying common-mode behavior. They treat a third-order voltage-controlled harmonic oscillator as a second-order non-controlled harmonic oscillator, e.g. the one in the previous example. Here, the order of an oscillator refers to the dimension of its core system's state vector. ▲

5.5 Oscillator perturbation analysis

The perturbation term $\varepsilon\mathbf{f}(\mathbf{x},\tau)$ drives the unperturbed core from one state to another. In order to unravel the details of this state-changing behavior, we study the structure of the waveforms $\mathbf{x}_\varepsilon(\tau)$ that solve the perturbed equations

$$\frac{\mathrm{d}}{\mathrm{d}\tau}\mathbf{c}(\mathbf{x}_\varepsilon) + \mathbf{g}(\mathbf{x}_\varepsilon) + \varepsilon\mathbf{f}(\mathbf{x}_\varepsilon,\tau) = 0\ . \tag{5.52}$$

Here, we use the notation $\mathbf{x}_\varepsilon(\tau)$ to indicate the fact that solutions of (5.52) in principle also depend on the perturbation variable ε. As a note aside, we mention that the theory that follows embeds the methods in [Dem00a, Dem00b, Kae90] as a special case.

5.5.1 Components of an oscillator's perturbed behavior

While all waveforms that solve the unperturbed equations (5.34) converge to the P-dimensional manifold M, the action of $\varepsilon\mathbf{f}(\mathbf{x},\tau)$ causes the $\mathbf{x}_\varepsilon(\tau)$ to deviate from the manifold. However, the attracting nature of M ensures that, for small perturbations, these deviations remain small. Hence, after some initial transients, any solution $\mathbf{x}_\varepsilon(\tau)$ will be located near M. They can therefore be written as

$$\mathbf{x}_\varepsilon(\tau) = \mathbf{x}_s(\tau,\mathbf{p}(\tau,\varepsilon)) + \varepsilon\Delta\mathbf{x}(\tau,\varepsilon)\ . \tag{5.53}$$

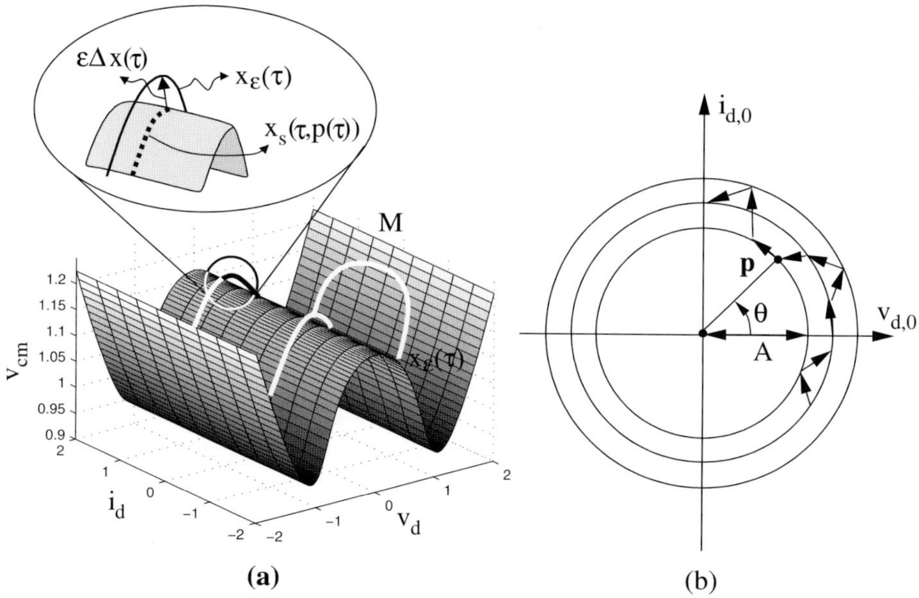

Figure 5.13: Perturbed harmonic oscillator behavior: (a) the solutions $\mathbf{x}_\varepsilon(\tau)$ of the perturbed harmonic oscillator in Fig. 5.11(a) are attracted to the neighborhood of the stable, 2-dimensional manifold M. These solutions contain two components: a motion $\mathbf{x}_s(\tau, \mathbf{p}(\tau))$ over the attracting manifold M and a small deviation $\varepsilon\Delta\mathbf{x}(\tau)$ from the manifold. (b) The motion over the manifold M can be represented as the resonant tank slipping from one state to another.

As indicated, a solution $\mathbf{x}_\varepsilon(\tau)$ contains two components: a motion $\mathbf{x}_s(\tau, \mathbf{p}(\tau, \varepsilon))$ over the manifold M and a deviation $\varepsilon\Delta\mathbf{x}(\tau, \varepsilon)$ from the manifold. The first component, $\mathbf{x}_s(\tau, \mathbf{p}(\tau, \varepsilon))$, corresponds to the state-changing behavior discussed in section 5.1.3: the perturbation term drives the core system from one state (steady-state solution) to another. Hence, the state vector $\mathbf{p}$ does no longer remain constant but changes with time. This time-varying behavior of the state vector $\mathbf{p}$ entirely characterizes the motion over M. The second component in (5.53), $\varepsilon\Delta\mathbf{x}(\tau, \varepsilon)$, describes a deviation from the manifold. Here, the manifold's stability guarantees that there exists a process $\mathbf{p}(\tau, \varepsilon)$ such that (5.53) is satisfied for a bounded (small) deviation $\varepsilon\Delta\mathbf{x}(\tau, \varepsilon)$. Note that the processes $\mathbf{p}(\tau, \varepsilon)$ and $\Delta\mathbf{x}(\tau, \varepsilon)$ in principle depend on both τ and ε. In what follows, for notational convenience, the dependence on ε will no longer be mentioned explicitly unless when relevant.

Example (Harmonic oscillator): Fig. 5.13 illustrates the components of the perturbed behavior of the harmonic oscillator from Fig. 5.11(a). Solutions $\mathbf{x}_\varepsilon(\tau)$ are attracted to the neighborhood of the stable, 2-dimensional manifold M obtained from the characterization of the oscillator's unperturbed core. These solutions contain two components: a motion $\mathbf{x}_s(\tau, \mathbf{p}(\tau))$ over the attracting manifold M and a small deviation $\varepsilon\Delta\mathbf{x}(\tau)$ from the manifold. Moreover, the motion over the manifold M can be represented as the oscillator's resonant tank (core system) slipping from one state to another. Hence, this implies that the state vector $\mathbf{p}(\tau) = \begin{bmatrix} A(\tau) & \theta(\tau) \end{bmatrix}$ (slowly) varies with time. ▲

Substituting the signal model (5.53) into (5.52) yields

$$\begin{aligned}\frac{\partial\mathbf{c}}{\partial\mathbf{x}}(\mathbf{x}_s(\tau,\mathbf{p}(\tau)))\cdot\frac{\partial\mathbf{x}_s}{\partial\mathbf{p}}(\tau,\mathbf{p}(\tau))\frac{\mathrm{d}\mathbf{p}}{\mathrm{d}\tau}+\varepsilon\frac{\mathrm{d}}{\mathrm{d}\tau}\left(\frac{\partial\mathbf{c}}{\partial\mathbf{x}}(\mathbf{x}_s(\tau,\mathbf{p}(\tau)))\cdot\Delta\mathbf{x}(\tau)\right)&\\ +\varepsilon\frac{\partial\mathbf{g}}{\partial\mathbf{x}}(\mathbf{x}_s(\tau,\mathbf{p}(\tau)))\Delta\mathbf{x}(\tau)+\varepsilon\mathbf{f}(\mathbf{x}_s(\tau,\mathbf{p}(\tau)),\tau)+O(\varepsilon^2)&=0\ .\end{aligned} \tag{5.54}$$

Here we used the fact that, for an arbitrary but fixed state vector $\mathbf{p}$, $\mathbf{x}_s(\tau,\mathbf{p})$ solves (5.52) for $\varepsilon = 0$. By projecting (5.54) on appropriate subspaces of $\mathbb{R}^N$, it is possible to compute the motion over M, i.e. the behavior of the state vector $\mathbf{p}(\tau)$, without needing the details of $\varepsilon\Delta\mathbf{x}(\tau)$. This is quite convenient. In most cases, the motion over M is of greatest interest as it captures the slow-varying components of the oscillator's dynamics, e.g. long-term amplitude and phase variations. Hence, proper projection helps us to avoid wasting time on the computation of the often irrelevant deviation $\varepsilon\Delta\mathbf{x}(\tau)$.

The next section discusses how to extract the equations governing the process $\mathbf{p}(\tau)$ from (5.54). The equations that govern $\varepsilon\Delta\mathbf{x}(\tau)$ are of less importance. Their derivation is, therefore, deferred to appendix D. Note that all analysis is performed up to first order in the perturbation variable ε. Furthermore, in what follows, we use

$$\mathbf{C}(\tau,\mathbf{p}) = \frac{\partial\mathbf{c}}{\partial\mathbf{x}}(\mathbf{x}_s(\tau,\mathbf{p})) \in \mathbb{R}^{N\times N} \tag{5.55}$$

$$\mathbf{G}(\tau,\mathbf{p}) = \frac{\partial\mathbf{g}}{\partial\mathbf{x}}(\mathbf{x}_s(\tau,\mathbf{p})) \in \mathbb{R}^{N\times N} \tag{5.56}$$

$$\mathbf{f}_p(\tau,\mathbf{p}) = \mathbf{f}(\mathbf{x}_s(\tau,\mathbf{p}),\tau) \in \mathbb{R}^{N} \tag{5.57}$$

as shorthand notations that help to simplify the algebra.

5.5.2 Motion $\mathbf{x}_s(\tau,\mathbf{p}(\tau))$ over the manifold M

The equations that characterize the process $\mathbf{p}(\tau)$ and, therefore, the oscillator's motion over the P-dimensional manifold M, are obtained by projecting (5.54) onto the manifold M. More specifically, at each point, we project them onto the *tangent space* of M. Hence, we focus on the behavior induced by those parts of the perturbation terms that are oriented along the surface M. Fig. 5.14 makes the analogy with a ball that is constrained to move on a curved rod (a 1-dimensional stable manifold). Since the ball cannot leave the rod, only the force components that are oriented along the rod can set the ball in motion (along the rod).

At each point of M, i.e. for each couple $(\tau,\mathbf{p}) \in \mathbb{R}\times\mathbb{R}^P$, the tangent space is spanned by the columns of

$$\mathbf{U}_1(\tau,\mathbf{p}) = \frac{\partial\mathbf{x}_s}{\partial\mathbf{p}}(\tau,\mathbf{p}) \in \mathbb{R}^{N\times P}\ . \tag{5.58}$$

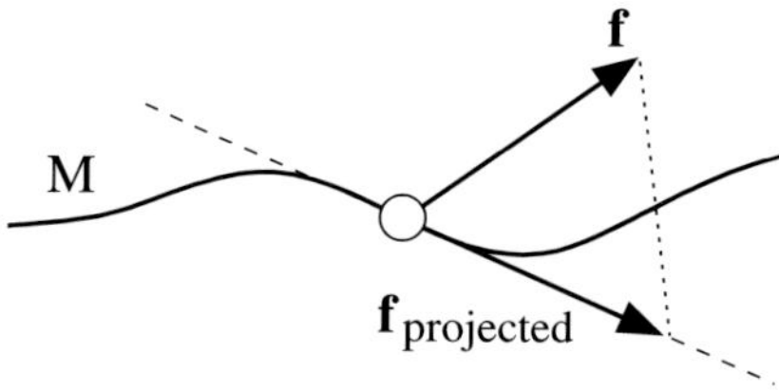

Figure 5.14: When a ball (a system) is constrained to move along a curved rod (a 1-dimensional surface), only the force components that are oriented along the rod can set the ball in motion. It hence makes sense to project the perturbation force on the tangent to the rod. As will be discussed later, this projection is not necessarily orthogonal.

Since, for a constant state vector $\mathbf{p}$, $\mathbf{x}_s(\tau,\mathbf{p})$ solves (5.52) for $\varepsilon = 0$, it is readily shown that all P columns of $\mathbf{U}_1(\tau,\mathbf{p})$ solve the linear N-dimensional homogeneous system of DAEs

$$\frac{\partial}{\partial\tau}\left(\mathbf{C}(\tau,\mathbf{p})\,\mathbf{y}(\tau)\right)+\mathbf{G}(\tau,\mathbf{p})\,\mathbf{y}(\tau)=0 \ . \tag{5.59}$$

Furthermore, if the system of DAEs (5.59) is index-1 [Dem00b, Marz95], then there exist matrices $\mathbf{U}_2(\tau,\mathbf{p}) \in \mathbb{R}^{N\times Q}$ and $\mathbf{U}_3(\tau,\mathbf{p}) \in \mathbb{R}^{N\times R}$, with $P+Q+R=N$, such that the columns of

$$\mathbf{U}(\tau,\mathbf{p}) = \left[\begin{array}{ccc} \mathbf{U}_1(\tau,\mathbf{p}) & \mathbf{U}_2(\tau,\mathbf{p}) & \mathbf{U}_3(\tau,\mathbf{p}) \end{array}\right] \in \mathbb{R}^{N\times N} \tag{5.60}$$

span the entire space $\mathbb{R}^N$. Here, the columns of $\mathbf{U}_2(\tau,\mathbf{p})$ correspond to the stable modes of (5.59). They solve (5.59) and satisfy

$$\lim_{\tau\to\infty}\|\mathbf{U}_2(\tau,\mathbf{p})\| = 0, \forall\mathbf{p}\in\mathbb{R}^P \ , \tag{5.61}$$

a property that expresses the stability of the manifold M. The columns of $\mathbf{U}_3(\tau,\mathbf{p})$ satisfy

$$\mathbf{C}(\tau,\mathbf{p})\mathbf{U}_3(\tau,\mathbf{p}) = 0 \ . \tag{5.62}$$

Hence, they span the null space of $\mathbf{C}(\tau,\mathbf{p})$.

In order to project (5.54) onto the tangent space of M, we need to determine a projector $\mathbf{V}_1^T(\tau,\mathbf{p}) \in \mathbb{R}^{P\times N}$ that satisfies

$$\mathbf{V}_1^T(\tau,\mathbf{p})\cdot\mathbf{C}(\tau,\mathbf{p})\cdot\mathbf{U}_1(\tau,\mathbf{p}) = \mathbf{I}_P, \forall(\tau,\mathbf{p})\in\mathbb{R}\times\mathbb{R}^P \ . \tag{5.63}$$

Here, $\mathbf{I}_P \in \mathbb{R}^{P\times P}$ is the P-dimensional unity matrix. This projector is a generalization the perturbation projection vector introduced in [Dem98] (see also section 5.7). The condition (5.63) is automatically satisfied for a projector $\mathbf{V}_1^T(\tau,\mathbf{p})$ that meets

$$\left(\frac{\partial}{\partial\tau}\mathbf{V}_1^T(\tau,\mathbf{p})\right)\cdot\mathbf{C}(\tau,\mathbf{p}) - \mathbf{V}_1^T(\tau,\mathbf{p})\cdot\mathbf{G}(\tau,\mathbf{p}) = 0 \tag{5.64}$$

$$\mathbf{V}_1^T(0,\mathbf{p})\cdot\mathbf{C}(0,\mathbf{p})\cdot\mathbf{U}_1(0,\mathbf{p}) = \mathbf{I}_P \ . \tag{5.65}$$

Indeed, combining (5.64) with the fact that $\mathbf{U}_1(\tau, \mathbf{p})$ solves (5.59) implies that the partial derivative with respect to τ of the left-hand side of (5.63) equals zero. The linear homogeneous system of DAEs (5.64) is called the system adjoint to (5.59). It is important to note that (5.64) and (5.65) do not uniquely determine $\mathbf{V}_1^T(\tau, \mathbf{p})$. There are still Q degrees of freedom left with Q the number of columns in $\mathbf{U}_2(\tau, \mathbf{p})$. As is discussed below, the extra conditions needed to fix $\mathbf{V}_1^T(\tau, \mathbf{p})$ stem from the requirement that the deviation $\varepsilon\Delta\mathbf{x}(\tau)$ must remain bounded.

By multiplying (5.54) with $\mathbf{V}_1^T(\tau, \mathbf{p})$ we obtain[9]

$$\begin{aligned} \frac{d\mathbf{p}}{d\tau} &= -\varepsilon\mathbf{V}_1^T(\tau, \mathbf{p})\left(\frac{d}{d\tau}(\mathbf{C}(\tau, \mathbf{p})\Delta\mathbf{x}(\tau)) + \mathbf{G}(\tau, \mathbf{p})\Delta\mathbf{x}(\tau) + \mathbf{f}_p(\tau, \mathbf{p})\right) + O(\varepsilon^2) \\ &= -\varepsilon\left(\frac{d}{d\tau}(\mathbf{V}_1^T(\tau, \mathbf{p})\mathbf{C}(\tau, \mathbf{p})\Delta\mathbf{x}(\tau)) + \mathbf{V}_1^T(\tau, \mathbf{p})\mathbf{f}_p(\tau, \mathbf{p})\right) + O(\varepsilon^2) \ . \end{aligned} \tag{5.66}$$

Here, we used (5.55)-(5.57) and (5.64)-(5.65). The ordinary differential equation (5.66) governs the behavior of the state vector $\mathbf{p}(\tau)$. Its right-hand side, however, depends on both $\mathbf{p}(\tau)$ and $\Delta\mathbf{x}(\tau)$. The dependency on $\Delta\mathbf{x}(\tau)$ can be removed by properly fixing the projector's remaining degrees of freedom. The extra conditions build on the requirement that $\Delta\mathbf{x}(\tau)$ must remain bounded. It can be shown that this requires $\Delta\mathbf{x}(\tau)$ to belong to the space spanned by the columns of $\left[\ \mathbf{U}_2(\tau, \mathbf{p}(\tau)) \quad \mathbf{U}_3(\tau, \mathbf{p}(\tau))\ \right]$ (see appendix D), i.e.

$$\Delta\mathbf{x}(\tau) = \left[\ \mathbf{U}_2(\tau, \mathbf{p}(\tau)) \quad \mathbf{U}_3(\tau, \mathbf{p}(\tau))\ \right] \left[\begin{array}{c} \alpha_2(\tau) \\ \alpha_3(\tau) \end{array} \right] \tag{5.67}$$

$$= \mathbf{U}_2(\tau, \mathbf{p}(\tau))\alpha_2(\tau) + \mathbf{U}_3(\tau, \mathbf{p}(\tau))\alpha_3(\tau) \ . \tag{5.68}$$

Using (5.68), $\Delta\mathbf{x}(\tau)$ is now removed from (5.66) by forcing the projector $\mathbf{V}_1^T(\tau, \mathbf{p})$ to satisfy

$$\mathbf{V}_1^T(\tau, \mathbf{p})\mathbf{C}(\tau, \mathbf{p})\mathbf{U}_2(\tau, \mathbf{p}) = 0, \forall(\tau, \mathbf{p}) \in \mathbb{R} \times \mathbb{R}^P \ . \tag{5.69}$$

With (5.62), it also automatically holds that $\mathbf{V}_1^T(\tau, \mathbf{p})\mathbf{C}(\tau, \mathbf{p})\mathbf{U}_3(\tau, \mathbf{p}) = 0$. Furthermore, since $\mathbf{V}_1^T(\tau, \mathbf{p})$ solves (5.64), it is sufficient that $\mathbf{V}_1^T(0, \mathbf{p})\mathbf{C}(0, \mathbf{p})\mathbf{U}_2(0, \mathbf{p}) = 0$ in order for (5.69) to hold. Using (5.57) and (5.66)-(5.69), we finally obtain

$$\frac{d\mathbf{p}}{d\tau} = -\varepsilon\mathbf{V}_1^T(\tau, \mathbf{p})\mathbf{f}(\mathbf{x}_s(\tau, \mathbf{p}), \tau) + O(\varepsilon^2) \tag{5.70}$$

as the P-dimensional system of equations that governs the behavior of the oscillator's state vector $\mathbf{p}$ and, therefore, the motion over the manifold M. Hence, changes in the state vector are driven the projected perturbation term $\varepsilon\mathbf{f}(\mathbf{x}_s(\tau, \mathbf{p}), \tau)$. Hereby, the condition (5.63) states that this projection is onto the hyperplane tangent to the manifold M. Moreover, (5.69) expresses that it runs, in a certain sense, in parallel to the hyperplane spanned by the stable modes of (5.59) augmented with the null space of $\mathbf{C}(\tau, \mathbf{p})$[10].

---

[9] In deriving (5.66), we interchanged the partial derivative $\partial\mathbf{V}_1^T/\partial\tau$ with the full derivative $d\mathbf{V}_1^T/d\tau = \partial\mathbf{V}_1^T/\partial\tau + (\partial\mathbf{V}_1^T/\partial\mathbf{p})\, d\mathbf{p}/d\tau$. This only introduces an error that is $O(\varepsilon^2)$ and, therefore, makes no difference for results up to first order in ε.

[10] Note that this projection is not necessarily done in a direction that is orthogonal to the hyperplane tangent to M. This actually only holds for some special cases.

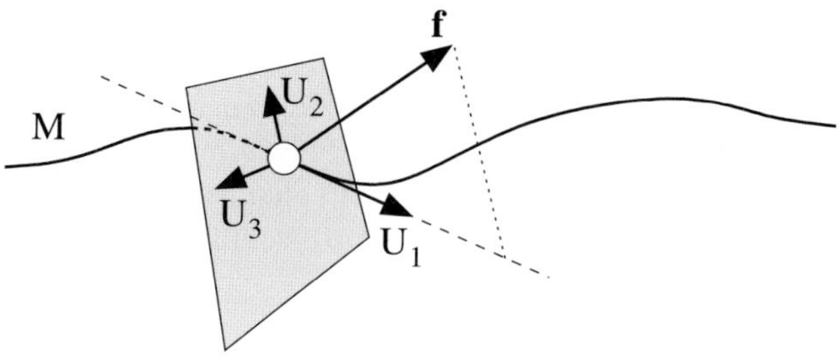

Figure 5.15: The motion over the manifold M is computed by projecting the perturbation terms onto the tangent space to M. This tangent space is spanned by the columns of $\mathbf{U}_1$. The projection runs, in a certain sense, parallel to the space spanned by the columns of $\mathbf{U}_2$ and $\mathbf{U}_3$. This projection is not necessarily orthogonal to the tangent space.

5.5.3 In summary

At this point, we recapitulate the main results so far. Up to first order in ε, the equation that governs the behavior of the oscillator's state vector $\mathbf{p}$ —and as such its motion over M— is

$$\frac{d\mathbf{p}}{d\tau} = -\varepsilon \mathbf{V}_1^T(\tau,\mathbf{p})\mathbf{f}(\mathbf{x}_s(\tau,\mathbf{p}),\tau) \ . \tag{5.71}$$

Hence, changes in the state vector are driven by the projected perturbation term. As illustrated in Fig. 5.15, the perturbation term $\varepsilon\mathbf{f}(\mathbf{x}_s(\tau,\mathbf{p}),\tau)$ is projected onto the tangent space of the manifold M. In most cases, the projection is a non-orthogonal one. The projector $\mathbf{V}_1^T(\tau,\mathbf{p})$ is calculated by solving the homogeneous linear system of DAEs

$$\left(\frac{\partial}{\partial\tau}\mathbf{V}_1^T(\tau,\mathbf{p})\right)\cdot\mathbf{C}(\tau,\mathbf{p}) - \mathbf{V}_1^T(\tau,\mathbf{p})\cdot\mathbf{G}(\tau,\mathbf{p}) = 0 \ . \tag{5.72}$$

Furthermore, its initial value $\mathbf{V}_1^T(0,\mathbf{p})$ satisfies

$$\mathbf{V}_1^T(0,\mathbf{p})\left[\begin{array}{ccc}\mathbf{C}(0,\mathbf{p})\mathbf{U}_1(0,\mathbf{p}) & \mathbf{C}(0,\mathbf{p})\mathbf{U}_2(0,\mathbf{p}) & \mathbf{G}(0,\mathbf{p})\mathbf{U}_3(0,\mathbf{p})\end{array}\right] = \left[\begin{array}{ccc}\mathbf{I}_P & 0 & 0\end{array}\right] \tag{5.73}$$

with $\mathbf{I}_P \in \mathbb{R}^{P\times P}$ the P-dimensional unity matrix. Here, $\mathbf{U}_1(\tau,\mathbf{p})$ and $\mathbf{U}_3(\tau,\mathbf{p})$ are respectively defined by (5.58) and (5.62). The columns of $\mathbf{U}_2(\tau,\mathbf{p})$ contain the stable modes of (5.59), i.e. the solutions that satisfy (5.61). The condition $\mathbf{V}_1^T(0,\mathbf{p})\mathbf{G}(0,\mathbf{p}) \times \mathbf{U}_3(0,\mathbf{p}) = 0$ is imposed in order for $\mathbf{V}_1^T(0,\mathbf{p})$ to be a valid initial condition for solving the system of DAEs (5.72).

Example (Voltage-controlled harmonic oscillator): For the voltage controlled harmonic oscillator in Fig. 5.9, we find, by means of (5.46)-(5.50) and with

$$\mathbf{p} = \left[\begin{array}{ccc} A & \theta & z\end{array}\right]^T \ , \tag{5.74}$$

that

$$\mathbf{U}_1(\tau,\mathbf{p}) = \begin{bmatrix} \cos(\tau+\theta) & -A\sin(\tau+\theta) & 0 \\ \sin(\tau+\theta) & A\cos(\tau+\theta) & 0 \\ 0 & 0 & 1 \\ 0 & 0 & 0 \\ 0 & 0 & 1 \end{bmatrix} . \tag{5.75}$$

Furthermore, with

$$\mathbf{C}(\tau,\mathbf{p}) = \begin{bmatrix} 1 & & & & \\ & 1 & 0 & & \\ & & 1 & & \\ & 0 & 0 & & \\ & & & & 0 \end{bmatrix}, \mathbf{G}(\tau,\mathbf{p}) = \begin{bmatrix} 0 & 1 & 0 & 0 & 0 \\ -1 & 0 & 0 & 0 & 0 \\ 0 & 0 & 0 & 0 & 0 \\ 0 & 0 & -1 & 0 & 1 \\ 0 & 0 & 0 & 1 & 0 \end{bmatrix} , \tag{5.76}$$

we obtain

$$\mathbf{U}_3(\tau,\mathbf{p}) = \begin{bmatrix} 0 & 0 \\ 0 & 0 \\ 0 & 0 \\ 1 & 0 \\ 0 & 1 \end{bmatrix} \tag{5.77}$$

while $\mathbf{U}_2(\tau,\mathbf{p})$ is empty. Using (5.72) and (5.73), the projector $\mathbf{V}_1^T(\tau,\mathbf{p})$ can be shown to equal

$$\mathbf{V}_1^T(\tau,\mathbf{p}) = \begin{bmatrix} \cos(\tau+\theta) & \sin(\tau+\theta) & 0 & 0 & 0 \\ -\frac{\sin(\tau+\theta)}{A} & \frac{\cos(\tau+\theta)}{A} & 0 & 0 & 0 \\ 0 & 0 & 1 & 0 & 0 \end{bmatrix} . \tag{5.78}$$

Hence, the equations that govern the changes of the state vector $\mathbf{p}$ are

$$\frac{dA}{d\tau} = -\cos(\tau+\theta)\frac{I_{M,d}(\tau+\theta,A,z)}{I_{osc}} + \frac{C_g}{2C}\frac{A}{2}\sin(2\tau+2\theta) - \frac{A\sin(\tau+\theta)^2}{Q} \tag{5.79}$$

$$\frac{d\theta}{d\tau} = \frac{\sin(\tau+\theta)}{A}\frac{I_{M,d}(\tau+\theta,A,z)}{I_{osc}} - \frac{C_g}{2C}\sin(\tau+\theta)^2 - \frac{\sin(2\tau+2\theta)}{2Q} \tag{5.80}$$

$$\frac{dz}{d\tau} = -\frac{I_{M,cm}(\tau+\theta,A,z)}{I_{osc}} + \frac{I_b}{2I_{osc}} . \tag{5.81}$$

These equations capture the dynamic behavior of the oscillation amplitude A, the phase θ and the common-mode level z. In (5.79)-(5.81), the differential transistor current $I_{M,d}(\tau+\theta,A,z)$ is computed by substituting the voltages $v_1(\tau) = V_{osc}(z+A\cos(\tau+\theta))$ and $v_1(\tau) = V_{osc}(z-A\cos(\tau+\theta))$ (see the equations (5.12)-(5.15), (5.27) and (5.46)-(5.50)) into (5.21). Using a likewise procedure, we can obtain the common mode transistor current $I_{M,cm}(\tau+\theta,A,z)$. Note that the phase θ always appears in conjunction with the time variable τ.

The equations (5.79)-(5.81) capture that part of the oscillator's dynamic behavior that is, for most practical purposes, of greatest interest. They describe the transient dynamics of the oscillator's amplitude, phase and common-mode behavior. Compared to the original system of five equations, i.e. (5.28)-(5.32), the number of equations needed

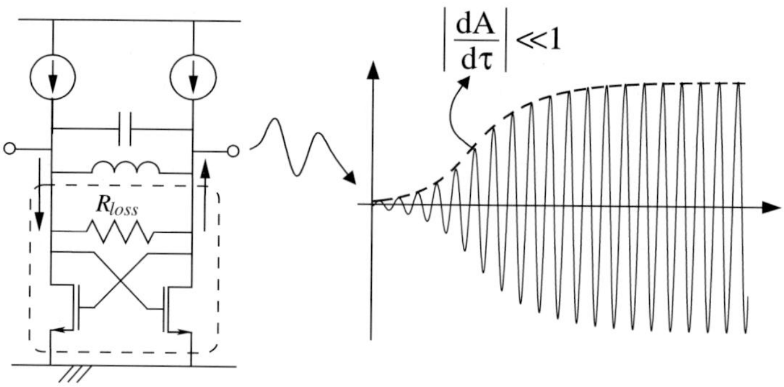

Figure 5.16: Startup behavior of a harmonic oscillator. Here, amplitude changes, i.e. changes in the value of the state vector **p**, are slow as compared to the oscillations.

to describe the oscillator's dynamic behavior has now been reduced to three. Hence, in a certain sense, the procedure above amounts to nonlinear model order reduction. Furthermore, as will be discussed below, the equations (5.79)-(5.81), when properly processed, can be solved using a time step that is much larger that the one needed to solve the original system of equations. ▲

5.6 Averaging

The ordinary differential equations (ODEs) (5.71) that govern the behavior of the state vector **p** belongs to the class of ODEs structured as

$$\frac{d\mathbf{p}}{d\tau} = \varepsilon \mathbf{h}(\mathbf{p},\tau) \tag{5.82}$$

where, in this case,

$$\mathbf{h}(\mathbf{p},\tau) = -\mathbf{V}_1^T(\tau,\mathbf{p})\,\mathbf{f}(\mathbf{x}_s(\tau,\mathbf{p}),\tau) \ . \tag{5.83}$$

Since the perturbation term $\varepsilon\mathbf{f}$ is small, so is the right-hand side of (5.82). Hence, the time derivative of the process $\mathbf{p}(\tau)$ is small in magnitude. This in turn implies that it takes a long time for the state vector $\mathbf{p}(\tau)$ to substantially change value. Apart from some small fluctuations, $\mathbf{p}(\tau)$ is therefore inherently a slow-varying process. It only changes significantly over a (normalized) time interval that is at least $1/\varepsilon$ in order of magnitude.

Example (Harmonic Oscillator): Fig. 5.16 shows the startup behavior of a harmonic oscillator. For the partitioning indicated in the figure, the state vector $\mathbf{p} = \begin{bmatrix} A & \theta \end{bmatrix}^T$ is 2-dimensional, containing amplitude and phase (since there is no control voltage, there is no common-mode component in **p**). It is observed that amplitude changes, i.e. changes in the value of the state vector **p**, occur on a time scale that is slow compared to that of the oscillations. Hence, with respect to normalized time we find that $|dA/d\tau| \ll 1$. This is in correspondence with (5.82). ▲

Equation (5.82) can be solved efficiently using averaging [Bogo61, Verhu96][11]. Averaging exploits the fact that solutions of (5.82) only vary substantially on a time scale proportional to $1/\varepsilon$. Hence, these solutions are dominated by a slow-varying component. Averaging techniques allow us to extract those parts of (5.82) that produce this dominant slow-varying component. This can be interpreted as "averaging the fast-varying components out of the equations". The averaged system of ODEs that results serves as an excellent starting point for noise and stability analysis, for efficient simulation and for extraction of compact behavioral models.

Technically, averaging involves the construction of a time-varying transformation of variables $\mathbf{p} \rightarrow \overline{\mathbf{p}}$. Hereby, the values of the original vector $\mathbf{p}$ and the newly introduced vector $\overline{\mathbf{p}}$ are related to each other by means of

$$\mathbf{p} = \overline{\mathbf{p}} + \varepsilon\tilde{\mathbf{h}}(\overline{\mathbf{p}}, \tau) + O\left(\varepsilon^2\right) \ . \tag{5.84}$$

Note that the difference $\mathbf{p} - \overline{\mathbf{p}}$ is a quantity that is $O(\varepsilon)$, i.e. small, in order of magnitude. Therefore, the transformation (5.84) is called a *near-identity* transformation. In this transformation, the term $\tilde{\mathbf{h}}(\overline{\mathbf{p}}, \tau)$ is chosen so that

1. the behavior of the transformed state vector $\overline{\mathbf{p}}(\tau)$ only contains slow-varying components.
2. the process $\varepsilon\tilde{\mathbf{h}}(\overline{\mathbf{p}}(\tau), \tau)$ captures the fast-varying fluctuations of $\mathbf{p}(\tau)$ while remaining bounded over the time interval of interest. The latter is, in our case, infinite.

By construction, the process $\overline{\mathbf{p}}(\tau)$ captures the slow-varying long-term characteristics of the state vector $\mathbf{p}(\tau)$. Therefore, it is called the *averaged process* corresponding to $\mathbf{p}(\tau)$. In many cases, this average is the quantity of main interest. The equations that govern this behavior are obtained by substituting the transformation (5.84) into the starting equation (5.82).

Example (Harmonic oscillator): Fig. 5.17(a) shows a typical harmonic oscillator amplitude startup characteristic. Here, $A(\tau)$ is solved from a set of equations structured as (5.82). These equations are obtained by means of the perturbation analysis outlined in the previous section. Fig. 5.17(b) illustrates the decomposition of $A(\tau)$ into a dominant slow-varying component $\overline{A}(\tau)$ and a small fast-varying fluctuation $\varepsilon h(\overline{A}(\tau), \tau)$. Note that, for sake of a clear illustration, the magnitude of the latter component has been exaggerated. In most practical cases, we are mainly interested in the behavior of the averaged amplitude $\overline{A}(\tau)$. The equations governing this averaged behavior are obtained by substituting, amongst others, the transformation of variables

$$A = \overline{A} + \varepsilon\tilde{h}(\overline{A}, \tau) \tag{5.85}$$

into the starting-point equations obtained from the perturbation analysis. ▲

[11] The theory of averaging can also be seen as a particular application of the theory of continuous transformation (Lie) groups [Cohn65, Firk00]

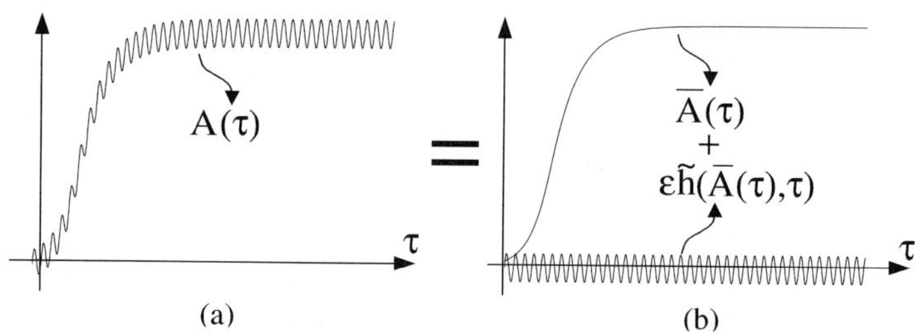

Figure 5.17: Typical harmonic oscillator startup characteristic. (a) The oscillation amplitude $A(\tau)$ solved from the starting-point equation (5.82). (b) The amplitude startup behavior can be decomposed into a dominant slow-varying component $\overline{A}(\tau)$ with a small fast-varying fluctuation $\varepsilon h(\overline{A}(\tau),\tau)$ on top of it.

By substituting (5.84) into (5.82), we obtain

$$\frac{\mathrm{d}\overline{\mathbf{p}}}{\mathrm{d}\tau} = \varepsilon\left(\mathbf{h}(\overline{\mathbf{p}},\tau) - \frac{\partial\tilde{\mathbf{h}}(\overline{\mathbf{p}},\tau)}{\partial\tau}\right) + O(\varepsilon^2) \ . \tag{5.86}$$

For $\overline{\mathbf{p}}$ to meet the conditions mentioned above, an averaging transformations imposes $\tilde{\mathbf{h}}(\overline{\mathbf{p}},t)$ to satisfy

$$\frac{\partial\tilde{\mathbf{h}}(\overline{\mathbf{p}},\tau)}{\partial\tau} = \mathbf{h}(\overline{\mathbf{p}},\tau) - M[\mathbf{h}](\overline{\mathbf{p}},\tau) \ . \tag{5.87}$$

Here, the operator $M[\cdot]$ is called an *averaging operator*. $M[\cdot]$ acts as a lowpass filter that removes the fast-varying *explicit* dependence on τ of $\mathbf{h}(\overline{\mathbf{p}},\tau)$. Hence, in computing $M[\mathbf{h}](\overline{\mathbf{p}},\tau)$, the value of $\overline{\mathbf{p}}$ is kept constant. With (5.87), (5.86) becomes

$$\frac{\mathrm{d}\overline{\mathbf{p}}}{\mathrm{d}\tau} = \varepsilon M[\mathbf{h}](\overline{\mathbf{p}},\tau) = \varepsilon\overline{\mathbf{h}}(\overline{\mathbf{p}},\tau) \ . \tag{5.88}$$

This relation is accurate up to first order in ε. By careful selection of the lowpass averaging operator $M[\cdot]$, the transformed equation (5.88) has a right-hand side that only contains slow-varying components. As a result, the same holds for the any process $\overline{\mathbf{p}}(\tau)$ that solves this equation. Equation (5.88) is called the *averaged equation* corresponding to (5.82). Because of the slow-varying nature of its right-hand side, it can be solved very efficiently using a large integration time step. Furthermore, it makes up a valuable starting point for further analysis or for the construction of compact behavioral models.

Selecting the averaging operator $M[\cdot]$ is a degree of freedom that can be used to optimize the properties of the averaged ODEs (5.88). This allows us to exploit particular properties of the right-hand side $\varepsilon\mathbf{h}(\mathbf{p},\tau)$ of the starting-point equation (5.82). Convenient averaging operators are

$$M_{traditional}[\mathbf{h}](\overline{\mathbf{p}},\tau) = \frac{1}{T_M}\int_{s-T_M/2}^{s+T_M/2}\mathbf{h}(\overline{\mathbf{p}},s)\mathrm{d}s \tag{5.89}$$

$$M_{ideal}[\mathbf{h}](\overline{\mathbf{p}},\tau) = \frac{1}{T_M}\int_{-\infty}^{+\infty}\frac{\sin(\pi(\tau-s)/T_M)}{\pi(\tau-s)/T_M}\mathbf{h}(\overline{\mathbf{p}},s)\mathrm{d}s \ . \tag{5.90}$$

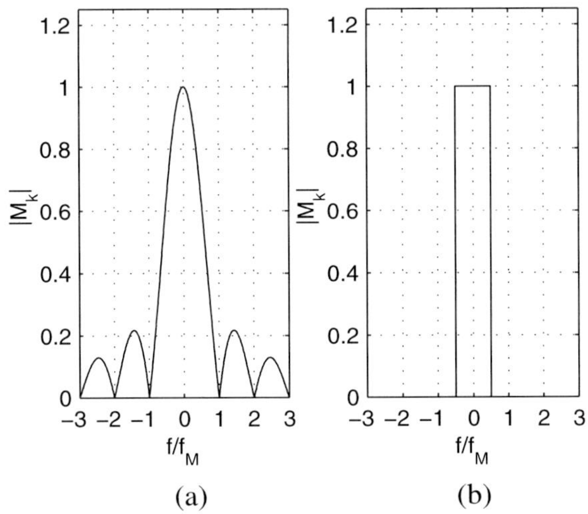

Figure 5.18: Frequency-domain characteristic corresponding to (a) the averaging operator (5.89) and (b) the averaging operator (5.90). Here, $f_M = 1/T_M$.

Here, in both cases, T_M is the time constant of the (lowpass) averaging operator (filter). The operator (5.89) corresponds to a time average over the time span T_M. This time average coincides with our intuitive notions on averaging. The operator (5.89) is the one that typically appears in traditional literature on averaging [Bogo61, Verhu96]. Fig. 5.18(a) shows its frequency-domain characteristic. The second operator (5.90) represents the ideal lowpass filtering operation over the range $f = [-1/2T_M, 1/2T_M]$. Its frequency characteristic is shown in Fig. 5.18(b). This operator is *ideal* in a sense that it realizes a perfect separation of slow- and fast-varying signal components. Numerically, however, (5.90) is sometimes more difficult or expensive to evaluate than (5.89).

We conclude this discussion with some notes on the term $\varepsilon\tilde{\mathbf{h}}(\overline{\mathbf{p}},\tau)$ in the transformation (5.84). This term captures the fast fluctuations in $\mathbf{p}(\tau)$ and is solved from (5.87). With $M[\cdot]$ chosen so that the right-hand side of (5.87) has no DC component with respect to the variable τ, $\varepsilon\tilde{\mathbf{h}}(\overline{\mathbf{p}},\tau)$ is guaranteed to remain bounded $\forall\tau$. However, since (5.87) is a partial differential equation, we need a boundary condition to fix $\varepsilon\tilde{\mathbf{h}}(\overline{\mathbf{p}},\tau)$ uniquely. Bearing in mind the interpretation of $\varepsilon\tilde{\mathbf{h}}(\overline{\mathbf{p}},\tau)$ as the small fast-varying fluctuations of $\mathbf{p}(\tau)$, it is most sensible to impose the term $\varepsilon\tilde{\mathbf{h}}(\overline{\mathbf{p}},\tau)$ to be void of a DC component with respect to the variable τ, or

$$\lim_{T\to\infty}\frac{1}{T}\int_{-T/2}^{+T/2}\tilde{\mathbf{h}}(\overline{\mathbf{p}},s)\mathrm{d}s = 0 \ . \tag{5.91}$$

Equations (5.87) and (5.91) uniquely fix the term $\varepsilon\tilde{\mathbf{h}}(\overline{\mathbf{p}},\tau)$ and, therefore, the averaging transformation (5.84). Fig. 5.19 summarizes the steps needed to compute the compo-

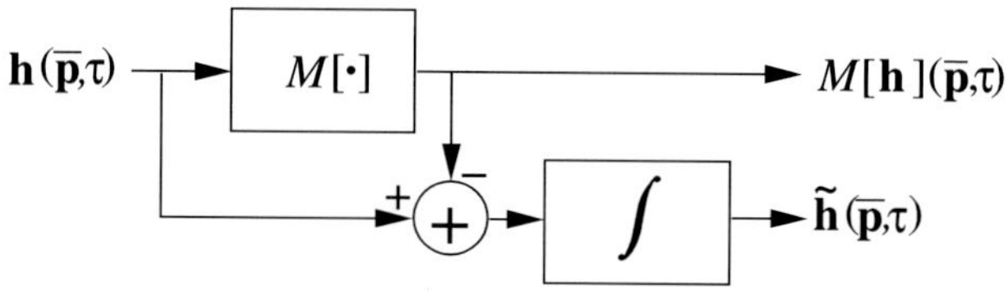

Figure 5.19: Steps needed to compute the components of the averaging transformation and the associated averaged equations.

nents of the averaging transformation and the associated averaged equations.

Example (Voltage-controlled harmonic oscillator): This example derives the averaged equations that describe the behavior of the amplitude, phase and common-mode level of the voltage-controlled oscillator in Fig. 5.9. Its intent is to illustrate the procedure to do so and to provide some numerical results that demonstrate the accuracy of the behavior predicted by these averaged equations. Note that further numerical results are presented in section 5.8.

In previous examples, the behavior of the oscillator's state vector $\mathbf{p} = \begin{bmatrix} A & \theta & z \end{bmatrix}^T$ was shown to be governed by (5.79)-(5.81). The right-hand sides of these equations are 2π-periodic in τ. Hence, applying the ideal lowpass averaging operator (5.90) corresponds to selecting the DC component of their Fourier series expansion with respect to τ. This is equivalent to the use of the averaging operator (5.89) for $T_M = 2\pi$. The averaged equations associated with (5.79)-(5.81) are then found to equal

$$\frac{d\overline{A}}{d\tau} = -\frac{1}{2\pi}\int_{-\pi}^{\pi}\cos(s)\frac{I_{M,d}(s,\overline{A},\overline{z})}{I_{osc}}ds - \frac{\overline{A}}{2Q} \quad (5.92)$$

$$\frac{d\overline{\theta}}{d\tau} = -\frac{C_g}{4C} \quad (5.93)$$

$$\frac{d\overline{z}}{d\tau} = -\frac{1}{2\pi}\int_{-\pi}^{\pi}\frac{I_{M,cm}(s,\overline{A},\overline{z})}{I_{osc}}ds + \frac{I_b}{2I_{osc}} \quad . \quad (5.94)$$

Note that none of the averaged right-hand sides depends on the phase θ. This is due to the periodic dependence on τ of all right-hand side terms in (5.92)-(5.94) combined with the fact that θ always appears in conjunction with τ as $\tau + \theta$.

Fig. 5.20 demonstrates the accuracy of the results obtained from (5.92)-(5.94) by comparing the predicted amplitude and common-mode startup behavior with results obtained by a direct (SPICE-like[12]) solution of the original circuit equations (5.28)-(5.32). The computations are performed for an oscillator with $Q \approx 8$. The results in Fig. 5.20 clearly show that (5.92)-(5.94) accurately predicts the behavior of the oscillator's averaged amplitude and common-mode behavior. Moreover, the lower plot in Fig. 5.20

[12] In this text, a "SPICE-like" solver corresponds either to a Runge-Kutta method or an algorithm based on numerical differentiation formulas.

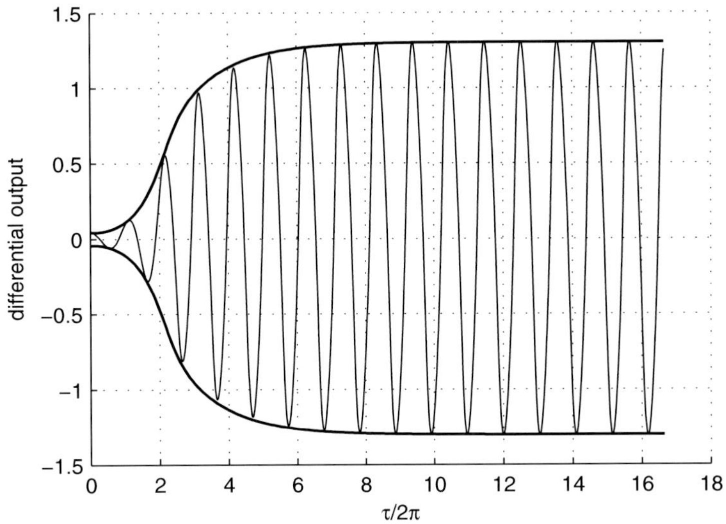

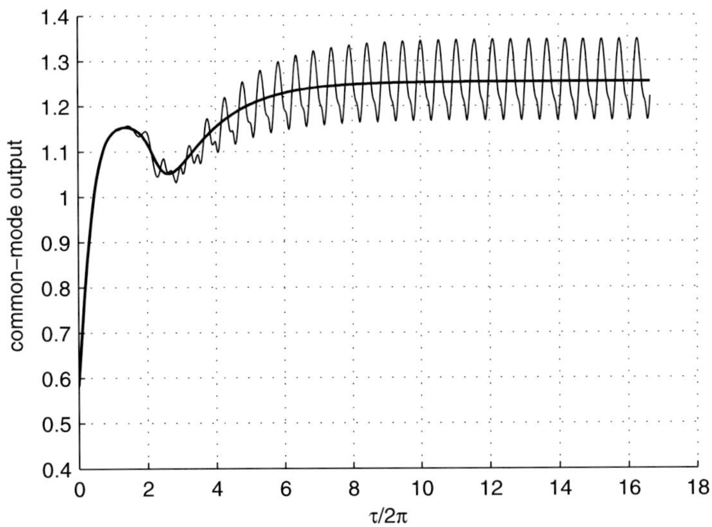

Figure 5.20: Oscillator startup behavior as respectively obtained from the averaged equations (5.92)-(5.94) (thick slow-varying envelopes/line) and from the original equations (5.28)-(5.32) (thin oscillating lines). The upper figure plots the differential output $x_d(\tau)$ while the lower one plots the common-mode output $x_{cm}(\tau)$.

emphasizes the fact that solutions to (5.92)-(5.94) average out the high-frequency components in the oscillator's dynamic behavior. With regard to computational complexity, (5.92)-(5.94) can be solved using a much larger timestep than the starting point equations (5.28)-(5.32) since they directly model the behavior of the dominant slow-varying (averaged) components of the oscillator's dynamics. As a consequence, for this example, solving (5.92)-(5.94) runs 10 times faster than solving the original equations (5.28)-(5.32).

The averaged equations (5.92)-(5.94) can be used for the computation of oscillator's operating point. The oscillator's (averaged) operating-point amplitude $\overline{A}_{op}$ and common-mode level $\overline{z}_{op}$ are obtained by equaling $\mathrm{d}\overline{A}/\mathrm{d}\tau$ and $\mathrm{d}\overline{z}/\mathrm{d}\tau$ to zero in (5.92) and (5.94). Hence, we solve $\overline{A}_{op}$ and $\overline{z}_{op}$ from

$$\frac{1}{2\pi}\int_{-\pi}^{\pi}\cos(s)\frac{I_{M,d}(s,\overline{A}_{op},\overline{z}_{op})}{I_{osc}}\mathrm{d}s+\frac{\overline{A}_{op}}{2Q} = 0 \tag{5.95}$$

$$\frac{1}{2\pi}\int_{-\pi}^{\pi}\frac{I_{M,cm}(s,\overline{A}_{op},\overline{z}_{op})}{I_{osc}}\mathrm{d}s-\frac{I_b}{2I_{osc}} = 0\ . \tag{5.96}$$

For the numerical values used to obtain the results in Fig. 5.20, solving (5.95)-(5.96) yields

$$\overline{A}_{op} = 1.30 \tag{5.97}$$

$$\overline{z}_{op} = 1.25 \tag{5.98}$$

By linearizing (5.92)-(5.94) in the neighborhood of these operating-point values, we obtain an affine system of differential equations that allows us to study the oscillator's local dynamic behavior, e.g. the stability of the operating point. As demonstrated in section 5.8, these linearized averaged equations can also serve as compact behavioral models useful for architectural-level exploration and verification.

In some cases, the small variations that come on top of the averaged behavior are also of interest. They are, for instance, responsible for the high-order harmonics in the oscillator's output spectrum. Up to first order, these variations are determined by the second term in the averaging transformation (5.84). This term is solved from (5.87) and (5.91). This yields, for instance,

$$z(t)-\overline{z}_{op} = -\sum_{n\neq 0}\frac{\sin(n\tau+n\theta)}{\pi n}\int_{-\pi}^{\pi}\cos(ns)\frac{I_{M,cm}(s,\overline{A}_{op},\overline{z}_{op})}{I_{osc}}\mathrm{d}s \tag{5.99}$$

as an expression for the oscillator's steady-state common-mode behavior. Fig. 5.21 compares the steady-state common-mode behavior predicted using (5.99) with that computed by solving (5.28)-(5.32). The figure illustrates a clear correspondence between both results. Deviations are due to perturbation effects of second and higher order that are not accounted for by the first-order analysis presented in this text. In magnitude, the RMS value of this error equals about $1/Q$ of the RMS value of the high-frequency steady-state common-mode variations $z(\tau)-\overline{z}_{op}$. Hence, in this example, due to the low Q-factor, use of first-order analysis introduces an error with an energy that lies about a factor 10 below that contained in the high-frequency variations.

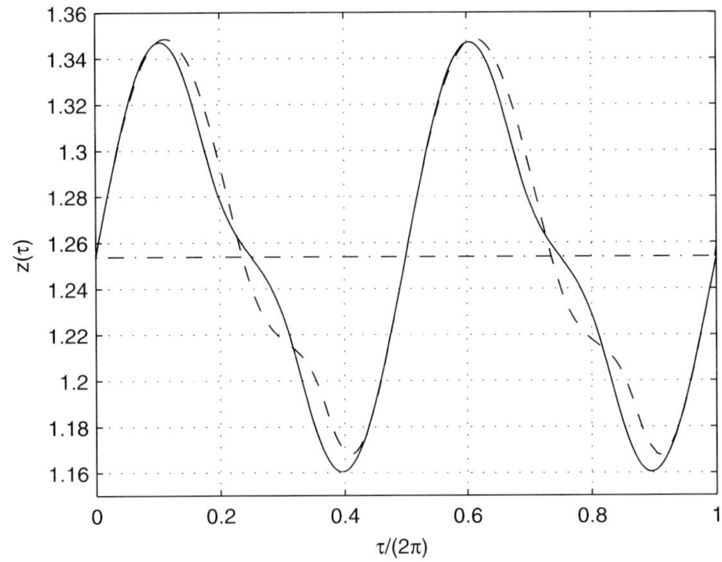

Figure 5.21: Common-mode steady-state variations respectively obtained using first-order averaging (solid line) and by solving (5.28)-(5.32) (dashed line). The dashdot line denotes the averaged operating-point common-mode level $\bar{z}_{op}$.

By taking a closer look at the harmonic content of the common-mode steady-state variations, it is observed that the error energy is mainly located at high frequencies. Table 5.1 lists the energy contained in the first couple of even harmonics (the odd ones are negligible). Values are given based on results respectively predicted by first-order averaging and SPICE-like evaluation of (5.28)-(5.32). These results clearly indicate good agreement for the low-order harmonics. Significant deviations only occur from the 6th harmonic on.

A final example that demonstrates the accuracy of the equations (5.92)-(5.94) follows from (5.93). This equation models the impact that perturbations have on the oscillator's phase and frequency. For example, the averaged unnormalized effective oscillation frequency equals

$$\omega_{effective} \approx \omega_0 \left(1 + \frac{d\bar{\theta}}{d\tau}\right) = \omega_0 \left(1 - \frac{C_g}{4C}\right) \tag{5.100}$$

where $\omega_0 = 1/\sqrt{LC}$. Up to first order, the small (transistor) capacitances C_g (see Fig. 5.9) decrease the oscillation frequency by $\omega_0 C_g/4C$. This corresponds with

$$\omega_{effective} = \frac{1}{\sqrt{L(C + C_g/2)}} = \frac{1}{\sqrt{LC}} \frac{1}{\sqrt{1 + C_g/2C}} \approx \omega_0 \left(1 - \frac{C_g}{4C}\right) \ . \tag{5.101}$$

Here, instead of being treated as perturbations, the C_g are taken into account as part of the resonant tank capacitance. ▲

Harmonic	Averaged	Spice-like
0	1.96 dB	1.98 dB
2	-21.56 dB	-21.58 dB
4	-34.04 dB	-35.26 dB
6	-45.92 dB	-43.44 dB
8	-62.76 dB	-54.47 dB

Table 5.1: Harmonic content of the steady-state common-mode waveform respectively obtained from first-order averaging results and SPICE-like simulation results. Due to higher-order perturbation components, results start to deviate significantly from the 6th harmonic on.

In summary, averaging is a powerful method for the analysis and modeling of an oscillator's dynamic behavior. It allows us to obtain numerical results as well as symbolic expressions. In the sections to come, we demonstrate its practical use for both the analysis of oscillator phase behavior and for behavioral model extraction. Furthermore, it is shown that averaging not only simplifies the analysis of an oscillator's deterministic dynamics, e.g. settling and stability, but also that of its stochastic behavior, e.g. phase noise.

5.7 Oscillator phase (noise) analysis

The methods outlined above can be used to model an oscillator's phase (noise) behavior [Dem00a, Dem00b, Dem02, Kae90, Haji98, Vana02d]. For this particular application, the perturbation analysis presented in section 5.5 reduces to the one in [Dem00a, Dem00b]. Hence, this work embeds existing theories on oscillator phase behavior within a more global framework.

Phase (noise) behavior concerns the prediction of oscillator phase shifts induced by the injection of small (noise) currents and/or voltages into the circuit. With regard to this particular application, Fig. 5.22(a) illustrates how to partition the oscillator circuit. The core system corresponds to the input-free oscillator while the perturbation terms are associated with the injected currents and/or voltages, e.g. noise. For a single external input $u(\tau) \in \mathbb{R}$, the (normalized) circuit equations corresponding to this setup are written as

$$\frac{\mathrm{d}}{\mathrm{d}\tau}\mathbf{c}(\mathbf{x}) + \mathbf{g}(\mathbf{x}) + \varepsilon\mathbf{b}(\mathbf{x})u(\tau) = 0 \ . \tag{5.102}$$

Here, $\mathbf{x} \in \mathbb{R}^N$ contains the circuit's state variables. The injected signal $u(\tau)$ can be either a deterministic signal or a noise process. The strength (standard deviation) of this signal (random process) is assumed to be well below that of the current and voltage levels in the input-free system (i.e. for $u(\tau) = 0$). If not, $u(\tau)$ can no longer be considered a perturbation. As usual, a perturbation variable ε is introduced to mark the perturbation terms, labeling them as being small.

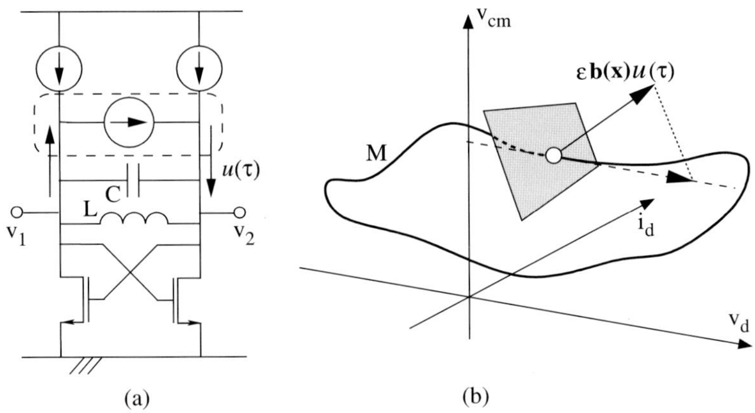

Figure 5.22: (a) With regard to the impact of a small external input signal $\varepsilon u(\tau)$ on the oscillator's phase behavior, we partition the system into the input-free oscillator (the core system) and the set of perturbing input signals. (b) The solutions of the input-free oscillator are attracted to a 1-dimensional orbit. The perturbing input signal causes the oscillator to (phase) shift along its own orbit.

5.7.1 Capturing oscillator phase behavior

If the input-free oscillator (the core system) is mono-stable, then all its steady-state solutions are phase-shifted copies of each other, i.e.

$$\mathbf{x}_s(\tau, \mathbf{p}) = \mathbf{x}_{osc}(\tau + \theta) \quad . \tag{5.103}$$

Here, $\mathbf{x}_{osc}(\tau)$ is a T-periodic solution of (5.102) for $\varepsilon = 0$. All other processes that solve this unperturbed equation are attracted to the 1-dimensional orbit

$$M = \{\mathbf{x}_{osc}(\tau) | \tau \in \mathbb{R}\} \tag{5.104}$$

traced by the steady-state solution $x_{osc}(\tau)$. In (5.103), $\mathbf{p} = [\theta]$ is the core system's 1-dimensional state vector. Hereby, θ represents the oscillator's phase which is, for instance, a measure for the positions in time of the oscillation signal's zero crossings.

As discussed in section 5.5, the perturbation term $\varepsilon \mathbf{b}(\mathbf{x})u(\tau)$ forces the oscillator to move over the manifold M. In this case, this is translated as: the oscillator starts to shift along its own orbit. This behavior is captured by means of slow-varying changes in the oscillator's phase $\theta(\tau)$. As illustrated in Fig. 5.22(b), the equation that governs the oscillator's phase behavior is obtained by projecting the perturbation term $\varepsilon \mathbf{b}(\mathbf{x})u(\tau)$ onto the tangent line to the orbit M. This yields

$$\frac{d\theta}{d\tau} = -\varepsilon \mathbf{v}_1^T(\tau + \theta)\, \mathbf{b}(\mathbf{x}_{osc}(\tau + \theta))\, u(\tau) \quad . \tag{5.105}$$

Here, $\mathbf{v}_1^T(\tau+\theta) \in \mathbb{R}^{1\times N}$ corresponds to the projector $\mathbf{V}_1^T(\tau,\mathbf{p})$ discussed in the sections 5.5.2 and 5.5.3. In the literature on phase noise analysis [Dem00a, Dem00b], $\mathbf{v}_1(\tau)$ is called the oscillator's *perturbation projection vector* (PPV). Furthermore, the function

$$\Gamma(\tau) = -\mathbf{v}_1^T(\tau)\,\mathbf{b}\left(\mathbf{x}_{osc}(\tau)\right) \tag{5.106}$$

has been termed the *impulse sensitivity function* (ISF) [Haji98] associated with the input source $u(\tau)$. This ISF is periodic with respect to τ. Using the ISF, (5.105) is written as

$$\frac{\mathrm{d}\theta}{\mathrm{d}\tau} = \varepsilon\Gamma(\tau+\theta)u(\tau) \ . \tag{5.107}$$

The averaged phase behavior, i.e. the dominant slow-varying component that often grows unbounded, is then determined by

$$\frac{\mathrm{d}\overline{\theta}}{\mathrm{d}\tau} = \varepsilon M_{ideal}\left[\Gamma(\tau+\overline{\theta})u(\tau)\right] \ . \tag{5.108}$$

Here, we applied the ideal averaging operator (5.90) which, in this case, is often the most convenient one to proceed with. In many cases, the equations (5.107) and (5.108) capture the oscillator behavior that is of main interest.

5.7.2 Practical application: oscillator injection locking

We illustrate the practical application of (5.107) and (5.108) by means of the study of an oscillator's injection locking behavior. To make the example more concrete, results are elaborated for the harmonic oscillator in Fig. 5.22(a). It is, however, straightforward to generalize them to other oscillator circuits. Moreover, all results that follow are described by means of symbolic expressions. This illustrates a powerful feature of the framework here presented: the ability to derive (approximate) symbolic expressions to describe the often complex nonlinear phenomena.

In what follows, we study what happens when the harmonic oscillator in Fig. 5.22(a) is injected with a weak sinusoidal current

$$u(t) = I_u \sin\left((\omega_0+\Delta\omega)t\right) = I_u \sin\left((1+\Delta\omega/\omega_0)\tau\right) \ . \tag{5.109}$$

Here, ω_0 is the oscillator's free-running frequency while, as usual, $\tau = \omega_0 t$ is the normalized time variable. I_u is the amplitude of the injected current. Furthermore, the frequency $\omega_0+\Delta\omega$ of the injected signal $u(t)$ is assumed to deviate only slightly from ω_0, i.e. $|\Delta\omega/\omega_0| \ll 1$. The SPICE simulation result in Fig. 5.23 illustrates the kind of behavior that occurs: the oscillator's output signal phase-locks onto the injected signal. In literature, this phenomenon is known as injection locking [Adl46, Jez74]: an oscillator locks, both in phase and frequency, onto an externally injected signal. Injection locking underlies a number of phase and frequency modulators. It is also the working principle for systems of coupled oscillators. In what follows, we check whether (5.107) correctly predicts injection locking to occur.

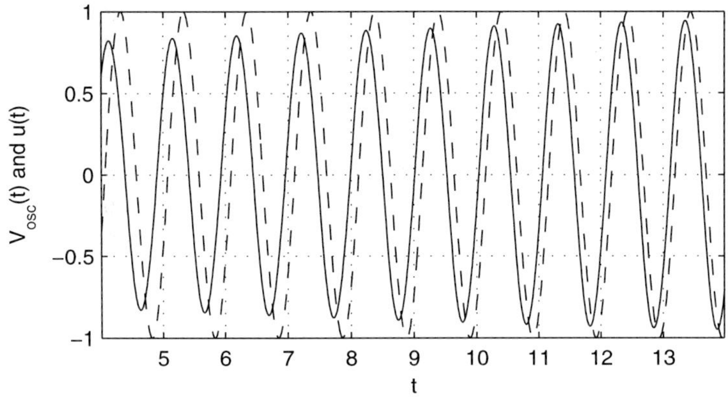

Figure 5.23: SPICE simulation result of a harmonic oscillator injected with a sinewave. The oscillator's output signal (solid line) phase-locks onto the injected signal (dashed line). Both the oscillator and the injected signal are normalized to have their amplitudes (nearly) equal to 1.

If the resonant tank of the harmonic oscillator in Fig. 5.22(a) has a sufficiently high Q-factor, then it is derived in appendix E that

$$v_d(\tau) \approx A\cos(\tau+\theta(\tau)) \tag{5.110}$$

$$\Gamma(\tau) \approx \frac{\sin(\tau)}{I_{osc}} \tag{5.111}$$

Here, $v_d = v_1 - v_2$ is the differential voltage over the resonant tank's capacitor while A is a is the oscillator's fixed operating point amplitude. Furthermore, $I_{osc} = CA\omega_0 = A/L\omega_0$ is the amplitude of the current circulating in the resonant tank. Substituting (5.109) and (5.111) into (5.107) yields

$$\frac{d\theta}{d\tau} = \frac{I_u}{I_{osc}} \sin(\tau+\theta)\sin\left((1+\Delta\omega/\omega_0)\tau\right) \ . \tag{5.112}$$

In order for the right-hand side of this equation to be small, i.e. in order for the results from the perturbation analysis to be valid, it must hold that $I_u/I_{osc} \ll 1/Q$ where Q is the quality factor of the resonant tank. Hence, the injected current must be much smaller than the tank current. If this is satisfied, then the equation (5.112) governs the behavior of the harmonic oscillator's phase changes induced by the injection of a sinusoidal current.

In order to solve the phase equation (5.112), we apply the averaging framework discussed in section 5.6. Use of the ideal lowpass averaging operator (5.90) and the assumption that $|\Delta\omega/\omega_0| \ll 1$, yields

$$\frac{d\bar{\theta}}{d\tau} = \frac{1}{2}\frac{I_u}{I_{osc}} \cos\left(\frac{\Delta\omega}{\omega_0}\tau - \bar{\theta}\right) \tag{5.113}$$

as the equation governing the averaged phase behavior $\overline{\theta}(\tau)$. This equation in further simplified by introducing

$$\psi(\tau) = \frac{\Delta\omega}{\omega_0}\tau - \overline{\theta}(\tau) \ . \tag{5.114}$$

If we substitute this into (5.114), then we obtain

$$\frac{d\psi}{d\tau} = \frac{\Delta\omega}{\omega_0} - \frac{I_u}{2I_{osc}}\cos(\psi) \ . \tag{5.115}$$

This equation models fairly accurately what happens when a harmonic oscillator is injected with a small sinusoidal current.

What about injection locking? Well, if

$$\left|\frac{\Delta\omega}{\omega_0}\right| < \frac{I_u}{2I_{osc}} \ , \tag{5.116}$$

then the right-hand side of (5.115) contains a stable fixed point

$$\psi_{op} = -\arccos\left(\frac{\Delta\omega}{\omega_0}\frac{2I_{osc}}{I_u}\right) \ . \tag{5.117}$$

By means of (5.114), we then find

$$\overline{\theta}_{op}(\tau) = \frac{\Delta\omega}{\omega_0}\tau - \arccos\left(\frac{\Delta\omega}{\omega_0}\frac{I_u}{2I_{osc}}\right) \ . \tag{5.118}$$

As a result, the oscillator's effective oscillation frequency equals

$$\omega_{effective} = \omega_0\left(1 + \frac{d\overline{\theta}_{op}}{d\tau}\right) = \omega_0 + \Delta\omega \ . \tag{5.119}$$

Hence, the effective (averaged) oscillation frequency locks onto that of the injected sinewave. Here, (5.116) provides an expression for the locking range, i.e. the range of input-frequencies for which frequency locking is accomplished. Also, when in lock, the phase difference between the oscillation signal and the injected signal remains constant at $\pi/2 - \arccos(2I_{osc}\Delta\omega/I_u\omega_0)$. Again, note the simplicity whereby these symbolic expressions are derived by means of the methods here presented.

Fig. 5.24 plots the averaged phase difference between the oscillation signal and the injected signal as solved from (5.113) for $\Delta\omega = 0$. The initial phase difference gradually disappears as the oscillator locks onto the injected sinewave.

5.7.3 Averaging in the presence of random perturbations

The previous example illustrates how averaging facilitates solving the phase equation (5.107) (or equivalently (5.105)) for deterministic input signals. Such a simplification is also possible when the injected signals are random in nature. In what follows, we discuss oscillator phase noise theory. Phase noise involves an oscillator's random phase shifts due to injection of noise in the circuit. As illustrated in Fig. 5.25, this results, for

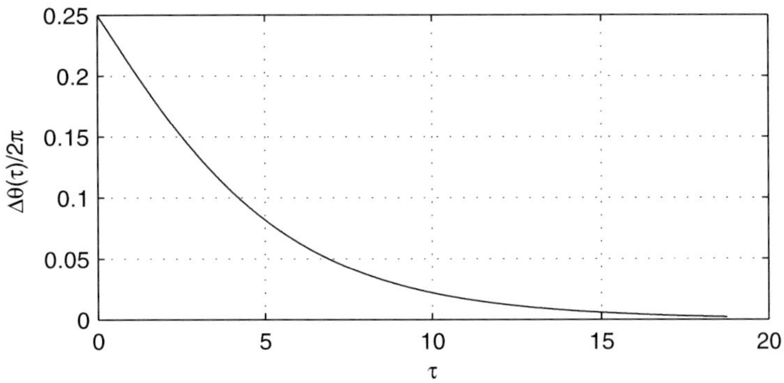

Figure 5.24: The (averaged) difference between the phase of the harmonic oscillator and that of the injected signal versus (normalized) time. Here, the injected signal runs at the oscillator's free-running frequency ($\Delta\omega = 0$).

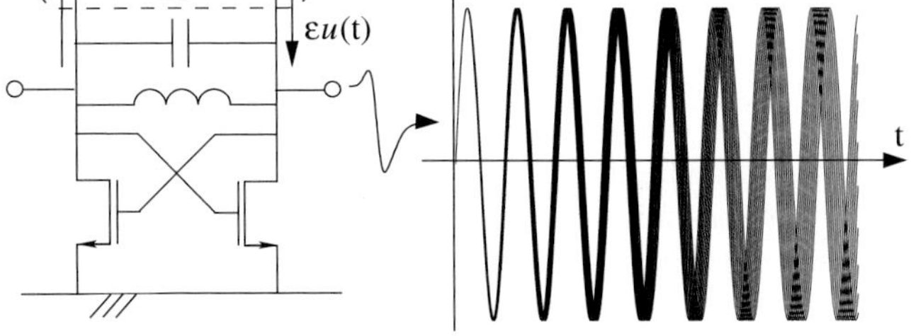

Figure 5.25: Phase noise involves an oscillator's random phase shifts due to injection of noise in the circuit. This results, for instance, in an ever increasing uncertainty on the positions of the zero crossings.

instance, in an ever increasing uncertainty on the positions of the zero crossings.

Let us be given a stationary zero-mean Gaussian noise process $u(\tau)$. This noise process is entirely characterized by its autocorrelation function

$$\Phi_{uu}(r) = \mathrm{E}\{u(\tau)u(\tau - r)\} \ . \tag{5.120}$$

By introducing

$$n(\tau, \theta) = \Gamma(\tau + \theta)u(\tau) \ , \tag{5.121}$$

we can write the oscillator's phase (noise) equation (5.107) as

$$\frac{\mathrm{d}\theta}{\mathrm{d}\tau} = \varepsilon n(\tau, \theta) \ . \tag{5.122}$$

Because of the random nature of $n(\tau, \theta)$, (5.122) is called a *stochastic differential equation* (SDE). *For constant* θ, the autocorrelation function of $n(\tau, \theta)$, i.e.

$$\Phi_{nn}(\tau, r) = \mathrm{E}\{n(\tau, \theta)n(\tau - r, \theta)\} = \Gamma(\tau + \theta)\Gamma(\tau + \theta - r)\Phi_{uu}(r) \ , \tag{5.123}$$

is periodic in τ. Hence, $n(\tau, \theta)$ is cyclostationary with respect to τ (see section 3.5.3 in chapter 3). This cyclostationarity together with the dependence of $n(\tau, \theta)$ on θ complicates solving (5.122).

Averaging reduces (5.122) to an SDE with a Gaussian stationary right-hand side that is independent of θ. Indeed, the application of the ideal lowpass averaging operator (5.90) to $n(\tau, \theta)$ results in an averaged Gaussian noise process $\overline{n}(\tau)$. Results from cyclostationary noise theory (again, see section 3.5.3 in chapter 3) allow us to prove that the autocorrelation function of $\overline{n}(\tau)$ is equal to

$$\Phi_{\overline{n}\overline{n}}(r) = \frac{1}{2\pi}\int_{-\pi}^{\pi} \mathrm{E}\{n(s)n(s - r)\}\,\mathrm{d}s \tag{5.124}$$

$$= \frac{\Phi_{uu}(r)}{2\pi}\int_{-\pi}^{\pi} \Gamma(s + \theta)\Gamma(s + \theta - r)\,\mathrm{d}s \tag{5.125}$$

$$= \frac{\Phi_{uu}(r)}{2\pi}\int_{-\pi}^{\pi} \Gamma(s)\Gamma(s - r)\,\mathrm{d}s \ . \tag{5.126}$$

The latter transition is a consequence of the periodicity of the ISF $\Gamma(\tau)$[13]. Equation (5.126) shows that $\overline{n}(\tau)$ is stationary and without explicit dependencies on θ. Hence, the averaged equation corresponding with (5.122) equals

$$\frac{\mathrm{d}\overline{\theta}}{\mathrm{d}\tau} = \varepsilon M_{ideal}[n](\tau) = \varepsilon\overline{n}(\tau) \ , \tag{5.127}$$

an SDE with a stationary right-hand side that is independent of the averaged phase $\overline{\theta}$.

[13] The equivalence of (5.125) and (5.126) is of major importance in resolving the difference between two widely publicized theories [Dem00a, Haji98] on oscillator phase noise behavior. In [Vana02d], it is shown that both theories yield similar results for stationary input signals $u(\tau)$. This is no longer true for non-stationary inputs —as is the case in the oscillator injection locking problem— with [Dem00a] being the one that yields correct results.

At this point, it is important to mention that the results above are not universally valid. Firstly, the input noise process $u(\tau)$ must be stationary. Furthermore, we require either that $\Phi_{uu}(r) \to 0$ on the time scale $1/\varepsilon$ or that most energy in the spectrum of $\Phi_{uu}(r)$ is contained at baseband. When these conditions are not satisfied, the stochastics of the right-hand side of (5.127) do start to depend on θ. Solving (5.127) then requires the construction of a Fokker-Planck equation [Dem02, Risk89].

Solving (5.127) is straightforward. Since $\overline{n}(\tau)$ is Gaussian, the same holds for the averaged phase $\overline{\theta}(\tau) = \varepsilon \int_0^\tau \overline{n}(s)\mathrm{d}s$ and the averaged phase differences $\overline{\theta}(\tau+r) - \overline{\theta}(\tau) = \varepsilon \int_\tau^{\tau+r} \overline{n}(s)\mathrm{d}s$. Hence, the variance

$$\sigma^2_{\overline{\theta\theta}}(r) = E\left\{\left(\overline{\theta}(\tau+r) - \overline{\theta}(\tau)\right)^2\right\}, \tag{5.128}$$

entirely characterizes the averaged phase behavior. With

$$\begin{aligned}
\frac{\mathrm{d}}{\mathrm{d}r}\sigma^2_{\overline{\theta\theta}}(r) &= \mathrm{E}\left\{2\left(\overline{\theta}(\tau+r) - \theta(\tau)\right)\frac{\mathrm{d}\overline{\theta}}{\mathrm{d}t}(\tau+r)\right\} && (5.129)\\
&= \varepsilon^2\mathrm{E}\left\{2\left(\int_\tau^{\tau+r} \overline{n}(s)\mathrm{d}s\right)\overline{n}(\tau+r)\right\} && (5.130)\\
&= \varepsilon^2 2\int_\tau^{\tau+r} \Phi_{\overline{nn}}(\tau+r-s)\,\mathrm{d}s && (5.131)\\
&= \varepsilon^2 2\int_0^{\tau} \Phi_{\overline{nn}}(s)\mathrm{d}s \ , && (5.132)
\end{aligned}$$

we find $\sigma^2_{\overline{\theta\theta}}(r)$ by solving

$$\frac{\mathrm{d}^2}{\mathrm{d}r^2}\sigma^2_{\overline{\theta\theta}}(r) = \varepsilon^2 2\Phi_{\overline{nn}}(r) \text{ with } \sigma^2_{\overline{\theta\theta}}(0) = 0,\ \frac{\mathrm{d}}{\mathrm{d}r}\sigma^2_{\overline{\theta\theta}}(0) = 0 \ . \tag{5.133}$$

This is in agreement with existing results [Dem98, Dem02]. Equation (5.133) is the starting point for computing, for instance, an oscillator's phase noise spectrum.

Example (White input noise): For a Gaussian stationary white input noise source $u(\tau)$, it holds that

$$\Phi_{uu}(r) = N_u\delta(r) \ . \tag{5.134}$$

Using (5.126) and (5.132), it is readily shown that the variance of the averaged phase behavior $\overline{\theta}(\tau)$ satisfies

$$\frac{\mathrm{d}\sigma^2_{\overline{\theta\theta}}}{\mathrm{d}r} = \frac{N_u}{2\pi}\int_{-\pi}^{\pi} \Gamma(s)^2\mathrm{d}s \ . \tag{5.135}$$

In the equations above, N_u is the power spectral density level of the white noise process $u(\tau)$. Equation (5.135) is solved by

$$\sigma^2_{\overline{\theta\theta}}(\tau) = \left(\frac{N_u}{2\pi}\int_{-\pi}^{\pi} \Gamma(s)^2\mathrm{d}s\right)\tau \ . \tag{5.136}$$

Hence, the variance of the averaged stochastic phase shift increases linearly with time.

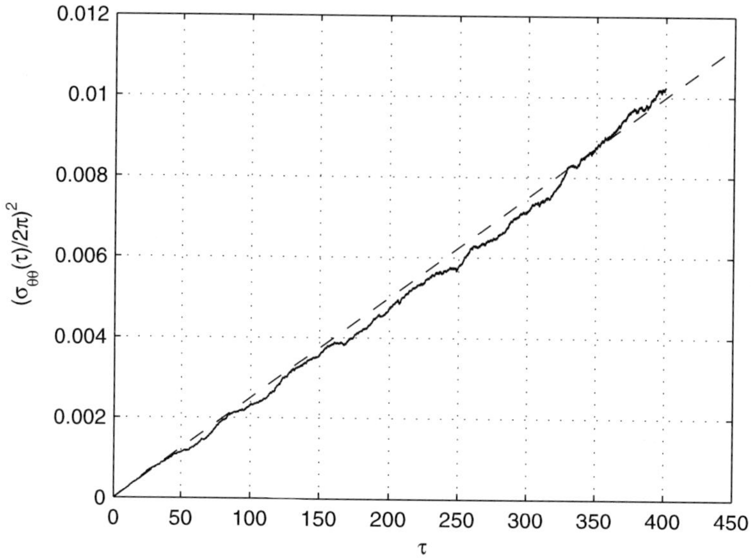

Figure 5.26: Variance $\sigma^2_{\theta\theta}(\tau)$ of an harmonic oscillator's phase $\theta(\tau)$ caused by the injection of a white noise current. The variance was computed by means of Monte-Carlo simulations. The dashed line marks the variance of the averaged phase shift as solved from (5.135).

In order to verify (5.136), the variance

$$\sigma^2_{\theta\theta}(\tau) = E\left\{\theta(\tau)^2\right\} \quad . \tag{5.137}$$

of the (non-averaged) phase shift $\theta(\tau)$ was computed numerically by means of Monte-Carlo simulations: (5.122) was solved for 500 randomly generated noise samples. The ISF $\Gamma(\tau)$ was set equal to that of the harmonic oscillator in Fig. 5.22. The numerically computed variance is shown in Fig. 5.26. It is in close agreement with the variance of the averaged phase shift predicted by (5.136). ▲

5.7.4 Practical application: computing oscillator phase noise spectra

The theory above relates the stochastics of an injected stationary noise process $u(\tau)$ to that of an oscillator's random phase behavior. In a next step, we need to figure out the impact this random phase behavior has on the PSD of the oscillator's output signal, i.e. on its *phase-noise spectrum*. The ultimate goal is to determine the relation between the PSD of the input noise process $u(\tau)$ and the phase-noise spectrum. In [Vana03c], we present a semi-analytic method to do so. For input noise processes that vary sufficiently fast over time, we derived an analytic expression. For cases that do not satisfy these conditions, we developed a numerical algorithm that computes corrections to this expression. These corrections mainly concern the shape of the very-close-in phase-noise spectrum. Either way, the computation of the oscillator's entire phase-noise spectrum is accomplished in only a few minutes. This is a huge improvement over methods that rely on lengthy simulations and a numerical discrete Fourier transformation. Moreover, our work in [Vana03c] allows the accurate computation of the very-close-in phase-noise spectrum, i.e. for frequency offsets down to a few Hz and below. In what follows, we summarize the main results. For details, we refer to [Vana03c].

The phase noise spectrum $S_{x,k}(\Delta f)$ near the k-th harmonic of an oscillator's output signal is related to the stochastic phase behavior described by (5.126) and (5.133) by means of the equation[14]

$$S_{x,k}(\Delta f) = \mathcal{F}\left\{ e^{-\frac{k^2\sigma^2_{\theta\theta}(2\pi f_0 t)}{2}} \right\} . \tag{5.138}$$

Here, Δf is the frequency offset to the k-th harmonic frequency kf_0. Furthermore, the Fourier transform in (5.138) is computed with respect to the unnormalized time variable t. If the averaged noise process $\overline{n}(\tau)$ varies sufficiently fast over time, then it is shown that [Vana03c]

$$S_{x,k}(\Delta f) \approx \varepsilon^2 \frac{k^2 \frac{f_0^2}{\Delta f^2} S_{\overline{nn}}(\Delta f)}{1 + \varepsilon^4 \pi^2 k^4 \frac{f_0^2}{\Delta f^2} f_0^2 S_{\overline{nn}}(0)^2} . \tag{5.139}$$

Here, $S_{\overline{nn}}(\Delta f) = \mathcal{F}\{\Phi_{\overline{nn}}(2\pi f_0 t)\}$ is the PSD of the averaged noise process $\overline{n}(\tau)$. Again, the Fourier transform is with respect to unnormalized time variable t. For sufficiently large Δf, (5.139) can be further approximated as

$$S_{x,k}(\Delta f) \approx \varepsilon^2 k^2 \frac{f_0^2}{\Delta f^2} S_{\overline{nn}}(\Delta f) . \tag{5.140}$$

This expression is consistent with the well-known $1/\Delta f^3 - 1/\Delta f^2$ characteristic. Moreover, (5.139) unifies the two distinct high- and low-frequency asymptotes presented in [Dem98] in a single expression.

For slow-varying ($1/f$) input noise sources, the validity of (5.139) —and therefore also that of the low-frequency asymptote in [Dem98]— breaks down for small frequency offsets Δf. For such cases, we propose (again, see [Vana03c]) a method that refines (5.139) by means of a routine based on exponential data fitting. In performing the numerical computations, the PSD of the averaged noise process $\overline{n}(\tau)$ is assumed to equal a combination of white and $1/f$ noise, i.e.

$$S_{\overline{nn}}(f) = N\left[1 + \frac{f_{1/f}}{|f|}\left(1 - \frac{2}{\pi}\arctan\left(\left|\frac{\gamma}{2\pi f}\right|\right)\right)\right] . \tag{5.141}$$

Here, $f_{1/f}$ is the input noise $1/f$ corner frequency, i.e. the frequency (in Hz) beyond which the PSD of the $1/f$ input noise drops below the white noise level. Furthermore

[14] This expression is somewhat different from the one presented in [Vana03c]. These differences arise from the fact that, contrary to [Vana03c], this dissertation performed computations with respect to the normalized time variable $\tau = 2\pi f_0 t$. We therefore need some de-normalizations to obtain the phase-noise spectrum.

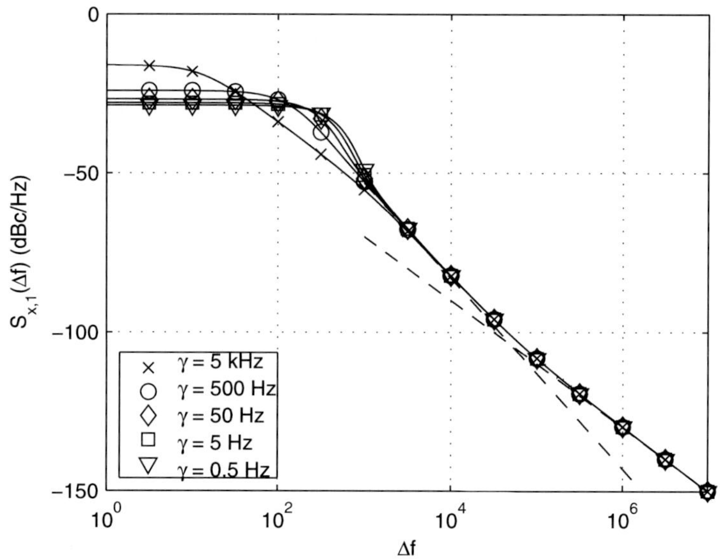

Figure 5.27: Phase noise spectrum $S_{x,1}(\Delta f)$ around carrier frequency. The spectrum is shown for different values of γ. For small values of γ, the very close-in phase noise spectrum flattens with a steep edge that makes the transition from the DC-level to the familiar $1/\Delta f^3 - 1/\Delta f^2$ characteristic.

the parameter γ represents the frequency whereby, for $f \to 0$, the $1/f$ characteristic of the second term in (5.141) goes over into a flat one with $S_{\overline{nn}}(0) = N\left[1 + 4f_{1/f}/\gamma\right]$. This $1/f$ noise model was borrowed from [Dem98, Kae90].

The algorithm mentioned above was used to compute $S_{x,1}(\Delta f)$, the phase-noise spectrum around the carrier frequency. The computation of a single spectrum finished in less than a minute. In this example, $f_0 = 1\,\text{GHz}$. The input noise PSD was modeled as in (5.141) with $f_{1/f}$ set to 50 kHz. The noise strength was chosen in order for the phase noise spectrum at 1 MHz to equal -130 dBc/Hz. These values correspond to typical (recent) designs, e.g. [Demu02]. Fig. 5.27 plots $S_{x,1}(\Delta f)$ for values of γ ranging from 0.5 Hz to 5 kHz. For small values of γ, i.e. for the second term in (5.141) becoming more $1/f$-like, the very-close-in phase noise spectrum flattens with a steep edge making the transition from the DC level to the familiar $1/\Delta f^3 - 1/\Delta f^2$ characteristic. In this example, this edge is located between 300 Hz and 1 kHz.

5.8 Harmonic oscillator behavioral modeling

We end this discussion on oscillators with a method for the construction of compact behavioral models for harmonic oscillators and sets of coupled harmonic oscillators [Vana03b]. The models are useful to speed up lengthy or repetitive system-level ver-

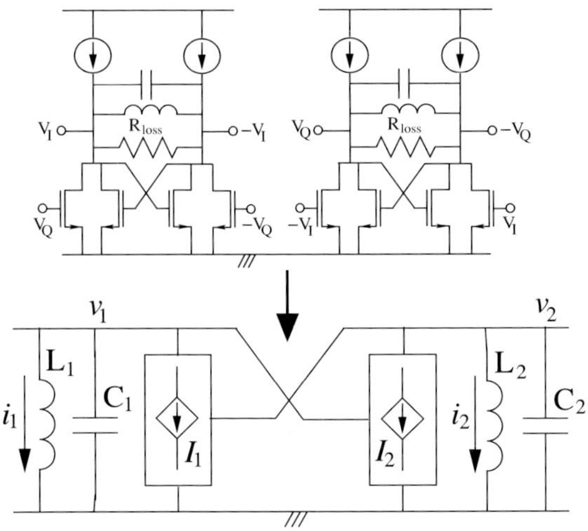

Figure 5.28: It is assumed that the oscillator circuit can be mapped onto a template consisting of K weakly interconnected lossless resonant tanks. The figure shows a template for $K = 2$.

ification and trade-off analysis. The model extraction procedure builds on the theory outlined in previous sections. The resulting models capture an oscillator's dynamic behavior by means of the averaged state-changing equations. The computational cost needed to evaluate them is well below that of solving the original circuit equations. The models also separate slow- and fast-varying behavior. As a result, they can be integrated using a large time step. Experimental results are presented for both a single harmonic oscillator and a harmonic quadrature oscillator.

5.8.1 Model extraction theory

The method assumes that the oscillator circuit can be mapped onto the template in Fig. 5.28. The template consists of K lossless resonant tanks and a number of controlled current sources. The current sources are assumed to be weak: the amount of energy they inject per oscillation period into the resonant tanks is small. Note that the template in Fig. 5.28 does not cover all possible oscillator topologies. For example, the oscillator circuit in Fig. 5.9 cannot be mapped onto it. However, it is straightforward to adapt the method to handle more general templates. In what follows, we briefly step through the theory of the sections 5.3 to 5.6 as it applies to a system of coupled harmonic oscillators.

5.8.1.1 Normalizing and partitioning the circuit equations

The equations that model a system of K coupled resonant tanks are given by

$$L_k \frac{\mathrm{d}i_k}{\mathrm{d}t} - v_k = 0 \tag{5.142}$$

$$C_k \frac{dv_k}{dt} + i_k + I_k\left(v_1, \frac{dv_1}{dt}, v_2, \frac{dv_2}{dt}, \ldots, t\right) = 0 \qquad (5.143)$$

for $k = 1, \ldots, K$. Here C_k and L_k respectively represent the effective capacitance and inductance of the k-th resonant tank. The currents $I_k\left(v_1, \frac{dv_1}{dt}, v_2, \frac{dv_2}{dt}, \ldots, t\right)$ model the contributions of the transistor feedback together with the inductor and capacitor losses.

In a first step, we normalize the system of equations (5.142). To do so, we change variables:

$$x_k = \frac{v_k}{V_k} \qquad (5.144)$$

$$y_k = \frac{i_k}{\omega_{0,k} C_k V_k} \qquad (5.145)$$

$$\omega_0 = \frac{1}{K}\sum_{k=1}^{K} \omega_{0,k} \qquad (5.146)$$

$$\tau = \omega_0 t \qquad (5.147)$$

$$f_k\left(x_1, \frac{dx_1}{d\tau}, \ldots, \tau\right) = \frac{I_k(V_1 x_1, V_1 \omega_0 \frac{dx_1}{d\tau}, \ldots, \tau/\omega_0)}{I_{k,max}} \qquad (5.148)$$

$$\varepsilon = \frac{I_{k,max}}{\omega_{0,k} C_k V_k} \ . \qquad (5.149)$$

Here, $\omega_{0,k} = 1/\sqrt{L_k C_k}$ is the resonant frequency of the k-th tank, V_k is an estimate for the magnitude of the voltage swing over the capacitor C_k and $I_{k,max}$ is a measure for the maximum feedback current delivered by the controlled current sources. The variable ε equals the ratio between the maximum feedback current $I_{k,max}$ and the current $I_{tank,k} = \omega_{0,k} C_k V_k = V_k/(\omega_{0,k} L_k)$ circulating in the resonant tank. The V_k should be chosen such that ε is the same for all tanks. Furthermore, it is assumed that all resonant frequencies $\omega_{0,k}$ cluster around the mean frequency ω_0, i.e.

$$\omega_{0,k} = \omega_0(1 + \varepsilon\delta_k) \qquad (5.150)$$

were $\varepsilon\delta_k$ is the (small) relative deviation of $\omega_{0,k}$ from ω_0.

With (5.144)-(5.149), the system (5.142)-(5.143) becomes

$$\frac{dy_k}{d\tau} - x_k - \varepsilon\delta_k \frac{dy_k}{d\tau} = 0 \qquad (5.151)$$

$$\frac{dx_k}{d\tau} + y_k + \varepsilon f_k\left(x_1, \frac{dx_1}{d\tau}, x_2, \frac{dx_2}{d\tau}, \ldots, \tau\right) - \varepsilon\delta_k \frac{dx_k}{d\tau} = 0 \ . \qquad (5.152)$$

The right-hand sides of (5.151)-(5.152) clearly fall into two components. The first two terms correspond to the behavior of the lossless resonant tanks. The terms preceded by an ε model the transistors currents, the tank losses and the resonant frequency mismatches. Since all latter terms are small, at least for well-matched resonant tanks with a sufficiently high Q-factor, they can be treated as perturbations. Hence, the (coupled) harmonic oscillator's core system is chosen to coincide with the K lossless resonant tanks.

5.8.1.2 Characterizing the oscillator's unperturbed core.

The oscillator's core system corresponds to K independent lossless resonant tanks. It is described by (5.151)-(5.152) for ε set to zero. The steady-state solutions to this system are given by

$$y_{s,k}(\tau) = A_k \sin(\tau + \theta_k) \tag{5.153}$$
$$x_{s,k}(\tau) = A_k \cos(\tau + \theta_k) \ . \tag{5.154}$$

Each tank resonates with an arbitrary but fixed amplitude A_k and phase θ_k. These resonant tank amplitudes and phases are combined in the core system's state vector

$$\mathbf{p} = \begin{bmatrix} A_1 & \theta_1 & A_2 & \theta_2 & \cdots & A_K & \theta_K \end{bmatrix}^T \in \mathbb{R}^{2K} \ . \tag{5.155}$$

Each particular value of the state vector identifies a possible state (steady-state solution) of the oscillator's core.

5.8.1.3 Oscillator perturbation analysis

For small but nonzero perturbations, i.e. for $0 < |\varepsilon| \ll 1$, the state vector $\mathbf{p}$ slowly changes with time. Up to first order in ε, the behavior of $\mathbf{p}$ is governed by

$$\frac{\mathrm{d}\mathbf{p}}{\mathrm{d}\tau} = -\varepsilon \mathbf{V}_1^T(\tau, \mathbf{p}) \mathbf{g}(\tau, \mathbf{p}) = \varepsilon \mathbf{h}_p(\mathbf{p}, t) \ . \tag{5.156}$$

Here,

$$\mathbf{g}(\tau,\mathbf{p}) = \begin{bmatrix} -\delta_k \frac{\partial y_{s,k}}{\partial \tau} \\ f_1\left(x_{s,1}, \frac{\partial x_{s,1}}{\partial \tau}, x_{s,2}, \frac{\partial x_{s,2}}{\partial \tau}, \ldots, \tau\right) - \delta_k \frac{\partial x_{s,1}}{\partial \tau} \\ \vdots \end{bmatrix} \tag{5.157}$$
$$= \begin{bmatrix} -\delta_1 A_1 \cos(\tau + \theta_1) \\ f_1\left(A_1 \cos(\tau+\theta_1), -A_1 \sin(\tau+\theta_1), \ldots, \tau\right) + \delta_1 A_1 \sin A_1 (\tau + \theta_1) \\ \vdots \end{bmatrix}$$

is the vector in $\mathbb{R}^{2K}$ containing the perturbation terms. The projector $\mathbf{V}_1^T(\tau,\mathbf{p}) \in \mathbb{R}^{2K \times 2K}$ can be shown to equal

$$\mathbf{V}_1^T(\tau,\mathbf{p}) = \begin{bmatrix} \sin(t+\theta_1) & \cos(t+\theta_1) & & & \\ \frac{\cos(t+\theta_1)}{A_1} & -\frac{\sin(t+\theta_1)}{A_1} & & 0 & \\ & & \sin(t+\theta_2) & \cos(t+\theta_2) & \\ & 0 & \frac{\cos(t+\theta_2)}{A_2} & -\frac{\sin(t+\theta_2)}{A_2} & \\ & & & & \ddots \end{bmatrix} , \tag{5.158}$$

a block-diagonal square matrix.

Since in most practical applications phase differences are of greater interest than absolute phase values, we again change variables. Equation (5.156) is reformulated in terms of one common-mode phase $\theta = \theta_1$ and the phase differences $\Delta\phi_k = \theta_k - \theta_1$ between the k-th resonant tank and the first one. In stable oscillators, these phase differences converge to a constant operating-point value. This property is exploited later on in our model extraction procedure. Introducing the vector

$$\mathbf{q} = \begin{bmatrix} A_1 & \cdots & A_K & \Delta\phi_2 & \cdots & \Delta\phi_K & \theta \end{bmatrix}^T \tag{5.159}$$

$$= \mathbf{T}_q \mathbf{p} \ , \tag{5.160}$$

where

$$\mathbf{T}_q = \begin{bmatrix} 1 & 0 & 0 & 0 & \cdots & 0 & 0 \\ 0 & 0 & 1 & 0 & \cdots & 0 & 0 \\ & \vdots & & & \ddots & & \vdots \\ 0 & 0 & 0 & 0 & \cdots & 1 & 0 \\ 0 & -1 & 0 & 1 & \cdots & 0 & 0 \\ & \vdots & & & \ddots & & \vdots \\ 0 & -1 & 0 & 0 & \cdots & 0 & 1 \end{bmatrix} \in \mathbb{R}^{2K\times 2K} \ , \tag{5.161}$$

the ODE (5.156) can be recast into

$$\frac{d\mathbf{q}}{d\tau} = \varepsilon \mathbf{T}_q \mathbf{h}_p \left(\mathbf{T}_q^{-1}\mathbf{q}, \tau\right) = \varepsilon \mathbf{h}_q(\mathbf{q}, \tau) \ . \tag{5.162}$$

Equation (5.162) captures the dynamic behavior of the oscillation amplitudes, the phase differences and the common-mode phase. It is accurate up to first order in ε.

5.8.1.4 Averaging

The presence of ε in the right-hand side of (5.162) implies a small time derivative of the process $\mathbf{q}(\tau)$. This ensures that substantial changes in $\mathbf{q}(\tau)$ only occur over a long period of time. Hence, the process $\mathbf{q}(\tau)$ contains a dominant low-frequency component. As outlined in section 5.6, the equation that governs the behavior of this component can be extracted by means of an averaging transformation

$$\mathbf{q} = \overline{\mathbf{q}} + \varepsilon \tilde{\mathbf{h}}_q(\overline{\mathbf{q}}, \tau) + O\left(\varepsilon^2\right) \ . \tag{5.163}$$

If we use the ideal lowpass averaging operator (5.90), then the averaged equation corresponding to (5.162) becomes

$$\frac{d\overline{\mathbf{q}}}{d\tau} = \varepsilon M_{ideal}\left[\mathbf{h}_q(\overline{\mathbf{q}}, \tau)\right] = \varepsilon \overline{\mathbf{h}}_q(\overline{\mathbf{q}}, \tau) \ . \tag{5.164}$$

As will be demonstrated experimentally in section 5.8.3, this averaged equation captures the essence of the dynamic behavior of a (system of coupled) harmonic oscillator(s). Furthermore, it can be solved very efficiently using traditional (SPICE-like) simulation algorithms. Hence, it serves as a compact behavioral harmonic oscillator model. Of course, in order to be useful in practice, the right-hand side of (5.164) must be stored in a way that can be evaluated efficiently while being easy to implement in environments for system-level simulation.

5.8.1.5 Storing the averaged equations

A first approach stores $\overline{\mathbf{h}}_q(\overline{\mathbf{q}},\tau)$ as a table of precomputed values. During simulation, model evaluations are performed using table lookup and interpolation in between. As illustrated in section 5.8.3.1, this approach works fine for modeling a single harmonic oscillator's behavior. However, since the storage space increases exponentially with the dimension of the vector $\overline{\mathbf{q}}$, it becomes cumbersome for sets of coupled oscillators.

For stable oscillators, a more elegant approach exists. Stability ensures that all amplitudes and phase differences converge to a neighborhood of a stable operating point. It is therefore possible to construct behavioral models by approximating the averaged equations near such an operating point. In order to put this idea to practice, we partition the state vector in two major components:

$$\begin{aligned} \mathbf{q} &= \left[\begin{array}{ccccccc} A_1 & \cdots & A_K & \Delta\phi_2 & \cdots & \Delta\phi_K & \theta \end{array}\right]^T \qquad (5.165)\\ &= \left[\begin{array}{cc} \mathbf{q}_1^T & \theta \end{array}\right]^T . \qquad (5.166)\end{aligned}$$

A first component $\mathbf{q}_1$ contains the amplitudes and phase differences. A second component contains the common-mode phase θ. Writing the system of ODEs (5.162), governing the behavior of the state vector $\mathbf{q} = \left[\begin{array}{cc} \mathbf{q}_1^T & \theta \end{array}\right]^T$, in terms of $\mathbf{q}_1$ and θ yields

$$\frac{\mathrm{d}}{\mathrm{d}\tau}\left[\begin{array}{c} \mathbf{q}_1 \\ \theta \end{array}\right] = \varepsilon\left(\mathbf{h}_{q,T}(\mathbf{q}_1,\tau+\theta) + \Delta\mathbf{h}_q\left(\left[\begin{array}{c} \mathbf{q} \\ \theta \end{array}\right],\tau\right)\right) . \qquad (5.167)$$

Here, we deliberately partitioned the right-hand side in two terms. The first one, $\mathbf{h}_{q,T}$, is typically largest in magnitude and captures the action of tank losses, non-stochastic transistor currents and resonant frequency mismatches. This component is periodic in the normalized time variable τ with a period $T = 2\pi$. The second component, $\Delta\mathbf{h}_q$, is typically much smaller than $\mathbf{h}_{q,T}$ and captures, for instance, (transistor) noise or other signal sources. The averaged ODEs corresponding to (5.167) are

$$\frac{\mathrm{d}}{\mathrm{d}\tau}\left[\begin{array}{c} \overline{\mathbf{q}}_1 \\ \theta \end{array}\right] = \varepsilon\left(\overline{\mathbf{h}}_{q,T}(\overline{\mathbf{q}}_1) + \overline{\Delta\mathbf{h}}_q\left(\left[\begin{array}{c} \overline{\mathbf{q}}_1 \\ \theta \end{array}\right],\tau\right)\right) . \qquad (5.168)$$

Since $\mathbf{h}_{q,T}(\mathbf{q}_1,\tau+\theta)$ is periodic in τ, the average of the first term does not explicitly depend on the common-mode phase θ.

For stable oscillators, the averaged oscillation amplitudes and phase differences, i.e. the elements of $\overline{\mathbf{q}}_1$, converge to the neighborhood of a constant operating point value $\overline{\mathbf{q}}_{1,op}$. This operating point is solved from

$$\overline{\mathbf{h}}_{q1,T}\left(\overline{\mathbf{q}}_{1,op}\right) = 0 . \qquad (5.169)$$

Here $\overline{\mathbf{h}}_{q1,T}$ contains the first $(2K-1)$ rows of $\overline{\mathbf{h}}_{q,T}$. Near the operating point, we capture the oscillator's dynamic behavior by means of approximations of the averaged equation (5.168). The simplest approximation is based on linearization. Section 5.8.3.2 illustrates the accuracy of these linearized models for an harmonic quadrature oscillator. More sophisticated approximators use multivariate polynomial regression, radial basis functions or support vector machines.

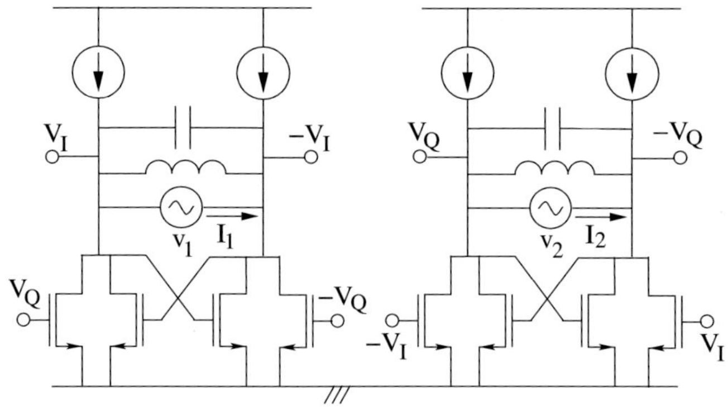

Figure 5.29: Harmonic quadrature oscillator measurement setup. Here, the voltage sources v_1 and v_2 are introduced for the purpose of extracting the resonant tank characteristics and the currents $I_k\left(v_1, \frac{dv_1}{dt}, v_2, \frac{dv_2}{dt}, \ldots, t\right)$ exchanged between the resonant tanks and the remainder of the circuit. The voltage sources should therefore not be considered as part of the oscillator circuit.

5.8.2 Numerical computations

Finally, we discuss the numerics that are needed to put the above theory to practice. Firstly, this requires us to map the oscillator's circuit topology on the template of Fig. 5.28. Furthermore, we need a way to evaluate the functions $\overline{\mathbf{h}}_q(\overline{\mathbf{q}}, \tau)$. Hereby, we focus on the dominant component $\overline{\mathbf{h}}_{q,T}(\overline{\mathbf{q}}_1)$. In what follows, all computational procedures are outlined for the harmonic quadrature oscillator shown in Fig. 5.29. Adapting the procedures to other types of (coupled) harmonic oscillators is straightforward. As presented, the procedures are easily implemented on top of existing simulation algorithms.

A resonant tank's effective capacitance and inductance values C_k and L_k are determined by fitting the admittance $G_{RLC}(j\omega)$ of an *RLC* tank onto the one simulated using the setup shown in Fig. 5.29. Here, we use voltage sources v_1 and v_2 to apply the in-phase and quadrature voltages

$$v_1(t) = \Delta v \cos(\omega t) \tag{5.170}$$

$$v_2(t) = 0 \tag{5.171}$$

with Δv a sufficiently small voltage. By measuring the resulting steady-state currents I_1 and I_2 through the voltage sources, we retrieve the information necessary to compute, for instance, $G_1(j\omega)$. This simulated admittance is fitted onto that of an *RLC* tank for frequencies ω in the neighborhood of the resonant frequency

$$\omega_{0,1} = \arg\min_{\omega} |G_1(j\omega)| \quad . \tag{5.172}$$

The setup in Fig. 5.29 guarantees that all fringe (transistor) capacitances are taken into account as part of the effective tank capacitances C_1. Accurate modeling also requires us to account for output loading due to subsequent blocks, e.g. buffers. This can be accomplished by adding extra capacitors at the output nodes that model the input impedance of these subsequent blocks.

Computing $\overline{\mathbf{h}}_{q,T}(\overline{\mathbf{q}}_1)$ for a given set of values for $\mathbf{q}_1$ requires the currents I_k. To obtain, for instance, $I_1\left(v_1, \frac{dv_1}{dt}, v_2, \frac{dv_2}{dt}, \ldots, t\right)$, we apply

$$v_1(t) = V_1\overline{A}_1\cos\left(\omega_0 t + \overline{\theta}\right) \quad (5.173)$$

$$v_2(t) = V_2\overline{A}_2\cos\left(\omega_0 t + \overline{\theta} + \overline{\Delta\phi_2}\right) \quad (5.174)$$

in the measurement setup of Fig. 5.29. Here, V_1 and V_2 are normalization constants introduced in (5.144). $I_1\left(v_1, \frac{dv_1}{dt}, v_2, \frac{dv_2}{dt}, \ldots, t\right)$ then equals the periodic steady-state current that flows through the voltage source v_1. These steady-state currents can be determined using a shooting or an harmonic balance method. Having computed the I_k, we use (5.144)-(5.162) to obtain the term $\mathbf{h}_{q,T}(\overline{\mathbf{q}}_1, \tau+\theta)$ in (5.162). Averaging this term yields $\overline{\mathbf{h}}_{q,T}(\overline{\mathbf{q}}_1)$. Here, averaging comes down to the selection of the DC component of the Fourier series expansion of $\mathbf{h}_{q,T}(\overline{\mathbf{q}}_1, \tau+\theta)$ with respect to τ.

5.8.3 Experimental results

The model extraction procedure was implemented in Matlab for both a single harmonic oscillator and an harmonic quadrature oscillator. In what follows, we compare the behavioral model's steady-state and transient behavior with results extracted from SPICE-like simulations. All computations were performed in Matlab running on a Pentium IV at 1.7 GHz.

5.8.3.1 Single harmonic oscillator

In a first example, we construct a behavioral model for the harmonic oscillator shown in Fig. 5.30. The behavioral model is constructed by precomputing the right-hand side of the averaged ODEs (5.168) for 31 different values of the oscillation amplitude. Model evaluations are based on table lookup and interpolation in between these 31 precomputed values. The oscillator is designed to have its oscillation frequency near 1 GHz. Its feedback transistors are sized such that

$$\gamma = \frac{G_{m,feedback}}{1/R_{loss}} = \frac{g_{m1}/2}{1/R_{loss}} \approx 2.5 \ . \quad (5.175)$$

The transistor's $(V_{GS} - V_{th})$ is set to 0.5 V. As shown in Fig. 5.30, the resonant tank model includes several parasitics for both the inductor and capacitor. The MOS transistors are modeled using the EKV equations [Enz95] for a 0.5μm CMOS technology.

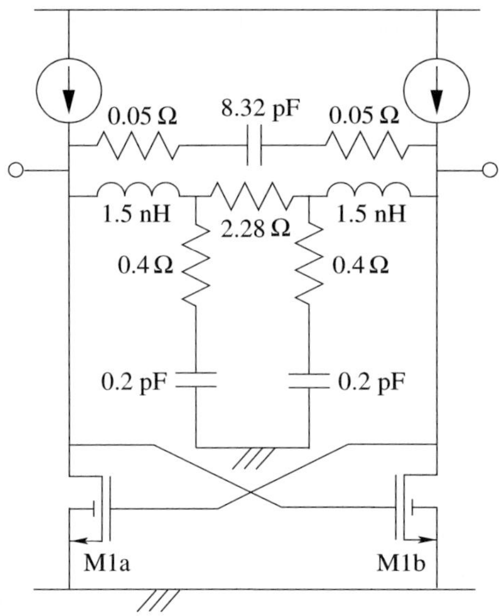

Figure 5.30: Harmonic oscillator with a single resonant tank. The tank includes several components that model parasitic behavior of the inductor and capacitor.

In a first step, we extract the characteristics of the resonant tank. Without transistor capacitances, the tank is characterized by:

$$f_0 = 1\,\text{GHz} \tag{5.176}$$

$$Q = \frac{2\pi f_0 C}{1/R_{loss}} = 8.04 \ . \tag{5.177}$$

Taking the transistor capacitances into account changes the resonance frequency to $f_0 = 0.986\,\text{GHz}$. Note that these capacitances are automatically accounted for if we use the computational procedure outlined in section 5.8.2.

Next, we construct the (normalized) equations that govern the behavior of the averaged state vector

$$\overline{\mathbf{q}} = [\ \overline{A} \quad \overline{\theta}\ ]^T \ . \tag{5.178}$$

This yields the system of ODEs

$$\frac{d\overline{A}}{d\tau} = \overline{h}_A(\overline{A}) \tag{5.179}$$

$$\frac{d\overline{\theta}}{d\tau} = \overline{h}_\theta(\overline{A}) \ . \tag{5.180}$$

Note that averaging removes the right-hand-side's dependency on the phase θ. Fig. 5.31 plots both $\overline{h}_A(\overline{A})$ and $\overline{h}_\theta(\overline{A})$ as a function of the averaged amplitude. The upper figure

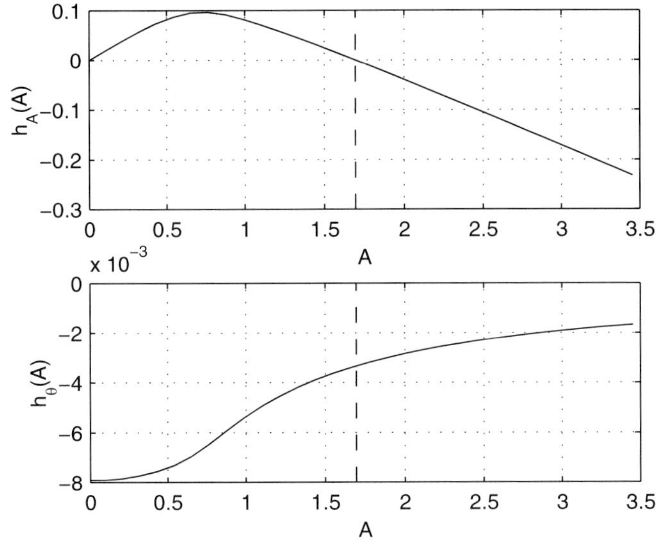

Figure 5.31: Right-hand sides of the ODEs (5.179) and (5.180) that govern the behavior of the resonant tank's averaged amplitude and phase. The vertical line indicates the position of the stable operating point.

shows $\overline{h}_A(\overline{A})$ to contain two operating points: an unstable one at $\overline{A}_{op} = 0$ and a stable one at $\overline{A}_{op} = 1.69$. Only the latter is of interest. It is indicated by the vertical line in Fig. 5.31. For this operating point, we notice from equation (5.180) that $\overline{\theta}(\tau) = \overline{h}_\theta(\overline{A}_{op})\tau$. Hence, the interactions between the resonant tank and the feedback elements cause a slight shift in oscillation frequency.

To verify the accuracy of the averaging method, we compare the oscillator's steady-state differential output voltage as computed using the (normalized) signal model

$$x_s(\tau) = A(\tau)\cos(\tau + \theta(\tau)) \tag{5.181}$$

and the averaging transformation

$$\begin{aligned} A(\tau) &= \overline{A}_{op} + \varepsilon\tilde{h}_A\left(\overline{A}_{op}, \overline{h}_\theta(\overline{A}_{op})\tau, \tau\right) && (5.182)\\ \theta(\tau) &= \overline{h}_\theta(\overline{A}_{op})\tau + \varepsilon\tilde{h}_\theta\left(\overline{A}_{op}, \overline{h}_\theta(\overline{A}_{op})\tau, \tau\right) && (5.183) \end{aligned}$$

with the solutions of the original circuit equations. The latter are obtained by means of a SPICE-like algorithm. The transformation (5.182)-(5.183) is evaluated at the stable operating point $\overline{A}_{op} = 1.69$. Table 5.2 lists the magnitudes of the fundamental tone and the first few odd harmonics (the even harmonics are negligible). The magnitude of the fundamental component is 1% (0.1 dB) accurate. The other components are 0.5-1.5 dB accurate. This shows that the steady-state behavior of the averaged equations corresponds with that of the circuit equations.

Harmonic	Averaged	SPICE-like
1	4.58 dB	4.68 dB
3	-33.2 dB	-32.1 dB
5	-46.8 dB	-46.4 dB
7	-60.6 dB	-59.0 dB

Table 5.2: Spectral components of the steady-state differential output voltage. These components are respectively obtained by means of averaging and via analysis of SPICE-like simulation results.

In a final step, we construct a behavioral model for the oscillator's transient behavior. This model implements the averaged equations (5.179) and (5.180). The right-hand sides $\overline{h}_A(\overline{A})$ and $\overline{h}_\theta(\overline{A})$ are stored as a table of 31 precomputed values. We evaluate them by means of table lookup and linear interpolation. Computing the 31 values requires 3 minutes of CPU time. Fig. 5.32 compares the oscillator startup behavior predicted by the behavioral model, which directly computes the oscillation amplitude, with the solution of the original circuit equations. The latter was obtained using a SPICE-like algorithm.As can be seen, there is excellent agreement (to within 1%) between the SPICE-like simulation results and the amplitude envelope predicted by the behavioral model.

As far as computational complexity is concerned, we observe that simulations based on the behavioral model only take 0.4 seconds to complete. Solving of the exact circuit equations takes 37.2 seconds. These numbers hold for simulations ranging over 20 periods of oscillation. The results demonstrate a significant simulation speedup whereby the behavioral model is solved up to a 100 times faster. The reasons for this speedup are twofold. Firstly, the behavioral model can be solved using a much larger time step. It "skips" fast-varying oscillations and directly computes amplitude and phase behavior.

The speedup mentioned above doesn't seem like a big deal. For a designer, there is little difference in having to wait 1 or 30 seconds. But the simulations above ran over a very limited simulation time span, covering only 20 periods of oscillation. Fig. 5.33 plots the simulation (CPU) time as a function of the simulation time span relative to the oscillation period. This is done for both SPICE-like transient simulations and for simulations performed with the model extracted by means of the methods here developed. As is observed from the figure, if the simulation time span grows large, then SPICE-like simulations tend to take minutes and even hours while the behavioral model still evaluates within seconds or minutes. Hence, use of the behavioral model becomes more and more advantageous when the total simulation time span grows large. Lengthy (or large numbers of short) simulations frequently occur in design practice: just think about system optimization methods that require us to run simulations in a loop, system-level verifications that run over many periods of the high-frequency carrier (oscillation) signal and the simulation of the startup behavior of oscillators with a really large settling time (e.g. when using a crystal resonant tank). As indicated by the dashed lines in Fig. 5.33: if, for the setup in this example, you want to avoid having to wait longer

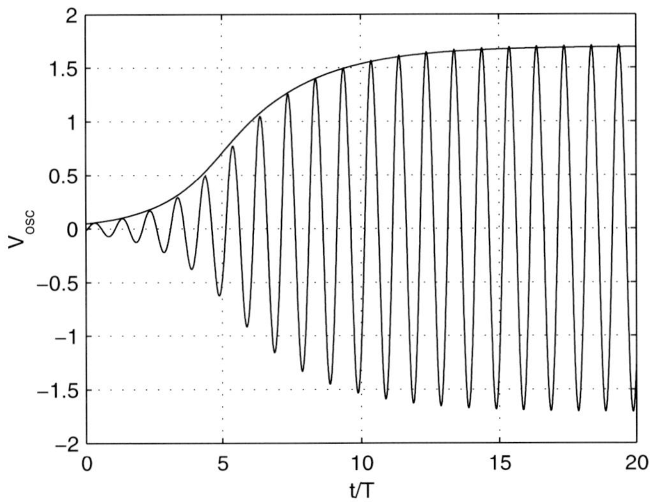

Figure 5.32: Harmonic oscillator startup behavior. The fast-oscillating line is the solution of the original circuit equations. The envelope solves the averaged system (5.179)-(5.180). The results agree to within 1%.

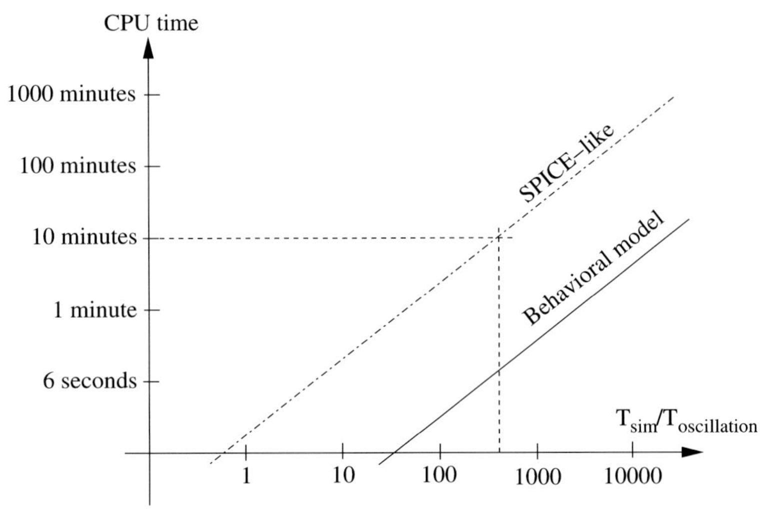

Figure 5.33: CPU time versus the normalized length of the simulation time span. For lengthy or repetitive simulations that run over a large number of oscillation periods, use of the behavioral model reduces simulation times from hours to minutes.

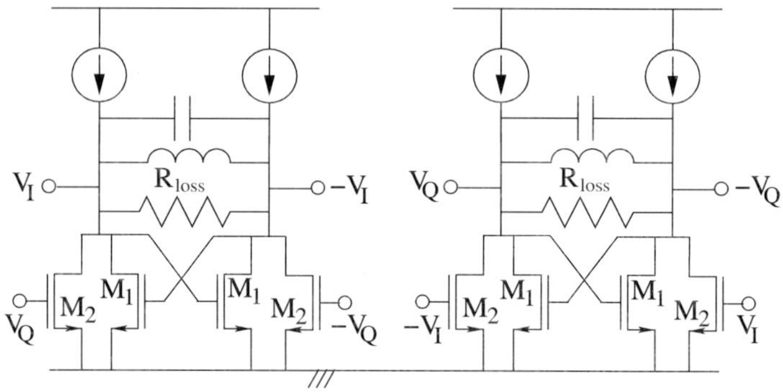

Figure 5.34: Harmonic quadrature oscillator topology. Tank losses are modeled by means of the parallel resistor R_{loss}.

than 10 minutes for the simulation results to be ready, then you should switch to the use of the behavioral model as soon as the simulation runs over a time span that exceeds 300 periods of the oscillation signal.

5.8.3.2 Harmonic quadrature oscillator

In a second example, we construct a behavioral model for the harmonic quadrature oscillator in Fig. 5.34. The model corresponds to the linearized averaged ODEs (5.168). In this example, the transistor models are based on the MOST level-1 equations. The resonant tanks have a Q-factor of about 10. Tank losses are modeled by putting a resistor R_{loss} in parallel with each resonant tank. The other parameters that specify the design of this quadrature oscillator are:

$$\gamma = \frac{G_{m,feedback}}{1/R_{loss}} = \frac{g_{m1}/2}{1/R_{loss}} = 2 \tag{5.184}$$

$$m = \frac{g_{m2}}{g_{m1}} = 0.3 \ . \tag{5.185}$$

Here, γ is the ratio between the small-signal transconductance of the active energy restoring feedback elements (as seen by the resonant tanks) and the loss conductance $G_{loss} = 1/R_{loss}$. The parameter m is the ratio of the transconductances of the transistors M_1 and M_2. It is a measure for the strength that couples the two resonant tanks.

The behavioral model corresponds to a linear approximation of the averaged ODEs that govern the state vector

$$\mathbf{q} = \left[\begin{array}{cccc} A_I & A_Q & \Delta\phi & \theta \end{array} \right]^T \ . \tag{5.186}$$

This state vector contains the amplitudes A_I and A_Q of the in-phase and quadrature oscillations, the phase difference $\Delta\phi$ between these oscillations and the common-mode phase θ. The averaged ODEs are linearized near the operating point values $\mathbf{q}_{1,op} = \left[\begin{array}{ccc} A_{I,op} & A_{Q,op} & \Delta\phi_{op} \end{array} \right]^T = \left[\begin{array}{ccc} 2.51 & 2.51 & \pi/2 \end{array} \right]^T$. The resulting behavioral model

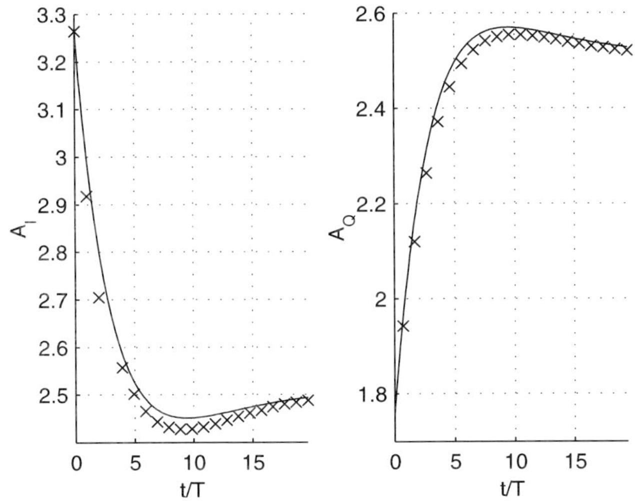

Figure 5.35: Amplitudes of the in-phase and quadrature oscillations obtained from the behavioral model (solid line) and extracted from the SPICE-like simulation of the original set of circuit equations (x-marks). The result are obtained for $\varepsilon = 0.2$, $\gamma = 0.5$ and $m = 0.3$.

is a system of linear time-invariant (LTI) differential equations. It is readily implemented in any environment for system-level simulation.

Fig. 5.35 compares the settling behavior predicted by the behavioral model

$$\frac{d}{d\tau}\Delta\mathbf{q}_1 = \mathbf{A}_{model}\Delta\mathbf{q}_1 \tag{5.187}$$

$$= \begin{bmatrix} -0.350 & -0.035 & -0.185 \\ -0.035 & -0.350 & 0.185 \\ 0.067 & -0.067 & -0.070 \end{bmatrix} \begin{bmatrix} \Delta A_I \\ \Delta A_Q \\ \Delta\Delta\phi \end{bmatrix} \tag{5.188}$$

with the envelopes extracted from the solutions of the original circuit equations. The latter are obtained by means of a SPICE-like algorithm. In (5.187)-(5.188), $\Delta\mathbf{q}_1 = \mathbf{q}_1 - \mathbf{q}_{1,op}$ with similar definitions for ΔA_I, ΔA_Q and $\Delta\Delta\phi$. The settling behavior is evaluated for initial values of the amplitudes A_I and A_Q deviate 30% from their operating-point values. The initial phase difference $\Delta\phi$ deviates 10% from its steady-state value. It is observed that the results predicted by the model are in good agreement with circuit-level behavior (to within 5%). Furthermore, simulations run about 30 times faster when using the behavioral model. Hence, use of the behavioral model significantly boosts simulation speed, even when compared to circuit simulations that use low-complexity transistor models (MOST level 1) instead of sophisticated models (BSIM4, MOS Model 11).

Finally, as an interesting application of the behavioral model (5.187)-(5.188) we men-

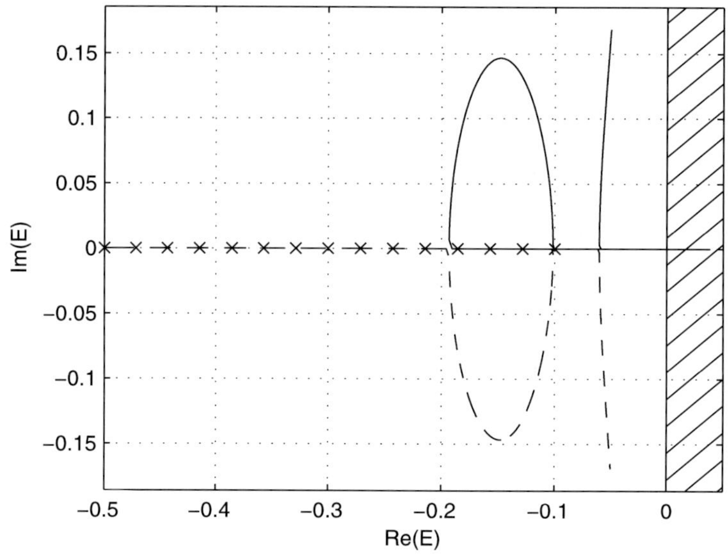

Figure 5.36: Loci of the oscillator eigenvalues (of the matrix $\mathbf{A}_{model}$) plotted in the complex plane for $\gamma = 0.5$ and $m = 0 \ldots 4$. The locus of the third eigenvalue, indicated by the crosses, lies entirely on the real axis. The hatched area marks the region of instability. If one of the eigenvalues enters this region, then the operating point around which linearization took place is unstable.

tion its use as an aid to optimize the design of an oscillator's dynamic behavior. With regard to this issue, it is particularly interesting to plot the loci of the eigenvalues of the matrix $\mathbf{A}_{model}$ in (5.187)-(5.188) [Vana01c]. As illustrated in Fig. 5.36, this is accomplished by computing $\mathbf{A}_{model}$ and its eigenvalues over a range of values for one of the design parameters Q, γ or m. The eigenvalues of $\mathbf{A}_{model}$ entirely characterize the oscillator's dynamics near the corresponding operating point. Their real values correspond to the oscillator's settling time while their complex parts, for instance, indicate the presence of ringing Furthermore, if one of the eigenvalues enters the right half of the complex plane (the hatced area in Fig. 5.36), then this indicates that the operating point around which linearization took place is unstable. Hence, loci like the ones in Fig. 5.36 are of great value in making design decisions, e.g. in determining a suitable value for the coupling parameter m. This value could, for example, be chosen in order to minimize the oscillator's settling time.

In summary, the behavioral modeling method here presented helps to derive oscillator models that contribute to boosting overall simulation efficiency. Moreover, these models can also be integrated in procedures or algorithms for making decisions on (optimal) design parameters as they help to focus on that part of the oscillator's dynamic behavior that is often of greatest interest.

5.9 Conclusions and directions for further research

This chapter has presented both the theory and algorithms to describe and model an oscillator's dynamic behavior. By proper partitioning of the circuit equations, it is possible to separately compute the periodic steady-state behavior and the transient dynamics. The theory builds on perturbation techniques and averaging. It provides a framework for the analysis of an oscillator's entire phase noise, injection locking and settling behavior. All these phenomena can be explained as consequences of the same underlying mechanism: perturbations drive the oscillator's core system from one state to another.

Two key applications of the theory above are oscillator phase-noise analysis and the extraction of compact behavioral models for (sets of coupled) harmonic oscillators. The work on phase noise culminated in a semi-analytic method to compute an oscillator's phase-noise spectrum in a few minutes time. Moreover, for the first time, a method was presented that enables us to accurately compute the very-close-in phase-noise spectrum. The harmonic oscillator behavioral models can be solved using a large simulation time step. As such, they are better suited for lengthy or repetitive (system-level) simulations than is the oscillator's original set of circuit equations. The accuracy and efficiency of the behavioral models was verified for both a single harmonic oscillator and an harmonic quadrature oscillator.

The work above is but the beginning of what promises to become a powerful theory and modeling technique for a wide variety of systems. Properly extended, it is applicable to virtually every system whose behavior is characterized by the presence of widely spaced time constants. Moreover, it is not required for these systems to be autonomous. In [Vana02e], the theory was extended to handle, for instance, PLL large-signal behavior.

Finally, it would be interesting to explore the link with the HTM framework discussed in the previous chapters. Projection of the perturbation terms requires us to solve systems of linear DAEs. In many cases, these systems turn out to be periodically time-varying. Hence, they can be solved using HTM-based techniques. A practical application is, for instance, the computation of the ISFs that are required for oscillator phase-noise computations.

Chapter 6

Conclusions

That's all folks! —*Woody Woodpecker*

This book has presented theory, methods and algorithms to support efficient and systematic modeling and analysis of telecommunication frontends and their building blocks. Focus was given to linear periodically time-varying (LPTV) systems and autonomous systems (oscillators). The resulting methods apply to a wide variety of systems and problems: the characterization of signal transfers in mixers, a PLL's time-varying stability behavior, noise folding in PLLs, oscillator injection-locking behavior, oscillator phase noise behavior, etc. Moreover, the development of all theory and methods was accompanied by the development of a sound intuitive understanding of the system behavior being studied: the story behind the math was considered as important as the math itself. As such, this work contributes to a better understanding of the behavior of telecom frontend building blocks and to the mathematical techniques necessary for systematic analysis of such systems.

6.1 Main achievements

The work presented in this book contributes to the state-of-the art in the analysis and modeling of telecommunication frontends and their building blocks. The two main subjects that it has dealt with are:

- **Linear periodically time-varying (LPTV) systems:** LPTV system behavior comes into play when system or circuit behavior is linearized in the neighborhood of a periodic (time-varying) operating point. This work has developed methods that allow us to manipulate LPTV systems in the frequency domain [Vana02a, Vana02c]. The methods build on the Harmonic Transfer Matrix (HTM) representation of an LPTV system [Maas88, Moll00, Vana02c, Were91a]. HTMs allow us to handle LPTV systems in a manner that is similar to dealing with LTI systems by means of (Laplace- or frequency-domain) transfer functions. This work has elaborated the theory of HTMs in full detail, including the frequency-domain characterization of signal transfers, LPTV stability analysis and LPTV noise analysis. The HTM framework has been applied to capture mixer and PLL behavior [Vana02a, Vana02c, Vana03a]. With regard to PLLs, our exact time-varying analysis gives solid mathematical underpinnings to a number of traditional (time-invariant) models while identifying their shortcomings.

- **Oscillator dynamics:** Oscillator behavior is often characterized by the presence of widely spaced time constants. While this often poses a bottleneck to traditional simulation algorithms, it enables the methods presented in this work to explicitly separate the oscillator's slow- and fast-varying signal components. This results in compact models that can be solved very efficiently using a large simulation time step . The modeling strategy is solidly grounded on the theory of dynamical systems, perturbation analysis and averaging [Bogo61, Kev81, Vana03b]. Applications of the theory involve oscillator phase noise analysis [Vana02d, Vana03c], injection locking behavior [Vana02d] and the construction of behavioral harmonic oscillator models [Vana01c, Vana02b, Vana03b]. Moreover, the theory here developed yields clear insights into the basic mechanisms that govern oscillator dynamic behavior (e.g. start-up behavior, etc.).

6.1.1 HTM-based LPTV system analysis

The HTM-based analysis of LPTV system behavior considers the signal content stored on different carrier waves, i.e. in different frequency bands, as independent input signals. In this way, a SISO LPTV system can be described as a limiting case of a MIMO LTI system. As a result, the behavior of an LPTV system can be captured by a matrix of transfer functions. Each transfer function models the transfer of signal content from a particular input frequency band to a particular output frequency band. The properties and the behavior of LPTV systems can then be analyzed using techniques for multi-variable LTI systems. This results in a powerful framework for handling LPTV system behavior, including the analysis of signal transfers, stability analysis and noise analysis.

A major bottleneck in making HTM-based techniques practical is the ∞-dimensional nature of a HTM. In principle, there are an infinite number of frequency bands that must be taken into account. When not handled carefully, this rapidly leads to computations that involve huge matrices [Moll00, Royc98, Royc99]. Although these matrices can be dealt with by means of sparse-matrix techniques, it is, in many cases, by far a suboptimal solution. This work has presented a number of methods that overcome these difficulties. These methods avoid tedious computations on huge matrices by exploiting both the structure inherent to HTMs and the properties of HTMs as they occur in design practice. In this way, they render HTM-based analysis of LPTV systems practical, both for obtaining numerical results and symbolic expressions.

The usefulness of the HTM framework was demonstrated by means of two applications: the analysis of PLL behavior and the SymbolicHTM algorithm. HTMs support an exact description of the time-varying aspects of a PLL's small-signal behavior. This time-varying behavior is inherent to all practical PLL implementations. Existing LTI techniques for PLL modeling were shown to coincide with specific approximations to the exact time-varying description. For the first time, these traditional LTI models were given solid mathematical groundings with their shortcomings clearly identified. As a second application, the SymbolicHTM algorithm has demonstrated that it is possible to (partially) automate the symbolic analysis of LPTV system behavior. It produces expressions for an LPTV system's input-output HTM elements when given the system's building block models. For a limited number of internal time-varying feedback loops, these symbolic expressions are quite accurate while only containing a limited number of terms. The algorithm complements existing algorithms for symbolic LTI network analysis [Fern98, Gie91, Lin91] and, for the first time, extends symbolic analysis towards (periodically) time-varying systems. SymbolicHTM is intended to support the introduction of HTM-based analysis in every day's circuit design practice.

6.1.2 Modeling oscillator dynamic behavior

By proper partitioning of the circuit equations, it has been shown possible to separate the computation of an oscillator's fast-varying periodic steady-state behavior and its slow-varying transient dynamics. The key to understanding oscillator dynamics lies in the observation that slow-varying transients are caused by small (perturbation) terms that push the oscillator's core system from one state to another. Here, a state of the oscillator's core corresponds to one of its fast-varying steady-state solutions. In this work, this understanding was translated into mathematics by means of perturbation techniques and averaging. This has resulted in a general theory that applies to a wide variety of problems involving oscillator dynamic behavior.

The practical applications addressed in this work concern the analysis of an oscillator's injection locking and phase noise behavior. Furthermore, we also presented an algorithm for the behavioral modeling of (sets of coupled) harmonic oscillators. All applications demonstrate the ease and efficiency with which the framework here developed allows us to handle complex nonlinear phenomena. With regard to oscillator phase noise, the theory here presented embeds existing work [Dem98, Dem00a] in a more global framework. Moreover, this work has resulted in a semi-analytic method to compute an oscillator's phase noise spectrum, including the details close to the oscillation frequency [Vana03c]. With regard to the behavioral modeling of harmonic oscillators, we obtained model equations that can be solved with a simulation time step that is much larger than would be possible for the (SPICE-like) solution of the original set of circuit equations. They were proven accurate and fast for both a single harmonic oscillator and a harmonic quadrature oscillator.

6.2 Leads for further work

It is the nature of research that the quest for knowledge tends to generate more questions than answers. This work should therefore be considered as a starting point for new research (by new researchers) rather than an as end point with nowhere to go from here. With regard to both linear system analysis and oscillator dynamics, there are still many challenges to be addressed. In what follows, some of them are listed.

In the area of linear system analysis, the topics to be explored concern both fundamental theory as practical applications:

- As presented in this text, the HTM-based frequency-domain analysis of LPTV systems can be generalized to other classes of LTV systems. These systems then transfer information between carrier waves other than the harmonic functions

$\{\psi_m(t) = e^{jm\omega_0 t}\}$ used in this work. Suggested directions for further research involve, for example, linear quasi-periodic time-varying systems and the use of principal component analysis for autodetecting relevant sets of carrier waves. The former is useful for heterodyne transceiver small-signal analysis while the latter can, for example, be applied to capture noise behavior in nonlinear systems.

- The systems to which HTM-based analysis applies, are by no means exhausted. As an example, we mention the computation of an oscillator's impulse sensitivity function by means of HTM-based techniques. The oscillator HTMs involved can be shown to have a rank-degenerate structure. This structure could be exploited to streamline computations.

- Having established methods that capture *linear* time-varying behavior, we can proceed by taking time-varying weak nonlinearities into account. Analyzing time-varying weakly-nonlinear behavior builds on perturbation techniques and linear time-varying system analysis. It thereby proceeds along the same lines of reasoning as time-invariant weakly-nonlinear system analysis [Wamb98a]. Developing methods for time-varying weakly-nonlinear system analysis would contribute yet another step towards a true top-down analysis (and hence design) of electronic systems and circuits.

With regard to the analysis of oscillator dynamic behavior, the work above is but the beginning of what promises to become a powerful technique to analyze and model a wide variety of systems:

- Properly adapted, it is applicable to virtually every system whose behavior is characterized by the presence of widely spaced time constants. Here, it is not required for these systems to be autonomous. In [Vana02e], the theory was adapted to handle, for instance, PLL large-signal behavior.

- A step that is of major importance to the methods here presented concerns the partitioning of the system equations, i.e. finding the core system. A general-purpose algorithm to do so would greatly contribute to the flexibility of the method and the scope of its applications.

Finally, it would be interesting to explore the link between both oscillator analysis and the HTM framework. As was discussed in chapter 5, projection of the perturbation terms requires us to solve systems of linear DAEs. In many cases, these systems turn out to be periodic. Hence, in principle, they can be handled using HTM-based techniques.

Appendix A

HTM norms and the comparison of HTMs

This appendix deals with HTM norms and the comparison of HTMs. Norms and criteria for comparing HTMs are especially useful for constructing approximations to HTMs. They allow us to quantify the notion of HTMs being "almost equal". Since HTMs represent LPTV operators (systems), the theory on HTM norms is closely related to the theory on operator norms.

A.1 Operator norms and the comparison of operators

Comparing two (LPTV) operators

$$H_1, H_2 \in S_i \rightarrow S_o \quad , \tag{A.1}$$

where S_i is the Banach space [Huts80] containing the input signals and S_o the one containing the output signals, requires us to define a notion of a distance $d(H_1,H_2)$ between them[1]. The most natural way to do so, is by comparing the outputs they produce for some relevant set S_i of test inputs. For continuous linear operators, this leads to [Huts80]

$$d(H_1,H_2) = ||H_1 - H_2|| \quad , \tag{A.2}$$

with

$$||H|| = \max_{u \in S_i} \frac{||H[u]||_o}{||u||_i} \quad . \tag{A.3}$$

Here, $||\cdot||_i$ is the norm defined over the space of input signals S_i and $||\cdot||_o$ is the norm defined over the space of output signals S_o. In what follows, we drop the subscripts $_i$ and $_o$. Which norm to use should be clear from the context.

The distance $||H_1 - H_2||$ can now be used to quantify the "degree of equality" of the two operators H_1 and H_2: stating that H_2 is close to H_1, or $H_2 \approx H_1$, implies that one expects $||H_1 - H_2||$ to be small. But what is "small"? The most sensible answer seems to be that, for each input u

$$||H_2[u] - H_1[u]|| \ll ||H_1[u]|| \, , \forall u \in S_i \quad , \tag{A.4}$$

[1] Note that S_i and S_o need not necessarily to be the same, e.g. in passing from the continuous-time to the discrete-time domain. In another example, S_i is the space of all signals with low-frequency content only, while the signals in S_o can have both low- and high-frequency content, i.e. $S_i \subset S_o$.

or

$$\max_{u \in S_i} \frac{\|H_2[u] - H_1[u]\|}{\|H_1[u]\|} \leq \varepsilon \ , \tag{A.5}$$

with $\varepsilon \ll 1$ some suitably chosen threshold level. However, there is still one matter open for debate. What if, for a certain $v \neq 0$, $H_1[v] = 0$? In that case, (A.4) becomes impossible to satisfy. To overcome this, we need to introduce the notion of a zero signal-level, i.e. there is a threshold level $\varepsilon_0 \|v\|$ for which

$$\|H[v]\| \approx 0 \Leftrightarrow \|H[v]\| \leq \varepsilon_0 \|v\| \ . \tag{A.6}$$

The threshold level is specified relative with respect to $\|v\|$ in order to have $\|H[v]\| \approx 0 \Rightarrow \|H[\alpha \cdot v]\| = |\alpha| \, \|H[v]\| \approx 0$. Taking this zero signal-level into account, (A.4) should be modified as

$$\|H_2[u] - H_1[u]\| \ll \|H_1[u]\| \vee \|H_2[u] - H_1[u]\| \leq \varepsilon_0 \|u\| \, , \forall u \in S_i \ . \tag{A.7}$$

This, approximately, is equivalent with

$$\max_{u \in S_i} \frac{\|H_2[u] - H_1[u]\|}{\|H_1[u]\| + (\varepsilon_0/\varepsilon)\|u\|} \leq \varepsilon \ . \tag{A.8}$$

A sufficient condition for (A.8) to hold)and one that is often much easier to compute) is given by

$$\frac{\max_{u \in S_i} \|H_2[u] - H_1[u]\| / \|u\|}{\min_{u \in S_i} \|H_1[u]\| / \|u\| + (\varepsilon_0/\varepsilon)} \leq \varepsilon \ . \tag{A.9}$$

This completes our discussion on operator norms and the comparison of operators.

A.2 Selecting the set of test inputs

Before continuing with the application of the theory outlined above to HTMs, we briefly discuss the selection of the space S_i of test input signals. As seen from (A.3), operators are compared by examining their response on signals chosen from S_i. In selecting this space, one should therefore be careful to select only those signals that are of greatest interest or that excite the modes of H_1 and H_2 that are of main importance. For example, if one is interested in comparing the response of H_1 and H_2 to narrow-banded signals centered around the frequency f_c, then it is often be sufficient to select $S_i = \{\cos(2\pi f_c t), \sin(2\pi f_c t)\}$. In this case, it is clearly useless to select test signals that have a large part of their frequency content outside the frequency band around f_c. Norms based on such a restricted test signal set S_i often simplify devising approximations for H_1 or H_2. Note, however, that these approximations are only valid when operating on input signals belonging to the restricted set S_i, e.g. they only hold for one particular frequency band.

A.3 Expressing LPTV operator norms in terms of the corresponding HTM elements

Computing the operator norm (A.3) and the related quantities needed for comparing operators comes down to computing the signal norms $\|u\|$ and $\|H[u]\|$. In what follows, we present expressions for these norms in terms of the equivalent baseband components of $u(t)$ and the elements of the HTM corresponding to $H[\cdot]$.

Let us be given a continuous-time signals $u(t)$ for which we define

$$\|u\|^2 = \int_{-\infty}^{+\infty} u(t)^* u(t) \mathrm{d}t \ , \tag{A.10}$$

i.e. the square of the signal norm equals the energy in $u(t)$. Given the signal's equivalent baseband components $u_m(t)$, where

$$u(t) = \sum_m u_m(t) e^{jm\omega_0 t} \ , \tag{A.11}$$

we find that

$$U(j\omega) = \mathcal{F}\left\{ \sum_{m=-\infty}^{+\infty} u_m(t) e^{jm\omega_0 (t)} \right\} = \sum_m U_m\left(j(\omega - m\omega_0)\right) \ . \tag{A.12}$$

Using Parseval's equality [Arf85], we then obtain

$$\|u\|^2 = \frac{1}{2\pi} \sum_{m=-\infty}^{+\infty} \sum_{n=-\infty}^{+\infty} \int_{-\infty}^{+\infty} U_m\left(j(\omega - m\omega_0)\right)^* U_n\left(j(\omega - n\omega_0)\right) \mathrm{d}\omega \ . \tag{A.13}$$

Note that when the $U_m(j\omega)$ are bandlimited within $[-\omega_0/2, \omega_0/2]$, (A.13) reduces to

$$\|u\|^2 = \frac{1}{2\pi} \sum_{m=-\infty}^{+\infty} \int_{-\omega_0/2}^{+\omega_0/2} U_m(j\omega)^* U_m(j\omega) \mathrm{d}\omega \tag{A.14}$$

$$= \frac{1}{2\pi} \int_{-\omega_0/2}^{+\omega_0/2} \mathbf{U}(j\omega)^* \mathbf{U}(j\omega) \mathrm{d}\omega \tag{A.15}$$

with $\mathbf{U}(j\omega) = \begin{bmatrix} \cdots & U_{-1}(j\omega) & U_0(j\omega) & U_1(j\omega) & \cdots \end{bmatrix}^T$ the vector containing the (bandlimited) equivalent baseband components. For $\|H[u]\|^2$ we find a similar expressions by replacing $\mathbf{U}(j\omega)$ with $\tilde{\mathbf{H}}(j\omega)\mathbf{U}(j\omega)$ in the expression above. For bandlimited equivalent baseband components $U_m(j\omega)$, this yields

$$\|H[u]\|^2 = \frac{1}{2\pi} \int_{-\omega_0/2}^{+\omega_0/2} \mathbf{U}(j\omega)^* \tilde{\mathbf{H}}(j\omega)^* \tilde{\mathbf{H}}(j\omega) \mathbf{U}(j\omega) \mathrm{d}\omega \ . \tag{A.16}$$

Using (A.16) and (A.3), we can now compute the norm of the operator H for any given set of input signals S_i. For instance, for $S_i = L^2([0, \infty[)$, i.e. the set of all square integrable functions, it is readily shown that

$$\|H\| = \max_{\omega \in [-\omega_0/2, \omega_0/2]} \left\|\tilde{\mathbf{H}}(j\omega)\right\|_2 \ .$$

For other sets of input signals, the corresponding operator norm can be computed in a similar manner.

A.4 Conclusions

This appendix has discussed the theory on HTM norms and their computation. HTM norms are a basic tool in comparing whether HTMs represent LPTV systems that show similar behavior. This is especially useful when trying to simplify a given HTM.

Appendix B

The Sherman-Morisson-Woodbury formula

The Shermann-Morisson-Woodbury formula [Gol96] is used to invert matrices structered as $\mathbf{A}+\mathbf{U}\mathbf{V}^T$, with $\mathbf{A} \in \mathbb{R}^{N\times N}$ and $\mathbf{U},\mathbf{V} \in \mathbb{R}^{N\times M}$. It goes as

$$\left(\mathbf{A}+\mathbf{U}\mathbf{V}^T\right)^{-1} = \mathbf{A}^{-1} - \mathbf{A}^{-1}\mathbf{U}\left(\mathbf{I}+\mathbf{V}^T\mathbf{A}^{-1}\mathbf{U}\right)^{-1}\mathbf{V}^T\mathbf{A}^{-1} \ . \tag{B.1}$$

The Shermann-Morisson-Woodbury formula is especially useful when $\mathbf{A}^{-1}$ can be computed at low cost (e.g. when $\mathbf{A}$ is diagonal, orthonormal or upper/lower triangular) and both $\mathbf{U}$ and $\mathbf{V}$ are of low rank, i.e. $M \ll N$. As a special case for $M = 1$, we find

$$\left(\mathbf{A}+\mathbf{u}\mathbf{v}^T\right)^{-1} = \mathbf{A}^{-1} - \frac{\mathbf{A}^{-1}\mathbf{u}\mathbf{v}^T\mathbf{A}^{-1}}{1+\mathbf{v}^T\mathbf{A}^{-1}\mathbf{u}} \ . \tag{B.2}$$

This is also known as the Sherman-Morisson formula.

Appendix C

HTM elements of the linear downconversion mixer

This appendix lists the full expressions for the HTM elements $\tilde{H}_{0,1}(s)$, $\tilde{H}_{0,0}(s)$ and $\tilde{H}_{0,2}(s)$ of the linear downconversion mixer topology discussed in section 4.3.7 of chapter 4. The expressions are presented for the SymbolicHTM algorithm performing computations up to first order in (the dummy variable) μ. In these expressions

$$L(s) = 1 - \tilde{G}_{0,0}(s) - \frac{H_c(s)\left(g_{m2,0} + sC_c\right)}{Y_o(s)} \tag{C.1}$$

with

$$\begin{cases} \tilde{G}_{0,0}(s) & = & \frac{H_i(s)\left(A_1(s)\left(g_{m2,0}+sC_c\right)+Y_{fb}(s)\right)}{Y_o(s)} \\ \tilde{G}_{k,0}(s) & = & \frac{H_i(s)A_1(s)g_{m2,k}}{Y_o(s+jk\omega_0)} \text{ for } k \neq 0 \ . \end{cases} \tag{C.2}$$

Furthermore, $g_{m,k}$ are the Fourier coefficients of the time-varying transconductance of the second OTA stage, as specified in (4.88), and the $V_{n1,k}$ are the Fourier coefficients corresponding to the time-varying operating point $V_{n1}(t)$ in equation (4.87).

The wanted signal transfer is described by

$$\begin{aligned} \tilde{H}_{0,1}(s) \quad = \quad & \frac{1}{2}\frac{\beta V_{osc}\tilde{G}_{0,0}(s)}{L(s)Y_{fb}(s)} \\ & + \frac{\gamma\tilde{G}_{-1,0}(s+j\omega_0)Y_{gs}(s+j\omega_0)}{L(s)Y_{fb}(s+j\omega_0)} \\ & + \frac{1}{2}\frac{\beta V_{osc}\tilde{G}_{-2,0}(s+2j\omega_0)}{L(s)Y_{fb}(s+2j\omega_0)} \\ & - \frac{\gamma^2\beta V_{n,-1}\tilde{G}_{0,0}(s)H_i(s+j\omega_0)Y_{gs}(s+j\omega_0)}{L(s)Y_{fb}(s)Y_{fb}(s+j\omega_0)} \\ & + \frac{\gamma\tilde{G}_{0,0}(s+j\omega_0)\tilde{G}_{-1,0}(s+j\omega_0)Y_{gs}(s+j\omega_0)}{L(s)L(s+j\omega_0)Y_{fb}(s+j\omega_0)} \\ & + \frac{\gamma g_{m,-1}\tilde{G}_{0,0}(s+j\omega_0)H_c(s+j\omega_0)Y_{gs}(s+j\omega_0)}{L(s)Y_o(s)L(s+j\omega_0)Y_{fb}(s+j\omega_0)} \\ & - \frac{\gamma^2\beta V_{n,-1}\tilde{G}_{0,0}(s)\tilde{G}_{0,0}(s+j\omega_0)H_i(s+j\omega_0)Y_{gs}(s+j\omega_0)}{L(s)Y_{fb}(s)L(s+j\omega_0)Y_{fb}(s+j\omega_0)} \end{aligned} \tag{C.3}$$

$$+\frac{1}{2}\frac{\beta V_{osc}\tilde{G}_{0,0}(s+2j\omega_0)\tilde{G}_{-2,0}(s+2j\omega_0)}{L(s)L(s+2j\omega_0)Y_{fb}(s+2j\omega_0)}$$
$$+\frac{1}{2}\frac{\beta V_{osc}g_{m,-2}\tilde{G}_{0,0}(s+2j\omega_0)H_c(s+2j\omega_0)}{L(s)Y_o(s)L(s+2j\omega_0)Y_{fb}(s+2j\omega_0)}$$

This expression is entirely dominated by the first term. The parasitic transfer from baseband to baseband is described by

$$\begin{aligned}
\tilde{H}_{0,0}(s) \quad = \quad & \frac{\gamma\tilde{G}_{0,0}(s)Y_{gs}(s)}{L(s)Y_{fb}(s)} \\
& +\frac{1}{2}\frac{\beta V_{osc}\tilde{G}_{0,0}(s-j\omega_0)\tilde{G}_{1,0}(s-j\omega_0)}{L(s)L(s-j\omega_0)Y_{fb}(s-j\omega_0)} \\
& +\frac{1}{2}\frac{\beta V_{osc}\tilde{G}_{0,0}(s+j\omega_0)\tilde{G}_{-1,0}(s+j\omega_0)}{L(s)L(s+j\omega_0)Y_{fb}(s+j\omega_0)} \\
& +\frac{1}{2}\frac{\beta V_{osc}g_{m,1}\tilde{G}_{0,0}(s-j\omega_0)H_c(s-j\omega_0)}{L(s)Y_o(s)L(s-j\omega_0)Y_{fb}(s-j\omega_0)} \\
& +\frac{1}{2}\frac{\beta V_{osc}g_{m,-1}\tilde{G}_{0,0}(s+j\omega_0)H_c(s+j\omega_0)}{L(s)Y_o(s)L(s+j\omega_0)Y_{fb}(s+j\omega_0)} \\
& -\frac{1}{2}\frac{\beta^2\gamma V_{osc}V_{n,1}\tilde{G}_{0,0}(s)\tilde{G}_{0,0}(s-j\omega_0)H_i(s-j\omega_0)}{L(s)Y_{fb}(s)L(s-j\omega_0)Y_{fb}(s-j\omega_0)} \\
& -\frac{1}{2}\frac{\beta^2\gamma V_{osc}V_{n,-1}\tilde{G}_{0,0}(s)\tilde{G}_{0,0}(s+j\omega_0)H_i(s+j\omega_0)}{L(s)Y_{fb}(s)L(s+j\omega_0)Y_{fb}(s+j\omega_0)} \\
& +\frac{1}{2}\frac{\beta V_{osc}\tilde{G}_{1,0}(s-j\omega_0)}{L(s)Y_{fb}(s-j\omega_0)} \\
& +\frac{1}{2}\frac{\beta V_{osc}\tilde{G}_{-1,0}(s+j\omega_0)}{L(s)Y_{fb}(s+j\omega_0)} \\
& -\frac{1}{2}\frac{\beta^2\gamma V_{osc}V_{n,1}\tilde{G}_{0,0}(s)H_i(s-j\omega_0)}{L(s)Y_{fb}(s)Y_{fb}(s-j\omega_0)} \\
& -\frac{1}{2}\frac{\beta^2\gamma V_{osc}V_{n,-1}\tilde{G}_{0,0}(s)H_i(s+j\omega_0)}{L(s)Y_{fb}(s)Y_{fb}(s+j\omega_0)}
\end{aligned} \tag{C.4}$$

while the parasitic transfer from $2f_0$ to baseband is described by

$$\begin{aligned}
\tilde{H}_{0,2}(s) \quad = \quad & \frac{1}{2}\frac{\beta V_{osc}\tilde{G}_{-1,0}(s+j\omega_0)}{L(s)Y_{fb}(s+j\omega_0)} \\
& +\frac{\gamma\tilde{G}_{-2,0}(s+2j\omega_0)Y_{gs}(s+2j\omega_0)}{L(s)Y_{fb}(s+2j\omega_0)} \\
& -\frac{1}{2}\frac{\beta^2\gamma V_{osc}V_{n,-1}H_i(s+j\omega_0)}{L(s)Y_{fb}(s)Y_{fb}(s+j\omega_0)} \\
& +\frac{1}{2}\frac{\beta V_{osc}\tilde{G}_{0,0}(s+j\omega_0)\tilde{G}_{-1,0}(s+j\omega_0)}{L(s)L(s+j\omega_0)Y_{fb}(s+j\omega_0)}
\end{aligned}$$

$$
\begin{aligned}
&+\frac{1}{2}\frac{\beta V_{osc} g_{m,-1}\tilde{G}_{0,0}(s+j\omega_0)H_c(s+j\omega_0)}{L(s)Y_o(s)L(s+j\omega_0)Y_{fb}(s+j\omega_0)} \\
&-\frac{1}{2}\frac{\beta^2\gamma V_{osc}V_{n,-1}\tilde{G}_{0,0}(s)\tilde{G}_{0,0}(s+j\omega_0)H_i(s+j\omega_0)}{L(s)Y_{fb}(s)L(s+j\omega_0)Y_{fb}(s+j\omega_0)} \\
&+\frac{\gamma\tilde{G}_{0,0}(s+2j\omega_0)\tilde{G}_{-2,0}(s+2j\omega_0)Y_{gs}(s+2j\omega_0)}{L(s)L(s+2j\omega_0)Y_{fb}(s+2j\omega_0)} \\
&+\frac{\gamma g_{m,-2}\tilde{G}_{0,0}(s+2j\omega_0)Y_{gs}(s+2j\omega_0)H_c(s+2j\omega_0)}{L(s)L(s+2j\omega_0)Y_{fb}(s+2j\omega_0)}\,.
\end{aligned}
\tag{C.5}
$$

It depends on the frequency interval to decide on which of the terms in the expressions (C.4) and (C.5) are dominant.

Appendix D

Oscillator dynamics: analysis of the deviation from the attracting manifold

In section 5.5 of chapter 5, we discussed the behavior of a weakly perturbed oscillator, i.e. the behavior of a core system interacting with weak disturbances. Mathematically, we were interested in the structure of the waveforms $\mathbf{x}_\varepsilon(\tau)$ that solve the system of DAEs

$$\frac{\mathrm{d}}{\mathrm{d}\tau}\mathbf{c}(\mathbf{x}_\varepsilon) + \mathbf{g}(\mathbf{x}_\varepsilon) + \varepsilon\mathbf{f}(\mathbf{x}_\varepsilon, \tau) \quad . \tag{D.1}$$

Here, as usual, the variable ε is called a perturbation variable and marks the disturbances. Solutions to (D.1) were shown to be structured as

$$\mathbf{x}_\varepsilon(\tau) = \mathbf{x}_s(\tau, \mathbf{p}(\tau, \varepsilon)) + \varepsilon\Delta\mathbf{x}(\tau, \varepsilon) \quad . \tag{D.2}$$

The first component of this solution models the motion over the attracting manifold M. This manifold consists of all steady-state solutions that solve the core system, i.e. the steady-state solutions to (D.1) for $\varepsilon = 0$. The motion over this manifold was discussed in depth in section 5.5 of chapter 5. This appendix deals with the second component in (D.2) which models the oscillator's deviation from the manifold M. In doing so, special interest is given as to prove that $\varepsilon\Delta\mathbf{x}(\tau)$ remains bounded. This study also yields a time constant that characterizes how long it takes for the manifold M to attract solutions in its neighborhood. Note that in what follows, we discuss the behavior of the normalized deviation $\Delta\mathbf{x}(\tau)$. The real deviation is obtained by multiplying this normalized deviation with the perturbation variable ε.

D.1 Components of the deviation $\Delta\mathbf{x}(\tau)$

By substituting the solution (D.2) into (D.1), it is obtained that

$$\begin{aligned} \mathbf{C}(\tau,\mathbf{p})\,\mathbf{U}_1(\tau,\mathbf{p})\frac{\mathrm{d}\mathbf{p}}{\mathrm{d}\tau} + \varepsilon\frac{\mathrm{d}}{\mathrm{d}\tau}\left(\mathbf{C}(\tau,\mathbf{p})\,\Delta\mathbf{x}(\tau)\right) & \\ +\varepsilon\mathbf{G}(\tau,\mathbf{p})\,\Delta\mathbf{x}(\tau) + \varepsilon\mathbf{f}_p(\tau,\mathbf{p}) + O\left(\varepsilon^2\right) &= 0 \ . \end{aligned} \tag{D.3}$$

Here, $\mathbf{C}(\tau,\mathbf{p}) \in \mathbb{R}^{N\times N}$,being $\mathbf{G}(\tau,\mathbf{p})$, $\mathbf{f}_p(\tau,\mathbf{p}) \in \mathbb{R}^N$ and $\mathbf{U}_1(\tau,\mathbf{p}) \in \mathbb{R}^{N\times P}$ are defined by (5.55)-(5.58) in section 5.5 of chapter 5. Furthermore, the process $\mathbf{p}(\tau)$ is solved from 5.66.

As stated in that same section, in order for $\Delta\mathbf{x}(\tau)$ to remain bounded, it must belong to the space spanned by the columns of the $[\mathbf{U}_2(\tau,\mathbf{p})\ \mathbf{U}_3(\tau,\mathbf{p})] \in \mathbb{R}^{N\times(Q+R)}$, or

$$\begin{aligned} \Delta\mathbf{x}(\tau) &= \left[\begin{array}{cc} \mathbf{U}_2(\tau,\mathbf{p}(\tau)) & \mathbf{U}_3(\tau,\mathbf{p}(\tau)) \end{array}\right] \left[\begin{array}{c} \alpha_2(\tau) \\ \alpha_3(\tau) \end{array}\right] && \text{(D.4)} \\ &= \mathbf{U}_2(\tau,\mathbf{p}(\tau))\alpha_2(\tau) + \mathbf{U}_3(\tau,\mathbf{p}(\tau))\alpha_3(\tau) && \text{(D.5)} \\ &= \Delta\mathbf{x}_2(\tau) + \Delta\mathbf{x}_3(\tau)\ . && \text{(D.6)} \end{aligned}$$

In the above, $\mathbf{U}_2(\tau,\mathbf{p}) \in \mathbb{R}^{N\times Q}$ spans the stable modes of the N-dimensional homogeneous system of DAEs

$$\frac{\partial}{\partial\tau}(\mathbf{C}(\tau,\mathbf{p})\mathbf{y}(\tau)) + \mathbf{G}(\tau,\mathbf{p})\mathbf{y}(\tau) = 0\ . \tag{D.7}$$

Moreover, $\mathbf{U}_3(\tau,\mathbf{p}) \in \mathbb{R}^{N\times R}$ is defined by $\mathbf{C}(\tau,\mathbf{p})\mathbf{U}_3(\tau,\mathbf{p}) = 0$, i.e. the directions stored in the columns of $\mathbf{U}_3(\tau,\mathbf{p})$ capture the deviation from the manifold defined by the constraints that are implicit to the DAEs (5.34) that model the core system. Note that $\mathbf{U}_3(\tau,\mathbf{p})$ can always be chosen to be normalized, i.e. $\|\mathbf{U}_3(\tau,\mathbf{p})\| = 1$. Having obtained these results, we continue with a discussion on both the behavior of $\Delta\mathbf{x}_2(\tau)$ and $\Delta\mathbf{x}_3(\tau)$.

D.2 Behavior of $\Delta\mathbf{x}_2(\tau)$

The first component of the (normalized) deviation is oriented along space spanned by the stable modes of the attracting manifold M. Since the manifold is stable along these directions, it is expected that the corresponding "attracting force" keeps the component $\Delta\mathbf{x}_2(\tau)$ bounded. In what follows, we first derive a zeroth-order approximation for $\Delta\mathbf{x}_2(\tau)$. Next, we address the issue of its boundedness.

An expression for $\Delta\mathbf{x}_2(\tau)$

By left multiplication of both sides of (D.3) with the projector $\mathbf{V}_2^T(\tau,\mathbf{p})$ satisfying

$$\begin{aligned} \left(\frac{\partial}{\partial\tau}\mathbf{V}_2^T(\tau,\mathbf{p})\right)\cdot\mathbf{C}(\tau,\mathbf{p}) - \mathbf{V}_2^T(\tau,\mathbf{p})\cdot\mathbf{G}(\tau,\mathbf{p}) &= 0 && \text{(D.8)} \\ \mathbf{V}_2^T(0,\mathbf{p})\cdot\mathbf{C}(0,\mathbf{p})\cdot\mathbf{U}_2(0,\mathbf{p}) &= \mathbf{I} && \text{(D.9)} \\ \mathbf{V}_2^T(0,\mathbf{p})\cdot\mathbf{C}(0,\mathbf{p})\cdot\mathbf{U}_1(0,\mathbf{p}) &= 0\ , && \text{(D.10)} \end{aligned}$$

with $\mathbf{I}$ the unity matrix, we obtain

$$\frac{\mathrm{d}}{\mathrm{d}\tau}\left(\mathbf{V}_2^T(\tau,\mathbf{p})\cdot\mathbf{C}(\tau,\mathbf{p})\cdot\Delta\mathbf{x}(\tau)\right) = -\mathbf{V}_2^T(\tau,\mathbf{p})\mathbf{f}_p(\tau,\mathbf{p}) + O(\varepsilon)\ . \tag{D.11}$$

Since $\mathbf{U}_3(\tau,\mathbf{p})$ satisfies $\mathbf{C}(\tau,\mathbf{p})\mathbf{U}_3(\tau,\mathbf{p}) = 0$ and with (D.5) and (D.8)-(D.10) it is then readily shown that (D.12) is equivalent to

$$\frac{\mathrm{d}\alpha_2}{\mathrm{d}\tau} = -\mathbf{V}_2^T(\tau,\mathbf{p}(\tau))\mathbf{C}(\tau,\mathbf{p}(\tau))\mathbf{f}_p(\tau,\mathbf{p}(\tau)) + O(\varepsilon)\ . \tag{D.12}$$

Hence, up to zeroth order in ε, we find

$$\begin{aligned}\Delta\mathbf{x}_2(\tau) &= \mathbf{U}_2(\tau,\mathbf{p}(\tau))\alpha_2(\tau) && \text{(D.13)}\\ &= \mathbf{U}_2(\tau,\mathbf{p}(\tau))\mathbf{V}_2^T(\tau_0,\mathbf{p}(\tau_0))\,\mathbf{C}(\tau_0,\mathbf{p}(\tau_0))\,\Delta\mathbf{x}_2(\tau_0)\\ &\quad -\int_{\tau_0}^{\tau}\mathbf{U}_2(\tau,\mathbf{p}(\tau))\mathbf{V}_2^T(s,\mathbf{p}(s))\,\mathbf{C}(s,\mathbf{p}(s))\,\mathbf{f}_p(s,\mathbf{p}(s))\,\mathrm{d}s && \text{(D.14)}\end{aligned}$$

as an expression for the first component of the oscillator's deviation from the manifold.

Boundedness of $\Delta\mathbf{x}_2(\tau)$

In order to prove that $\Delta\mathbf{x}_2(\tau)$ remains bounded, it is sufficient to show that the homogeneous part in (D.14) (the first term) is uniformly exponentially stable [Cor91], i.e. $\forall\tau,\tau_0$ and for an arbitrary $\Delta\mathbf{x}_2(\tau_0)$ with $||\Delta\mathbf{x}_2(\tau_0)|| = 1$, it holds

$$\left|\left|\Delta\mathbf{x}_{2,homogeneous}(\tau)\right|\right| < Ke^{-\beta(\tau-\tau_0)} \tag{D.15}$$

with $K,\beta > 0$. Since the initial condition $\Delta\mathbf{x}_2(\tau_0)$ is arbitrary, (D.15) is equivalent with

$$\left|\left|\mathbf{U}_2(\tau,\mathbf{p}(\tau))\mathbf{V}_2^T(\tau_0,\mathbf{p}(\tau_0))\,\mathbf{C}(\tau_0,\mathbf{p}(\tau_0))\right|\right| < Ke^{-\beta(\tau-\tau_0)},\ \forall\tau,\tau_0\ . \tag{D.16}$$

Using (D.16), it is readily shown that

$$\begin{aligned}||\Delta\mathbf{x}_2(\tau)|| &\leq Ke^{-\beta(\tau-\tau_0)}\,||\Delta\mathbf{x}_2(\tau_0)||\\ &\quad +\frac{K\cdot L}{\beta}\left(1-e^{-\beta(\tau-\tau_0)}\right)\ , && \text{(D.17)}\end{aligned}$$

with

$$L = \max_{\tau,\mathbf{p}}||\mathbf{f}_p(\tau,\mathbf{p})||\ , \tag{D.18}$$

is an upper bound for $||\Delta\mathbf{x}_2(\tau)||$. For $\tau \gg \tau_0$, this yields

$$||\Delta\mathbf{x}_2(\tau)|| \leq \frac{K\cdot L}{\beta}\ . \tag{D.19}$$

Hence, the deviation grows larger when the strength of the perturbing force $\mathbf{f}_p$ grows large —especially the strength of the components oriented along the space spanned by $\mathbf{U}_2$— and when the time constant $1/\beta$ increases. This time constant is a measure for how quick the manifold attracts solutions in its neighborhood. It remains of course to be answered whether it is possible to determine suitable values for the constants K,β. In what follows, we outline an argument for (D.7) being a system of DAEs that is periodic in τ.

If $\mathbf{p}$ is considered constant and assuming the system of linear DAEs (D.7) to be T-periodic in τ and stable, Floquet theory [Lam97] shows that we can write

$$\begin{aligned}\mathbf{U}_2(\tau,\mathbf{p}) &= \mathbf{U}_{p,2}(\tau,\mathbf{p})e^{\Lambda(\mathbf{p})t} && \text{(D.20)}\\ \mathbf{V}_2^T(\tau,\mathbf{p}) &= e^{-\Lambda(\mathbf{p})t}\mathbf{V}_{p,2}^T(\tau,\mathbf{p})\ . && \text{(D.21)}\end{aligned}$$

Here, both $\mathbf{U}_{p,2}(\tau,\mathbf{p})$ and $\mathbf{V}_{p,2}^T(\tau,\mathbf{p})$ are periodic in τ with their norm bounded $\forall\tau$. Furthermore

$$\Lambda(\mathbf{p}) = \begin{bmatrix} \mu_1(\mathbf{p}) & & & \\ & \mu_2(\mathbf{p}) & & 0 \\ 0 & & \ddots & \\ & & & \mu_Q(\mathbf{p}) \end{bmatrix} \in \mathbb{R}^{Q\times Q} \tag{D.22}$$

is a diagonal matrix containing the nonzero Floquet exponents corresponding to (D.7). Since the system is assumed to be stable, we can order the exponents μ_k such that

$$\mathrm{Re}\{\mu_Q\} \leq \mathrm{Re}\{\mu_{Q-1}\} \leq \cdots \leq \mathrm{Re}\{\mu_1\} < 0 \ . \tag{D.23}$$

Using (D.20)-(D.23), it is readily shown that

$$\left\|\mathbf{U}_2(\tau,\mathbf{p})\mathbf{V}_2^T(\tau_0,\mathbf{p})\,\mathbf{C}(\tau_0,\mathbf{p})\right\| \leq K(\mathbf{p})e^{-\beta(\mathbf{p})(\tau-\tau_0)} \tag{D.24}$$

with

$$\beta(\mathbf{p}) = |\mathrm{Re}\{\mu_1(\mathbf{p})\}| \tag{D.25}$$

$$K(\mathbf{p}) = \left(\max_{\tau\in[0,T]} \|\mathbf{U}_{p,2}(\tau,\mathbf{p})\|\right)\left(\max_{\tau\in[0,T]} \|\mathbf{V}_{p,2}(\tau,\mathbf{p})\mathbf{C}(\tau,\mathbf{p})\|\right) \ . \tag{D.26}$$

Substituting these results in (D.19), we then obtain

$$\|\Delta\mathbf{x}_2(\tau)\| \leq L\frac{K(\mathbf{p})}{\beta(\mathbf{p})} \ . \tag{D.27}$$

as an upper bound on the deviation.

The reasoning above, however, does not take into account that the state vector $\mathbf{p}$ also varies with time. Fortunately, (D.27) holds as long as the time constant $1/\beta(\mathbf{p})$, which characterizes the time it takes to attract solutions to the manifold M, is much smaller than the time scale T/ε over which the process $\mathbf{p}(\tau)$ is changing. The inequality (D.27) can, therefore, be seen as a kind of quasi-static approximation, expressing a slowly changing upper bound on the magnitude of $\Delta\mathbf{x}_2(\tau)$.

D.3 The behavior of $\Delta\mathbf{x}_3(\tau)$

The second component of $\Delta\mathbf{x}(\tau) = \Delta\mathbf{x}_2(\tau) + \Delta\mathbf{x}_3(\tau)$ models the deviation from the manifold defined by the constraints implicit to the system of the DAEs (5.34) that model the core system. It is found by left multiplication of (D.3) with $\mathbf{V}_3^T(\tau,\mathbf{p})$ where

$$\mathbf{V}_3^T(\tau,\mathbf{p})\mathbf{C}(\tau,\mathbf{p}) = 0 \tag{D.28}$$

and

$$\mathbf{V}_3^T(\tau,\mathbf{p})\mathbf{G}(\tau,\mathbf{p})\mathbf{U}_3(\tau,\mathbf{p}) = \mathbf{I} \ . \tag{D.29}$$

Hence, the rows of $\mathbf{V}_3^T(\tau,\mathbf{p})$ span the left null space of $\mathbf{C}(\tau,\mathbf{p})$. After a little algebra, it is then obtained that

$$\Delta\mathbf{x}_3(\tau) = -\mathbf{U}_3(\tau,\mathbf{p})\left(\mathbf{V}_3^T(\tau,\mathbf{p})\mathbf{G}(\tau,\mathbf{p})\Delta\mathbf{x}_2(\tau) + \mathbf{V}_3^T(\tau,\mathbf{p})\mathbf{f}_p(\tau,\mathbf{p})\right) \ . \tag{D.30}$$

If we assume that $\mathbf{C}(\tau,\mathbf{p})$ and $\mathbf{G}(\tau,\mathbf{p})$ are continuous in their arguments, and, therefore, also $\mathbf{V}_3^T(\tau,\mathbf{p})$, and since we assumed the system to be T-periodic (and therefore also $\mathbf{V}_3^T(\tau,\mathbf{p})$ and $\mathbf{G}(\tau,\mathbf{p})$), all matrices and vectors in the right-hand side of (D.30) are bounded. Hence, so is $\Delta\mathbf{x}_3(\tau)$.

D.4 Conclusions

In this appendix, we presented expressions for the two components of the (normalized) deviation $\Delta\mathbf{x}(\tau) = \Delta\mathbf{x}_2(\tau) + \Delta\mathbf{x}_3(\tau)$. The first component is directed along the space spanned by the stable modes of the attracting manifold. The second component models the deviation from the manifold defined by the constraints implicit to the system of the DAEs that model the core system. Both components were shown to remain bounded.

Conclusions

Appendix E

Analysis of a harmonic oscillator

This appendix deals with the analysis of the harmonic oscillator topology shown in Fig. E.1. Special interest is given to oscillator's operating-point phase behavior. This results in the expression for the impulse sensitivity function (ISF) corresponding to the source $u(\tau)$. This ISF was, amongst others, used in the analysis of the injection locking problem treated in section 5.7.2 of chapter 5.

E.1 Determining the oscillator's averaged dynamics

The normalized circuit equations corresponding to the topology shown in Fig. E.1 are easily shown to equal[1]

$$\frac{\mathrm{d}x_d}{\mathrm{d}\tau} + y_d + \left[\frac{x_d}{Q} + \frac{I_{M,d}(x_d, x_{cm})}{I_{osc}} + \frac{u(\tau)}{I_{osc}}\right] = 0 \tag{E.1}$$

$$\frac{\mathrm{d}y_d}{\mathrm{d}\tau} - x_d = 0 \tag{E.2}$$

$$I_{M,cm}(x_d, x_{cm}) - I_b = 0 \tag{E.3}$$

with

$$x_d = \frac{v_1 - v_2}{2V_{osc}} \tag{E.4}$$

$$y_d = \frac{i_L}{I_{osc}} \tag{E.5}$$

$$x_{cm} = \frac{v_1 + v_2}{2V_{osc}} \tag{E.6}$$

$$\tau = \omega_0 t \tag{E.7}$$

$$\omega_0 = 1/\sqrt{LC} \tag{E.8}$$

$$Q = \frac{\omega_0 C}{1/R_{loss}} = \frac{R_{loss}}{\omega_0 L} \tag{E.9}$$

$$I_{osc} = \omega_0 C V_{osc} = V_{osc}/\omega_0 L\ . \tag{E.10}$$

[1] Note that, in this appendix, we neglect all transistor capacitances. This introduces only small errors as far as the dynamics induced by $u(\tau)$ is concerned. However, when interested in the upconversion of the $1/f$ noise coming from the bias-current sources I_b, it is of great importance to take the transistor capacitances into account. These capacitances induce (a slight) skewing of the common-mode voltage which can be shown the main responsible for the upconversion of $1/f$ noise coming from the bias currents. With some extra effort, the reasoning outlined in this appendix can be extended to include the effects of the transistor capacitances as well.

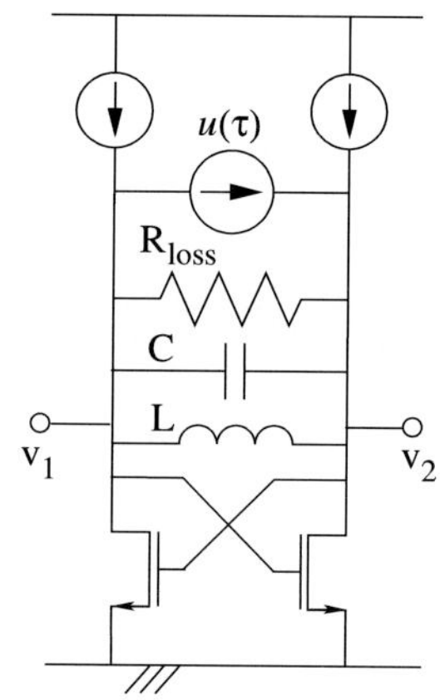

Figure E.1: Harmonic oscillator topology injected with a (small) current $u(\tau)$.

In the above, i_L is the current flowing through the inductor while V_{osc} is an estimate for the oscillation amplitude. Furthermore, $I_{M,d}$ is the differential transistor current while $I_{M,cm}$ equals the common-mode transistor current. As usual, the terms of (E.1)-(E.3) in square brackets denote perturbation terms while the others correspond to the oscillators core system.

The set of all steady-state solutions to the equations corresponding to the oscillator's core system behavior, i.e. (E.1)-(E.3) with the terms between the square brackets removed, is given by

$$\mathbf{x}_s(\tau,\mathbf{p}) = \begin{bmatrix} x_{d,s}(\tau,\mathbf{p}) \\ y_{d,s}(\tau,\mathbf{p}) \\ x_{cm,s}(\tau,\mathbf{p}) \end{bmatrix} = \begin{bmatrix} A\cos(\tau+\theta) \\ A\sin(\tau+\theta) \\ f(A\cos(\tau+\theta)) \end{bmatrix} \tag{E.11}$$

with $\mathbf{p} = \begin{bmatrix} A & \theta \end{bmatrix}^T$ being the oscillator's state vector. In (E.11), the function $x_{cm} = f(x_d)$ is determined by solving (E.3) for x_{cm}.

Next, using (E.11) we find

$$\mathbf{U}_1(\tau,\mathbf{p}) = \frac{\partial \mathbf{x}_s}{\partial \mathbf{p}} = \begin{bmatrix} \cos(\tau+\theta) & -A\sin(\tau+\theta) \\ \sin(\tau+\theta) & A\cos(\tau+\theta) \\ \frac{\mathrm{d}f}{\mathrm{d}x_d}\cos(\tau+\theta) & \frac{\mathrm{d}f}{\mathrm{d}x_d}A\sin(\tau+\theta) \end{bmatrix}. \tag{E.12}$$

Using the theory outlined in section 5.5.2 of chapter 5, we find that the corresponding projector $\mathbf{V}_1(\tau,\mathbf{p})$ equals the solution to

$$\left(\frac{\partial}{\partial\tau}\mathbf{V}_1^T(\tau,\mathbf{p})\right)\mathbf{C}(\tau,\mathbf{p}) - \mathbf{V}_1^T(\tau,\mathbf{p})\mathbf{G}(\tau,\mathbf{p}) = 0 \tag{E.13}$$

$$\mathbf{V}_1^T(\tau,\mathbf{p})\mathbf{C}(\tau,\mathbf{p})\mathbf{U}(\tau,\mathbf{p}) = \mathbf{I} \tag{E.14}$$

where $\mathbf{I} \in \mathbb{R}^{2\times 2}$ is the unit matrix and

$$\mathbf{C}(\tau,\mathbf{p}) = \begin{bmatrix} 1 & 0 & 0 \\ 0 & 1 & 0 \\ 0 & 0 & 0 \end{bmatrix} \tag{E.15}$$

$$\mathbf{G}(\tau,\mathbf{p}) = \begin{bmatrix} 0 & 1 & 0 \\ -1 & 0 & 0 \\ \frac{\partial I_{cm}}{\partial x_d}(\tau,\mathbf{p}) & 0 & \frac{\partial I_{cm}}{\partial x_{cm}}(\tau,\mathbf{p}) \end{bmatrix}. \tag{E.16}$$

This solution is found to be

$$\mathbf{V}_1^T(\tau,\mathbf{p}) = \begin{bmatrix} \cos(\tau+\theta) & \sin(\tau+\theta) & 0 \\ -\frac{\sin(\tau+\theta)}{A} & \frac{\cos(\tau+\theta)}{A} & 0 \end{bmatrix}. \tag{E.17}$$

Hence, using (5.70) from section 5.5.2 in chapter 5, the oscillator's amplitude and phase behavior is seen to be governed by

$$\frac{dA}{d\tau} = -\frac{A\cos(\tau+\theta)^2}{Q} - \cos(\tau+\theta)g(A,\tau+\theta) - \cos(\tau+\theta)\frac{u(\tau)}{I_{osc}} \tag{E.18}$$

$$\frac{d\theta}{d\tau} = \frac{A\sin(\tau+\theta)\cos(\tau+\theta)}{A} + \frac{\sin(\tau+\theta)}{A}g(A,\tau+\theta) + \frac{\sin(\tau+\theta)}{A\cdot I_{osc}}u(\tau). \tag{E.19}$$

In the above,

$$g(A,\tau) = \frac{I_{M,d}(A\cos(\tau), f(A\cos(\tau)))}{I_{osc}}. \tag{E.20}$$

Since we neglected all transistor capacitances, both $I_{M,d}(\cdot,\cdot)$ and $f(\cdot)$ are memoryless operators. Hence, $g(A,\tau)$ can be expanded as

$$g(A,\tau) = \sum_{k=0}^{\infty} g_k(A)\cos(k\tau) \tag{E.21}$$

with

$$g_k(A) = \frac{1}{\pi(1+\delta(k))}\int_{-\pi}^{\pi} g(A,s)\cos(ks)ds . \tag{E.22}$$

Here, $\delta(k)$ represents the Kronecker delta function.

Given (E.18)-(E.19), we extract the equations that govern the (dominant) slow-varying changes of A and θ. This is accomplished by applying the ideal lowpass averaging operator $M_{ideal}[\cdot]$ (see (5.90) in section 5.6 of chapter 5) to both right-hand sides in (E.18) and (E.19). This yields

$$\frac{d\bar{A}}{d\tau} = -\frac{\bar{A}}{2Q} - \frac{g_1(\bar{A})}{2} - M_{ideal}\left[\cos(\tau+\theta)\frac{u(\tau)}{I_{osc}}\right] \tag{E.23}$$

$$\frac{d\bar{\theta}}{d\tau} = M_{ideal}\left[\frac{\sin(\tau+\theta)}{\bar{A}\cdot I_{osc}}u(\tau)\right] \tag{E.24}$$

as the equations governing the oscillator's averaged (slow-varying) dynamics.

E.2 Phase behavior near operating point

In many practical problems, the magnitude of the the injected current $u(\tau)$ lies well below that of the currents flowing through the transistors, e.g. when $u(\tau)$ models some injected noise current. In that case, the right-hand side of (E.23) is dominated by the first two terms. If the oscillator is stable, these terms determine a (stable) operating-point amplitude $\overline{A}_{op}$, equal to the (non-zero) solution of

$$\frac{\overline{A}_{op}}{2Q} + \frac{g_1(\overline{A}_{op})}{2} = 0 \ . \tag{E.25}$$

Because of the stability of this operating point, disturbances, like the current $u(\tau)$, only cause small amplitude deviations to occur. Hence, for most practical purposes we can take $A(\tau) \approx \overline{A}_{op}$. The phase behavior near this operating point is then (approximately) governed by

$$\frac{d\overline{\theta}}{d\tau} = M_{ideal} \left[\frac{\sin(\tau + \theta)}{I_{osc,op}} u(\tau) \right] \tag{E.26}$$

where $I_{osc,op} = \overline{A}_{op} \cdot I_{osc} = (V_{osc}\overline{A}_{op})\omega_0 C$ is the amplitude of the operating-point current circulating in the resonant tank. If we now compare (E.26) to (5.108) in section 5.7 of chapter 5, we find that the ISF corresponding to the source $u(\tau)$ (approximately) equals

$$\Gamma(\tau) \approx \frac{\sin(\tau)}{I_{osc,op}} \ . \tag{E.27}$$

This concludes our analysis of the harmonic oscillator in Fig. E.1.

E.3 Conclusions

This appendix has presented expressions for the averaged amplitude and phase behavior of the harmonic oscillator in Fig. E.1. This, amongst others, resulted in an approximate expression for the oscillator's ISF with respect to a disturbing signal $u(\tau)$ that is directly injected into the resonant tank

Bibliography

[Abra88] R. Abraham, J. Marsden and T. Ratiu, *Manifolds, Tensors, Analysis and Applications*, 2nd. edition, Springer-Verlag, 1988

[Adl46] R. Adler, "A study of locking phenomena in oscillators", *Proc. IRE*, vol. 34, pp. 351-357, June 1946

[Arf85] G. Arfken, *Mathematical Methods for Physicists*, 3rd edition, Academic Press Inc., 1985

[Bam92] B.A. Bamieh and J.B. Pearson, "A General Framework for Linear Periodic Systems with Applications to H^∞ Sampled-Data Control", In *Transactions on Automatic Control*, vol. 37, no. 4, pp. 418-435, April 1992

[Barr94] R. Barret, M. Berry et al., *Templates for the Solution of Linear Systems: Building blocks for Iterative Methods*, SIAM, 1994.

[Blah87] R.E. Blahut, *Principles and practice of information theory*, Addison-Wesley Reading, 1987

[Bode45] H.W. Bode, *Network Analysis and Feedback Amplifier Design*, Van Nostrand, 1945

[Bogo61] N.N. Bogoliubov and Y.A. Mitropolsky, *Asymptotic methods in the theory of non-linear oscillations*, Hindustan Pulishing Corp., Delhi, 1961

[Bra96] H. Brachtendorf, G. Welsch, R. Laur and A. Bunse-Gerstner, "Numerical steady state analysis of electronic circuits driven by multi-tone signals", Electrical Engineering series, vol. 79, pp. 103-112, Springer-Verlag, 1996.Delhi, 1961

[Buo02] A. Buonomo and A. Lo Schiavo, "Analyzing the Dynamic Behavior of RF Oscillators", In *IEEE Transactions on Circuits and Systems-I: Fundamental Theory and Applications*, vol. 49, no. 11, November 2002

[Chua92] L.O. Chua, "The genesis of Chua's circuit," In *Archiv fur Elektronik und Ubertragungstechnik*, vol. 46, no. 4, pp. 250-257, 1992

[Cohn65] P.M. Cohn, *Lie groups*, Cambridge University Press, 1965

[Cor91] C. Corduneanu, *Integral equations and applications*, Cambridge University Press, 1991

[Cran98] J. Craninckx and M. Steyaert, *Wireless CMOS Frequency Synthesizer Design*, Kluwer Academic Publishers, 1998

[Crol95] J. Crols and M. Steyaert, "A 1.5 GHz Highly Linear CMOS Downconversion Mixer", In *IEEE Journal of Solid State Circuits*, vol. 30, no. 7, pp. 736-742, July 1995

[Crol97] J. Crols, *CMOS Wireless Tranceiver Design*, Kluwer Academic Publishers, 1997

[Dae99] W. Daems, W. Verhaegen, P. Wambacq, G. Gielen and W. Sansen, "Evaluation of error-control strategies for the linear symbolic analysis of analog integrated circuits", In *IEEE Transactions Circuits and Systems-I: Fundamental Theory and Applications*, vol. 46, no. 5, pp. 594-606, May, 1999.

[Dae02] W. Daems, G. Gielen and W. Sansen, "Circuit Simplification for Symbolic Analysis of Analog Integrated Circuits", In *IEEE Transactions on Computer-Aided Design of Integrated Circuits and Systems*, vol. 21, no. 4, pp. 395-407, April, 2002

[Dae03] W. Daems, G. Gielen and W. Sansen, "Simulation-based Generation of Posynomial Performance Models for the Sizing of Analog Integrated Circuits", In *IEEE Transactions on Computer-Aided Design of Integrated Circuits and Systems*, vol. 22, no. 5, May 2003

[Dem98] A. Demir, "Phase Noise in Oscillators: DAEs and Colored Noise Sources", In *Proceedings of IEEE/ACM International Conference on Computer-Aided Design*, pp. 170-177, 1998

[Dem00a] A. Demir, A. Mehrotra and J. Roychowdhury, "Phase Noise in Oscillators: A Unifying Theory and Numerical Methods for Characterization", In *IEEE Transactions on Circuits and Systems-I: Fundamental Theory and Applications*, vol. 47, no. 5, pp. 655-674, May 2000

[Dem00b] A. Demir, "Floquet theory and non-linear perturbation analysis for oscillators with differential-algebraic equations", In *International Journal of Circuit Theory and Applications*, vol. 28, pp. 163-185, March-April 2000

[Dem02] A. Demir, "Phase Noise and Timing Jitter in Oscillators with Colored Noise Sources", In *IEEE Transactions on Circuits and Systems and Systems-I: Fundamental Theory and Applications*, vol. 49, no. 12, pp. 1782-1791, December 2002

[Dem03] A. Demir and J. Roychowdhury, "A Reliable and Efficient Procedure for Oscillator PPV Computation, With Phase Noise and Macromodeling Applications", In *IEEE Transactions on Computer-Aided Design of Integrated Circuits and Systems*, vol. 22, no. 2, pp. 188-197 Febraury 2003

[Demu02] B. Demuer, *Monolithic CMOS fractional-N frequency Synthesizers*, Ph.D. thesis, Katholieke Universiteit Leuven, Department of Electrical Engineering, November 2002

[Deso80] C. Desoer and Y. Wang, "On the Generalized Nyquist Stability Criterion", In *IEEE Transactions on Automatic Control*, vol. AC-25, no. 2, pp. 187-196, April 1980

[Donn98] S. Donnay, *Analog High-Level Design Automation in Mixed-Signal Asics*, Ph.D. dissertation, Katholieke Universiteit Leuven, December 1998

[Enz95] C. Enz, F. Krummenacher and E. Vittoz, "An analytical MOS transistor model valid in all regions of Operation and dedicated to low-voltage and low-current applications", In *Journal of Analog Integrated Circuits and Signal Processsing*, Kluwer Academic Publishers, vol. 26, pp. 83-114, July 1995

[Fern98] F. Fernandez, A. Rodríguez-Vázquez, L. Huertas and G. Gielen, *Symbolic analysis techniques-Applications to Analog Design Automation*, IEEE Press, 1998.

[Firk00] F. Firk, *Introduction to Groups, Invariants and Particles*, http://www.physicsforfree.com/, 2000

[Floq83] G. Floquet, "Sur les équations différentielles linéaires a coefficients périodques", Annales de l'Ecole Normale Supérieure, vol. 12, pp. 47-89, 1883.

[Fran94] G. Franklin, J. Powell, and A. Emami-Naeini, *Feedback Control of Dynamic Systems*, 3rd edition, Addison-Wesley, 1994.

[Freu95] R. Freund and P. Feldmann, "Numerical Simulation of Electronic Circuits: State-of-the-Art Techniques and Challenges", short course, SIAM Annual Meeting, 1995

[Gard79] F.M. Gardner, *Phaselock Techniques*, 2nd. ed., John Wiley & Sons Inc., 1979

[Gard80] F.M. Gardner, "Charge-Pump Phase-Lock Loops", In *IEEE Transactions on Communications*, vol. com-28, no. 11, November 1980

[Gelb68] A. Gelb and W. Vander Velde, *Multiple-input Describing Functions and Nonlinear System Design*, McGraw-Hill Book Company, 1968

[Gie91] G. Gielen and W. Sansen, *Symbolic Analysis for Automated Design of Analog Integrated Circuits*, Kluwer Academic Publishers, 1991

[Gie94] G. Gielen, P. Wambacq, W. Sansen, "Symbolic analysis methods and applications for analog circuits: a tutorial overview", In *Proceedings of the IEEE*, vol. 82, no. 2, pp. 287-304, February, 1994

[Gol96] G.H. Golub and C.F. Van Loan, *Matrix Computations*, 3rd edition, The John Hopkins University Press, 1996

[Guck80] J. Guckenheimer, "Dynamics of the Van der Pol Equation", In *IEEE Transactions on Circuits and Systems*, vol. 27, no. 11, November 1980

[Haji98] A. Hajimiri and T.H. Lee, "A General Theory of Phase Noise in Electrical Oscillators", In *IEEE Journal of Solid-State Circuits*, vol. 33, no. 2, pp. 179-194, February 1998

[Ham01] D. Ham and A. Hajimiri, "Concepts and Methods in Optimization of Integrated LC VCOs", In *Journal of Solid-State Circuits*, vol. 36, no. 6, June 2001

[Hays76] J.D. Hays, J. Imbrie and N.J. Shackleton "Variations in the Earth's Orbit: Pacemaker of the Ice Ages", In *Science*, vol. 194, pp. 1121-1132, 1976.

[Hein88] J.P. Hein and J.W. Scott, "z-Domain Model for Discrete-Time PLL's", In *IEEE Transactions on Circuits and Systems*, vol. 35, no. 11, November 1988

[Hest87] D. Hestenes, *New foundations for classical mechanics*, Kluwer Academic Publishers, 1987

[Huts80] V. Hutson and J.S. Pym, *Applications of Functional Analysis and Operator Theory*, Math. in Science and Engineering, vol. 146, Academic Press, 1980

[Jayn03] E.T. Jaynes, *Probability theory: the logic of science*, Cambridge University Press, 2003

[Jez74] M.T. Jezewski, "An Approach to the Analysis of Injection-Locked Oscillators", In *IEEE Transactions on Circuits and Syst*ems, vol. 21, no. 3, pp. 395-401, May 1974

[Kae90] F.X. Kaertner, "Analysis of White and $f^{-\alpha}$ Noise in Oscillators", In *International Journal of Circuit Theory and Applications*, vol. 18, pp. 485-519, 1990

[Kesh82] M. Keshner, "$1/f$ Noise", In *Proceedings of the IEEE*, vol. 70, no. 3, pp. 212-218, 1982

[Kev81] J. Kevorkian and J.D. Cole, *Perturbation Methods in Applied Mathematics*, Springer-Verlag, New York, 1981

[Kha96] H.K. Khalil, *Nonlinear systems,* 2nd ed., Prentice Hal, 1996

[Kolm92] A.N. Kolmogorov, A.N. Shiryayev, G. Linduquist, *Selected Works of A.N.Kolmogorov : Probability Theory and Mathematical Statistics*, vol. 2, Kluwer Academic Publishers, 1992

[Kryl47] N. Kryloff and N. Bogoliubov, *Introduction to nonlinear mechanics, Princeton University Press*, New York, 1947

[Kund86] K. Kundert and A. Sangiovanni-Vincentelli, "Simulation of Nonlinear Circuits in the Frequency Domain", *IEEE Transactions on Computer-Aided Design of Integrated Circuits and Systems*, vol. cad-5, no. 4, pp. 521-535, October 1986

[Kund90] K. Kundert, J. White, and A. Sangiovanni-Vincentelli, *Steady-State Methods for Simulating Analog And Microwave Circuits*, Kluwer Academic Publishers, 1990

[Kund97] K. Kundert, "Simulation Methods for RF Integrated Circuits", In *Proceedings IEEE/ACM International Conference on Computer-Aided Design*, pp.752-765, 1997

[Lam97] R. Lamour, R. März, R. Winkler, "How floquet theory applies to differential-algebraic equations", Preprint, 1997

[Lee98] T.H. Lee, *The design of CMOS radio-frequency integrated circuits*, Cambridge University Press, 1998

[Lees66] D. Leeson, "A Simple Model of Feedback Oscillator Noise Spectrum", In *Proceedings of the IEEE*, vol. 54, pp. 329-330, February 1966

[Leyn01] F. Leyn, W. Sansen, and G. Gielen, "Analog Small-Signal Modeling—Part I: Behavioral Signal Path Modeling for Analog Integrated Circuits", In *IEEE Transactions on Circuits and Sytems-II: Analog and Digital Signal Processing*, vol. 48, no. 7, pp. 701-711, july 2001

[Lin91] P.M. Lin, *Symbolic Network Analysis*, Elsevier, 1991

[Maas88] S.A. Maas, *Nonlinear microwave circuits*, Norwood: Artech House, 1988

[Mac89] J.M. Maciejowski, *Multivariable feedback design*, Addison-Wesley Publishers Ltd., 1989

[Marz95] R. März, "On linear differential-algebraic equations and linearizations", In *Applied Numerical Mathematics*, vol. 18, pp. 267-292, 1995

[Mid87] D. Middleton, *Statistical Communication Theory*, Peninsula Publishing, Los Altos, California, 1987

[Mind02] D.A. Mindell, "Opening Black's Box", In *Between Human and Machine*, Johns Hopkins, 2002

[Moll00] E. Möllerstedt and B. Bernhardsson, "Out of Control Because of Harmonics", In *IEEE Control Systems Magazine*, pp. 70-81, 2000

[Murd88] J. Murdock, "Qualitative Theory of Nonlinear Resonance by Averaging and Dynamical Systems Methods", *Dynamics Reported* (U. Kirchgraber and H. Walther), vol. 1, pp. 91-172, Wiley, 1988

[Nag75] L. Nagel, "SPICE2: A computer program to simulate semiconductor circuits", Electron. Res. Lab., Univ. California at Berkeley, UCB/ERL M520, 1975.

[Nar99] O. Narayan and J. Roychowdhury, "Multi-Time Simulation of Voltage-Controlled Oscillators", In *Proceedings IEEE/ACM Design Automation Conference*, Session 36, 1999

[Ngo96] E. Ngoya and R. Larcheveque, "Envelope transient analysis: a new method for the steady state analysis of microwave communication circuits and systems", In *IEEE MTT Symposium Digest*, pp. 1365-1368, June 1996

[Odab97] A. Odabasioglu, M. Celik and L.T. Pileggi, "PRIMA: Passive Reduced-Order Interconnect Macromodeling Algorithm", In *Proceedings IEEE/ACM International Conference on Computer-Aided Design*, Session 1D, 1997

[Perr02] M.H. Perrot, "Fast and Accurate Behavioral Simulation of Fractional-N Frequency Synthesizers and other PLL/DLL Circuits", In *Proceedings IEEE/ACM Design Automation Conference*, Session 33, June 2002

[Phil98] J. Philips, "Model Reduction of Time-Varying Linear systems using Approximate Multipoint Krylov-subspace Projectors", In *Proceedings IEEE/ACM International Conference on Computer-Aided Design*, pp. 96-102, November 1998

[Pop34] K. Popper, *Logik Der Forschung* (*The Logic of Scientific Discovery*), Springer, 1934

[Proa01] J.G. Proakis, *Digital Communications*, McGraw-Hill, 2001

[Rand03] R.H. Rand, *Lecture Notes on Nonliner Vibrations*, http://www.tam.cornell.edu/randdocs/, 2003

[Risk89] H. Risken, *The Fokker-Planck Equation*, Springer-Verlag, 1989

[Riz96] V. Rizzoli, A. Neri and F. Mastri, "A modulation-oriented piecewise harmonic-balance technique suitable for transient analysis and digitally modulated signals", In *Proceedings European Microwave Conference*, pp. 546-550, Prague, September 1996

[Rob83] C. Robinson, "Second-Order Averaging of Forced and Coupled Nonlinear Oscillators", In *IEEE Transactions on Circuits and Systems*, vol. 30, no. 8, August 1983

[Royc98] J. Roychowdhury, D. Long and P. Feldmann, "Cyclostationary noise analysis of large RF circuits with multitone excitations", In *IEEE Journal of Solid State Cir*cuits, vol. 33, no. 3, pp. 324-336, March 1998

[Royc99] J. Roychowdhury, "Reduced-order modeling of time-varying systems", In *IEEE Transactions on Circuits and Systemts-II: Analog and Digital Signal Processing,* vol. 46, no. 10, pp. 1273-1288, October 1999

[Royc01] J. Roychowdhury, "Analyzing Circuits with Widely Separated Time Scales using Numerical PDE Methods", In *IEEE Transactions on Circuits and Systems-I: Fundamental Theory and Applications*, vol. 48, pp. 578-594, May 2001

[Saad92] Y. Saad, *Numerical Methods for Large Eigenvalue Problems*, Halstead Press, 1992

[Saad96] Y. Saad, *Iterative Methods for Sparse Linear Systems*, PWS Publishing, 1996

[Shan48a] C.E. Shannon, "A Mathematical Theory of Communication, Part I and II", In *The Bell System Technical Journal*, vol. 27, pp. 379-423, July 1948

[Shan48b] C.E. Shannon, "A Mathematical Theory of Communication, Part III", In *The Bell System Technical Journal*, vol. 27, pp. 623-656, October 1948

[Sore00] D. Sorensen and R. Lehoucq, "Implicitly Restarted Lanczos Method", In *Templates for the Solution of Algebraic Eigenvalue Problems: A Practical Guide*, pp. 67-80, 166-184, SIAM, 2000

[Star80] J. Starzyk and E. Sliwa, "Hierarchic decomposition method for the topological analysis of electronic networks", In *International Journal of Circuit Theory and Applications*, vol. 8, pp. 407-417, 1980

[Stav00] A. van Staveren, C.J.M. Verhoeven, A.H.M. van Roermund, *Structured Electronic Design: High-Performance Harmonic Oscillators and Bandgap References*, Kluwer Academic Publishers, 2000

[Sto50] J.J. Stoker, *Nonlinear Vibrations in Mechanical and Electrical Systems*, Wiley, New York, 1950

[Strom77] T. Ström and S. Signell, "Analysis of Periodically Switched Linear Circuits", In *IEEE Transactions on Circuits Systems*, vol. 24, no. 10, pp. 531-541, October 1977

[Suz96] J. Suzuki, *A History in the stability problem of celestial mechanics from Newton to Laplace*, Ph.D. Dissertation, Boston University, 1996

[Teli96] R. Telichevesky, K. Kundert and J. White, "Receiver characterization using periodic small-signal analysis", In *Proceedings IEEE Custom Integrated Circuits Conference*, May 1996

[Thul92] K. Thulasiraman and M. Swamy, *Graphs: theory and algorithms*, John Wiley & Sons Inc., 1992

[Vana01a] P. Vanassche, G. Gielen and W. Sansen, "Efficient Time-Domain Simulation of Telecom Frontends Using a Complex Damped Exponential Signal Model", *In Proceedings Design Automation and Test in Europe*, Session 4A, Munchen, 2001

[Vana01b] P. Vanassche, G. Gielen and W. Sansen, "Efficient High-Level Simulation of Analog Telecom Frontends", *In Low-power design techniques and CAD tools for analog and RF integrated circuits*, Kluwer Academic Publishers, 2001

[Vana01c] P. Vanassche, G. Gielen and W. Sansen, "Modeling Harmonic Oscillators for Use in High-Level Simulations", *In Proceedings IEEE ProRISC*, Veldhoven, 2001

[Vana02a] P. Vanassche, G. Gielen and W. Sansen, "Constructing Symbolic Models for the Input/Output Behavior of Periodically Time-Varying Systems using Harmonic Transfer Matrices", *In Proceedings Design Automation and Test in Europe*, Session 3C, Paris, 2002

[Vana02b] P. Vanassche, G. Gielen and W. Sansen, "Behavioral Modeling of (Coupled) Harmonic Oscillators", *In Proc. IEEE/ACM DAC,* Session 35, New Orleans, 2002

[Vana02c] P. Vanassche, G .Gielen and W. Sansen, "Symbolic modeling of periodically time-varying systems using harmonic transfer matrices", In *IEEE Transactions on Computer-Aided Design of Integrated Circuits and Systems,* vol. 21, nr. 9, pp. 1011-1024, September 2002

[Vana02d] P. Vanassche, G. Gielen and W. Sansen, "On the Difference Between Two Widely Publicized Methods for Analyzing Oscillator Phase Behavior", *In Proceedings IEEE/ACM International Conference on Computer-Aided Design*, Session 4A, San Jose, 2002

[Vana02e] P. Vanassche, G. Gielen and W. Sansen, "Analyzing Slowly Modulated Oscillator Behavior", *In Proceedings IEEE ProRISC*, pp. 507-515, Veldhoven, 2002

[Vana03a] P. Vanassche, G. Gielen and W. Sansen, "Time-Varying, Frequency-Domain Modeling and Analysis of Phase-Locked Loops with Sampling Phase-Frequency Detectors", *In Proceedings Design Automation and Test in Europe*, Session 3F, Munchen, 2003

[Vana03b] P. Vanassche, G. Gielen and W. Sansen, "Behavioral Modeling of (Coupled) Harmonic Oscillators", In *IEEE Transactions on Computer-Aided Design of Integrated Circuits and Systems*, vol. 22, nr. 8, August 2003

[Vana03c] P. Vanassche, G. Gielen and W. Sansen, "A Generalized Method for Computing Oscillator Phase Noise Spectra", To appear in *Proceedings IEEE/ACM International Conference on Computer-Aided Design*, San Jose, 2003

[Vand20] B. van der Pol, *Radio Reviews*, vol. 1, pp. 704-754, 1920

[Verha01] W. Verhaegen and G. Gielen, "Symbolic Distortion Analysis of Analog Integrated Circuits", In *Proceedings ECCTD*, pp. I.21-I.24, August 2001

[Verhu96] F. Verhulst, *Nonlinear Differential Equations and Dynamical Systems*, 2nd ed., Springer 1996

[Vlad94] A. Vladimirescu, *The Spice Book*, John Wiley & Sons, 1994

[Wamb98a] P. Wambacq and W. Sansen, *Distortion analysis of analog integrated circuits*, Kluwer Academic Publishers, 1998

[Wamb98b] P. Wambacq, G. Gielen and W. Sansen, "Symbolic network analysis methods for practical analog integrated circuits: a survey", In *IEEE Transactions on Circuits and Systems-II: Analog and Digital Signal Processing*, vol. 45, no. 10, pp. 1331-1341, October 1998.

[Were91a] N. Wereley, *Analysis and Control of Linear Periodically Time Varying Systems*, PhD thesis, Department of Aeronautics and Astronautics, MIT, 1991

[Were91b] N. Wereley and S. Hall, "Linear time periodic systems: Transfer function, poles, transmission zeros and directional properties", In *Proceedings American Control Conference*, Boston, 1991

[Zad61] L.A. Zadeh, "Time-Varying Networks, I", In *Proceedings of the IRE*, 49, pp. 1488-1503, October 1961.

[VHDL] http://www.vhdl.org/analog, http://www.vhdl-ams.com

[VERI] http://www.eda.org/verilog-ams